# Inorganic Chemistry: Reactions, Structures and Mechanisms

# Inorganic Chemistry: Reactions, Structures and Mechanisms

Editor: Bridget Kent

NY RESEARCH PRESS

New York

Published by NY Research Press
118-35 Queens Blvd., Suite 400,
Forest Hills, NY 11375, USA
www.nyresearchpress.com

Inorganic Chemistry: Reactions, Structures and Mechanisms
Edited by Bridget Kent

International Standard Book Number: 978-1-63238-666-3 (Hardback)

**Cataloging-in-Publication Data**

   Inorganic chemistry : reactions, structures and mechanisms / edited by Bridget Kent.
       p. cm.
   Includes bibliographical references and index.
   ISBN 978-1-63238-666-3
   1. Chemistry, Inorganic.  2. Reaction mechanisms (Chemistry).
   3. Chemical structure. I. Kent, Bridget.
QD151.3 .I56 2019
546--dc23

# Contents

# Preface

Every book is a source of knowledge and this one is no exception. The idea that led to the conceptualization of this book was the fact that the world is advancing rapidly; which makes it crucial to document the progress in every field. I am aware that a lot of data is already available, yet, there is a lot more to learn. Hence, I accepted the responsibility of editing this book and contributing my knowledge to the community.

Inorganic chemistry is that branch of chemistry that studies the properties and synthesis of inorganic and organometallic compounds. Such compounds can be classed into several groups based on their properties such as coordination compounds, transition metal compounds, organometallic compounds, bioinorganic compounds, etc. Some of the technological tools generally applied in inorganic chemistry are X-ray crystallography, electrochemistry, dual polarization interferometry and spectroscopy. The study of inorganic chemistry is of relevance in material science, coating, medicine, agriculture, the chemical industry, etc. This book elucidates new techniques and applications of inorganic chemistry in a multidisciplinary manner. Some of the diverse topics covered in this book address the varied aspects of reactions, structures and mechanisms that fall under this discipline. In this book, using case studies and examples, constant effort has been made to make the understanding of the difficult concepts of inorganic chemistry, as easy and informative as possible, for the readers.

While editing this book, I had multiple visions for it. Then I finally narrowed down to make every chapter a sole standing text explaining a particular topic, so that they can be used independently. However, the umbrella subject sinews them into a common theme. This makes the book a unique platform of knowledge.

I would like to give the major credit of this book to the experts from every corner of the world, who took the time to share their expertise with us. Also, I owe the completion of this book to the never-ending support of my family, who supported me throughout the project.

**Editor**

# A Structural and Spectroscopic Study of the First Uranyl Selenocyanate, [Et$_4$N]$_3$[UO$_2$(NCSe)$_5$]

**Stefano Nuzzo, Michelle P. Browne, Brendan Twamley, Michael E. G. Lyons and Robert J. Baker ***

School of Chemistry, Trinity College, University of Dublin, 2 Dublin, Ireland; nuzzos@tcd.ie (S.N.); Brownm6@tcd.ie (M.P.B.); twamleyb@tcd.ie (B.T.); melyons@tcd.ie (M.E.G.L.)
* Correspondence: bakerrj@tcd.ie

Academic Editors: Stephen Mansell and Steve Liddle

**Abstract:** The first example of a uranyl selenocyanate compound is reported. The compound [Et$_4$N]$_3$[UO$_2$(NCSe)$_5$] has been synthesized and fully characterized by vibrational and multinuclear ($^1$H, $^{13}$C{$^1$H} and $^{77}$Se{$^1$H}) NMR spectroscopy. The photophysical properties have also been recorded and trends in a series of uranyl pseudohalides discussed. Spectroscopic evidence shows that the U–NCSe bonding is principally ionic. An electrochemical study revealed that the reduced uranyl(V) species is unstable to disproportionation and a ligand based oxidation is also observed. The structure of [Et$_4$N]$_4$[UO$_2$(NCSe)$_5$][NCSe] is also presented and Se$\cdots$H–C hydrogen bonding and Se$\cdots$Se chalcogen–chalcogen interactions are seen.

**Keywords:** uranyl; structural determination; photophysics

---

## 1. Introduction

The chemistry of uranium in its highest oxidation state has held scientists fascination for a long period of time. The uranyl moiety, [UO$_2$]$^{2+}$, is well studied in aqueous phases due, in part, to relevance in the nuclear waste treatment. Moreover, the photophysical properties of uranyl were first used in ancient roman times in colored glass [1], whilst comprehensive understanding of the bonding, and therefore photophysical properties, has come from both experiment and theory. An authoritative review by Denning summarizes these fundamental developments [2], and further reviews cover recent results [3–6]. The photophysical properties of the uranyl ion have been elucidated from these studies and the optical properties are due to a ligand-to-metal charge transfer (LMCT) transition involving promotion of an electron from a bonding –yl oxygen orbital ($\sigma_u$, $\sigma_g$, $\pi_u$ and $\pi_g$) to a non-bonding 5f$_\delta$ and 5f$_\phi$ orbital on uranium. De-excitation of this $^3\Pi_u$ triplet excited state causes the characteristic green emission at *ca.* 500 nm. Visible on the absorption and emission bands are the vibronic progression arising from strong coupling of the ground state Raman active symmetric vibrational O=U=O ($\nu_1$) mode with the $^3\Pi_u$ electronic triplet excited state. Time resolved studies allow sometimes complex speciation in water to be deconvoluted [7], whilst in non-aqueous media the positions of the emission maxima and lifetimes can be used as electronic and structural probes. For instance in the family of complexes *trans*-[UO$_2$X$_2$(O=PPh$_3$)$_2$] (X = Cl, Br, I) the photoluminescent properties do not vary [8], but for the compounds *trans*-[UO$_2$Cl$_2$L$_2$] (L = Ph$_3$P=NH, Ph$_3$P=O and Ph$_3$As=O) a red shift in the O$_{yl}$→U LMCT band is observed, in line with the increased donor strength of the ligand [9]. A further interesting photophysical property of certain uranyl compounds are thermochromic effects. Thus the compound [C$_4$mim]$_3$[UO$_2$(NCS)$_5$] (C$_4$mim = 1-butyl-3-methylimidazolium) is thermochromic in ionic liquids [10] but in organic solvents [Et$_4$N]$_3$[UO$_2$(NCS)$_5$] is not [11]. We have reported on the latter compound recently and now extend our study to the selenocyanate [NCSe]$^-$ derivatives which have not been reported. Indeed there is only one structurally characterized U–NCSe complex, *viz.* $[Pr_4N]_4[U(NCSe)_8]$ [12]. In this work we have synthesized $[Et_4N]_3[UO_2(NCSe)_5]$ and have characterized this by vibrational and multinuclear NMR spectroscopy and a photophysical investigation. X-ray diffraction of the compound $[Et_4N]_4[UO_2(NCSe)_5][NCSe]$ is also reported.

## 2. Results and Discussion

The synthesis of $[Et_4N]_3[UO_2(NCSe)_5]$, **1**, was conducted in a comparable way to that for the thiocyanate derivatives. Thus uranyl nitrate was treated with five equivalents of K[NCSe] followed by three equivalents of $Et_4NCl$ in acetonitrile. A yellow precipitate was formed which was soluble in dichloromethane or acetone. We have noted that whilst this compound is air and moisture stable, it is somewhat light sensitive so reactions were conducted in the dark; the uranium(IV) compound $[^nPr_4N]_4[U(NCSe)_8]$ was also reported to be light sensitive [12]. Decomposition to red selenium powder was sometimes observed but the fate of the uranium was not determined. An alternative route to this compound was to treat a THF solution of $[UO_2Cl_2(THF)_3]$ sequentially with K[NCSe] and $Et_4NCl$. **1** was characterized by spectroscopic methods and single crystals were grown from slow evaporation of an acetonitrile solution. Unfortunately, crystals grown from different solvents always proved to be twinned so refinement to a satisfactory standard was not possible, however it did prove atom connectivity (Figure S1). During the course of one experiment, a few single crystals which had a different morphology were observed; these were separated by hand and the structure was solved to be $[Et_4N]_4[UO_2(NCSe)_5][NCSe]$, **2**.

The solid state structure of **2** is shown in Figure 1 and the packing shown in Figure 2. The structure of **2** contains disorder in two of the $Et_4N^+$ cations and the uncoordinated $[NCSe]^-$ anion which were modelled with restraints and constraints. The geometry around the uranyl in **2** are a typical pentagonal bipyramid with linear NCSe fragments and the N$\cdots$N intramolecular distances are similar to that seen in $[Et_4N]_3[UO_2(NCS)]_5$ (**2**: 2.89 Å; NCS: 2.87 Å) [13]. The U=O bond length is 1.771(2) Å and average U–N, N=C and C=Se bond lengths of 2.459 Å, 1.149 Å and 1.794 Å respectively can be compared to the uncoordinated $[N=C=Se]^-$ ion (N=C: 1.081(14) Å and C=Se: 1.846(7) Å) in **2**. Upon coordination to the uranyl ion the N=C bond lengthens slightly and the C=Se bond shortens slightly, suggesting a reorganization in the $\pi$-framework of the ligand; this effect has also been observed in uranyl thiocyanates experimentally and theoretically [11,14]. The average U–N bond in $[Et_4N]_3[UO_2(NCS)]_5$ is 2.443 Å [13], whilst in a suite of [pyridinium]$[UO_2(NCS)_4(H_2O)]$ compounds the U–N bond lengths are 2.454(3) Å and 2.437(4) Å [15]. As has been previously described for $[Et_4N]_4[An(NCS)_8]$ (An = Th, U, Pu) [16], the lack of perturbation of the $\pi$-system in the $[NCS(e)]^-$ ligands suggests no $\pi$-overlap in the U–N bond.

The packing diagram (Figure 2) shows that the structure is a layer type where the cationic components sit between layers of uranyl ions. Hydrogen bonds between the U=O and H–C of the cations link these layers ($d_{C\cdots O}$ = 3.175–3.300Å), as now commonly observed [17]. There are also number of Se$\cdots$H–C short contacts. The most recent IUPAC definition of a hydrogen bond states that *"in most cases, the distance between H and Y are found to be less than the sum of their van der Waals radii"* [18]. According to this criterion, and using van der Walls radii taken from reference [19], H$\cdots$Se distances of less than 3.02 Å are classed as hydrogen bonds. These form a link between the layers via a C–H of an ethyl group and a Se atom in the coordinated and non-coordinated [NCSe] anion ($d_{C\cdots Se}$ = 3.687(16)–3.856(10); C–H$\cdots$Se = 142°–155°) [20]. Also present in the structure are close contacts between a coordinated selenium atom and the selenium of the non-coordinated $[NCSe]^-$ (3.427(1) Å) that are shorter than the van der Waals radii (3.64 Å) [19]. Chalcogen–chalcogen interactions have been studied both experimentally [21,22], and in the case of $[NCS]^-$ also for uranyl (S$\cdots$S = 3.536(2) Å) [15], and theoretically [23,24]. This may explain the difficulty in growing single crystals of **1** as these weak interactions may be important. Further studies are underway in our laboratory and will be reported on in due course.

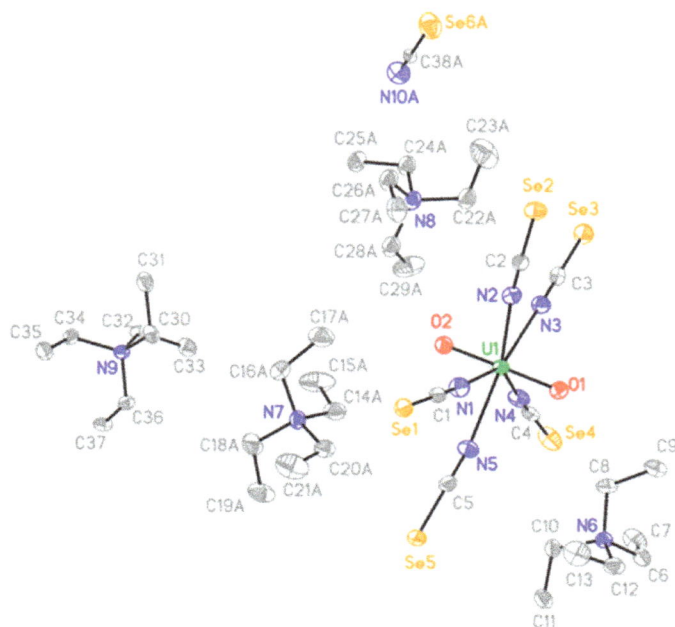

**Figure 1.** ORTEP plot of the structure of **2** refined with 70; 65; 55% occupancy for C14a–C21a; C22a–C29a; N10a–Se6a respectively. Thermal displacement shown at 50% occupancy and hydrogen atoms omitted for clarity. Selected bond lengths (Å): U(1)–O(1): 1.771(2); U(1)–N(1): 2.448(3); U(1)–N(2): 2.466(3); U(1)–N(3): 2.474(3); U(1)–N(4): 2.440(2); U(1)–N(5): 2.468(3); N(1)–C(1): 1.158(4); N(2)–C(2): 1.161(5); N(3)–C(3): 1.154(5); N(4)–C(4): 1.120(5); N(5)–C(5): 1.151(6); C(1)–Se(1): 1.791(3); C(2)–Se(2): 1.782(4); C(3)–Se(3): 1.798(4); C(4)–Se(4): 1.805(4); C(5)–Se(5): 1.794(4); N(10a)–C(38a): 1.081(14); C(38a)–Se(6a): 1.846(7).

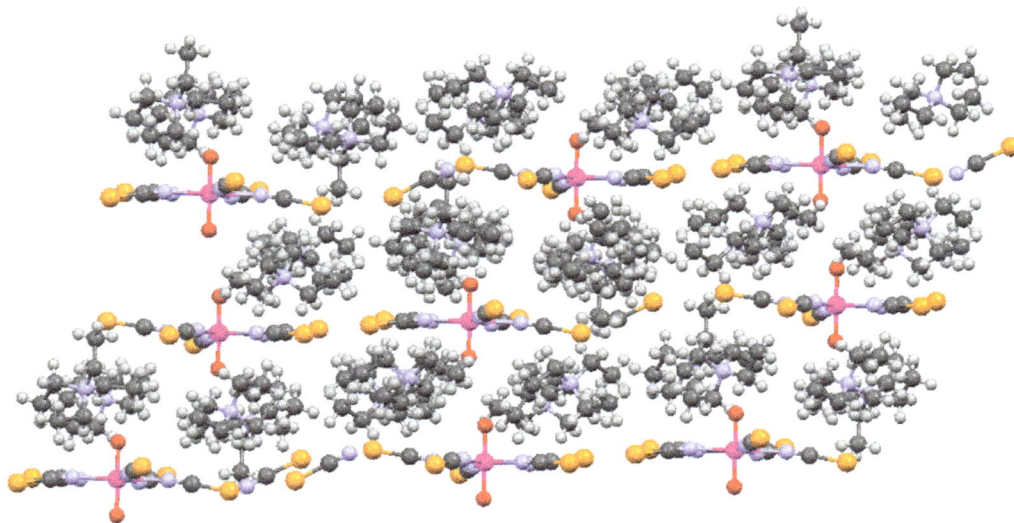

**Figure 2.** Packing diagram of **2** viewed down the *a* axis. Color code: U—pink; N—blue; C—Grey; S—yellow; O—red.

**1** has been spectroscopically characterized, whilst for **2** there was not enough material. The uranyl group has characteristic vibrations in both the infrared and Raman spectra (Figures S2 and S3). For **1** these bands occur at 921 cm$^{-1}$ (IR) and 845 cm$^{-1}$ (R) comparable to the thiocyanate analogue 924 (IR) and 849 cm$^{-1}$ (R) respectively. The N=C stretching frequency at 2056 cm$^{-1}$ (IR) and 2051, 2060, 2091 cm$^{-1}$ (R) are also similar to the NCS compound [2063 cm$^{-1}$ (IR) and 2088, 2058, 2044

$cm^{-1}$ (R)] [11], whilst the C=Se stretch of **1** is visible in the Raman spectrum at 635 and 672 $cm^{-1}$. $^1$H NMR spectroscopy was uninformative (Figure S4). $^{13}$C{$^1$H} NMR spectroscopy shows the resonance attributable to the selenocyanate at 117.4 ppm whilst a single peak is observed at −342.4 ppm in the $^{77}$Se{$^1$H} NMR spectrum. For comparison, in our hands these peaks occur in K[NCSe].at 119.2 and −314.2 ppm respectively. Therefore, on the basis of the metric parameters from the X-ray structure, vibrational and NMR spectroscopic data we suggest that the bonding in these compounds are ionic with little perturbation of the [NCSe]$^-$ anionic fragment upon coordination. Our recent theoretical study of the [NCS]$^-$ compound suggested a predominantly ionic interaction [11].

The photophysics of this compound has also been investigated (Figure 3). The electronic absorption spectrum of **1** (Figure 3a) shows a broad featureless band at 320 nm ($\varepsilon$ = 1,132 $mol^{-1} \cdot cm^{-1}$) assigned to transitions due to the [NCSe]$^-$ fragment and a weak vibronically coupled band at 460 nm ($\varepsilon \sim 100$ $mol^{-1} \cdot cm^{-1}$) due to the LMCT uranyl band. Excitation at 340 nm gives an emission spectrum typical for a uranyl moiety (Figure 3b). Pertinent properties are recorded in Table 1, along with a comparison for the uranyl thiocyanate and other pseudohalides. The average vibronic progression of the emission bands are coupled to the Raman active vibrational modes, which at 861 $cm^{-1}$ is in close agreement with that measured in the Raman spectrum (849 $cm^{-1}$). The individual spacing (828, 868 and 888 $cm^{-1}$) reflect the transition of the vibronic parabola from harmonic to anharmonic. The luminescence lifetime of **1** was determined by the correlated single photon counting on the microsecond scale following excitation at 372 nm with a nanoLED (Table 1). The kinetic decay profile was fitted to a mono-exponential decay and the luminescence lifetime for **1** was measured to be 1.30 ± 0.02 µs. No significant change in lifetime was observed for the different pseudohalide systems given in Table 1. Given the ionicity of the U–N bond, ligand exchange processes may be faster than the lifetime of the uranyl excited state and so contributes to the shorter lifetime [25].

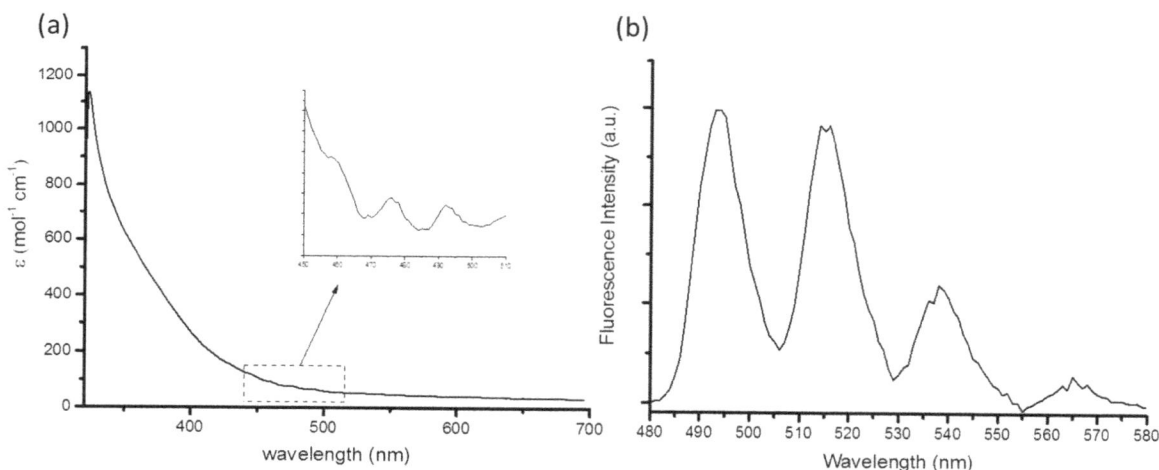

**Figure 3.** (**a**) UV-Vis absorption spectrum of **1** in acetone; (**b**) emission spectrum of **1** in acetone ($\lambda_{ex}$ = 340 nm).

**Table 1.** Comparison of photophysical properties of selected uranyl halides and pseudohalides.

| Compound (Solvent) | $\lambda_{abs}$ U=O (nm) | $\lambda_{em}$ (nm) | $E_{0-0}$ $(cm^{-1})$ | $\tau$ (µs) | $\chi^2$ | Ref. |
|---|---|---|---|---|---|---|
| **1** (MeCN) | 460 | 514 | 20,267 | 1.30 | 1.40 | This work |
| [Et$_4$N]$_3$[UO$_2$(NCS)$_5$] (MeCN) | 440 | 520 | 20,072 | 1.40 | 1.02 | [11] |
| [UO$_2$Cl$_2$(OPPh$_3$)$_2$] (MeCN) | 440 | 515 | 20,325 | 1.08 | 1.07 | [8] |
| [UO$_2$Cl$_4$]$^{2-}$ (MeBu$_3$N[Tf$_2$N]) | | 509 | 20,329 | 0.7 | | [26] |

We have also briefly examined the electrochemistry of **1** (Figure 4). Cyclic voltammetry of a solution of **1** in acetonitrile containing 0.1 M [$^n$Bu$_4$N][BPh$_4$] shows an irreversible cathodic wave at $E_{p,c} = -0.95$ V (vs. Fc/Fc$^+$) ascribed to the unstable $[UO_2]^{2+}/[UO_2]^+$ redox couple, in line with known formal redox potentials of U(VI)/U(V) reduction. For comparison the uranyl [NCS]$^-$ analogue displayed the reduction at $-1.45$ V [11]. The putative 1 e$^-$ reduced uranyl(V) species $[Et_4N]_4[UO_2(NCSe)_5]$ would be predicted to be quite unstable as it is now established that good $\pi$-donors and/or sterically bulky groups in the equatorial plane are required for stabilization of this unusual oxidation state [27,28]; although, there is evidence for the kinetic stabilization of the $[UO_2]^+$ ion in ionic liquids [29–31]. Any instability would manifest itself in an irreversible reduction, which is indeed what is observed. Also observed in this voltammogram is a broad, poorly defined irreversible oxidation at $E_{p,a} = +0.09$ V (vs. Fc/Fc$^+$) which is not observed at low scan rates, indicating the instability of this species. Given that the metal is in its highest oxidation state, it can be assigned as ligand based; we have observed similar behavior in the uranyl thiocyanate analogue ($E_{p,a} = +0.30$ V vs. Fc/Fc$^+$) and extends the family of uranyl coordinated to redox non-innocent ligands [11,32].

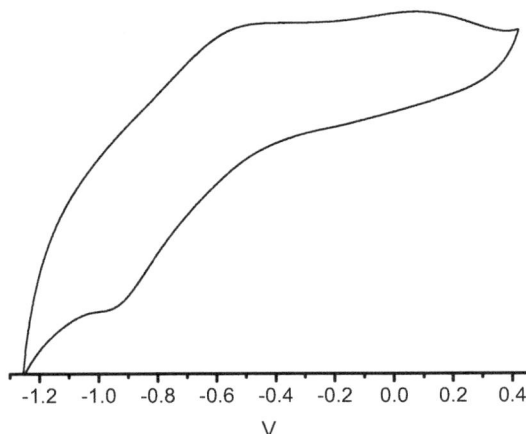

**Figure 4.** Cyclic Voltammogram of [1] vs. Fc/Fc$^+$ in MeCN at 293 K, with 0.1 M [Bu$_4$N][BPh$_4$] as a supporting electrolyte (scan rate = 0.1 V·s$^{-1}$).

## 3. Experimental Section

*Caution!* Although depleted uranium was used during the course of the experimental work, as well as the radiological hazards uranium is a toxic metal and care should be taken with all manipulations. Experiments were carried out using pre-set radiological safety precautions in accordance with the local rules of Trinity College Dublin.

$^1$H, $^{13}$C{$^1$H} and $^{77}$Se{$^1$H} NMR spectra were recorded on an AV400 spectrometer (Bruker, Karlsruhe, Germany) operating at 400.23, 155.54 and 76.33 MHz respectively, and were referenced to the residual $^1$H resonances of the solvent used or external Me$_2$Se. IR spectra were recorded on a Spectrum One spectrometer (Perkin Elmer, Norwalk, CT, USA) with attenuated total reflectance (ATR) accessory. Raman spectra were obtained using 785-nm excitation on a 1000 micro-Raman system (Renishaw, Wotton-under-Edge, UK) in sealed capillaries. X-ray crystallography data were measured on an Apex diffractometer (Bruker). The structures were solved by direct methods and refined on F2 by full matrix least squares (SHELX97) using all unique data. CCDC 1424467 contains the supplementary crystallographic data for this paper. This data can be obtained free of charge from The Cambridge Crystallographic Data Centre via www.ccdc.cam.ac.uk/data_request/cif. UV-Vis measurements were made on a Lambda 1050 spectrophotometer (Perkin Elmer USA), using fused silica cells with a path length of 1 cm. Steady-state photoluminescence spectra were recorded on a Fluorolog-3 spectrofluorimeter (Horiba–Jobin–Yvon, Stanmore, UK). Luminescence lifetime data were recorded following 372 nm excitation, using time-correlated single-photon counting (a PCS900 plug-in PC card for fast photon counting). Lifetimes were obtained by tail fit on the data obtained, and the quality of fit was judged by minimization of reduced chi-squared and residuals squared.

Cyclic voltammetry measurements were conducted in a standard three-electrode cell using a high performance digital potentiostat (CH model 1760 D Bi-potentiostat system monitored using CH1760D electrochemical workstation beta software). All solutions were degassed for 15 min before commencing analysis. A platinum electrode with a diameter of 2 mm was employed as the working electrode, a platinum rod (together with internal referencing *vs.* Fc/Fc$^+$) was used as a reference electrode and a platinum wire electrode as counter electrode. The electrolyte was a solution of 0.1 M [$^n$Bu$_4$N][BPh$_4$] in CH$_3$CN. [UO$_2$Cl$_2$(THF)$_3$] was prepared via the literature procedure [33] whilst all other reagents and solvents were obtained from commercial sources.

*Synthesis of 1*

**Method 1.** To a solution of UO$_2$(NO$_3$)$_2\cdot$6H$_2$O (400 mg, 0.80 mmol) in acetonitrile (30 cm$^3$) were added KNCSe (576 mg, 4.0 mmol) and Et$_4$NCl (328 mg, 2.4 mmol). The solution mixture was stirred at room temperature for 60 min. The resulting orange solution was filtered and the solvent was reduced in volume. After 48 h at room temperature, the orange solution deposited orange-yellow crystals suitable for X-ray diffraction (408 mg, 0.34 mmol, yield = 43%).

**Method 2.** Under an atmosphere of high purity dry argon, to a solution of UO$_2$Cl$_2$THF$_3$ (50 mg, 0.090 mmol) in dry THF (20 cm$^3$) were added sequentially KNCSe (65 mg, 0.45 mmol) and Et$_4$NCl (45 mg, 0.27 mmol). The solution was stirred at room temperature for 60 min. The resulting orange solution was filtered and the solvent was reduced in volume. Placement at $-20\,^\circ$C overnight yielded an orange powder (33 mg, 0.036 mmol, yield = 40%).

IR ($v$/cm$^{-1}$): 784 (C=Se), 921 (U=O), 2056 (C=N); Raman($v$/cm$^{-1}$): 635 and 672 (C=Se), 845 (U=O), 2051, 2060 and 2091 (C=N); $\delta_H$ (CD$_3$CN/ppm): 3.21 (q, 2H, $^3$J$_{H-H}$ = 7.28 Hz, CH$_2$), 1.25 (t, 3H, $^3$J$_{H-H}$ = 7.32 Hz, CH$_3$); $\delta_C$ (CD$_3$CN/ppm): 117.4 (N=C=Se), 29.9 (CH$_2$), 6.8 (CH$_3$); $\delta_{Se}$ (d-CH$_3$CN/ppm): $-342.2$ (N=C=Se).

## 4. Conclusions

To summarize, we have prepared and structurally characterized the first uranyl complexes of a selenocyanate ligand, which feature some unusual Se$\cdots$Se chalcogenide interactions and Se$\cdots$H–C hydrogen bonding. Vibrational and structural data suggest that the U–N bond is ionic and there is little perturbation of the [NCSe]$^-$ fragment compared to K[NCSe]. A photophysical investigation has shown that there is a small shift in the positions of the bands compared to the analogous [UO$_2$(NCS)$_5$]$^{3-}$ compound and the lifetime of the emission does not vary significantly with the nature of the pseudohalide. Finally, an electrochemical investigation revealed that the putative uranyl(V) compound is unstable with respect to disproportionation whilst there is a ligand based oxidation, similar to that observed in the [UO$_2$(NCS)$_5$]$^{3-}$ analogue.

**Acknowledgments:** We thank Trinity College Dublin for funding this work.

**Author Contributions:** Steffano Nuzzo conducted the synthesis and data analysis, Brendan Twamley conducted the X-ray crystallography, Michelle P. Browne and Michael E. G. Lyons did the electrochemistry measurements, Robert J. Baker conceived the experiments and wrote the manuscript.

## References

1.   Günther, R.T.; Manley, J.J. A mural glass mosaic from the Imperial Roman Villa near Naples. *Archaeologia* **1912**, *63*, 99–108. [CrossRef]

2.   Denning, R.G. Electronic structure and bonding in actinyl ions and their analogs. *J. Phys. Chem. A* **2007**, *111*, 4125–4143. [CrossRef] [PubMed]

3.   Liddle, S.T. The renaissance of non-aqueous uranium chemistry. *Angew. Chem. Int. Ed.* **2015**, *54*, 8604–8641. [CrossRef] [PubMed]

4.  Jones, M.B.; Gaunt, A.J. Recent developments in synthesis and structural chemistry of nonaqueous actinide complexes. *Chem. Rev.* **2013**, *113*, 1137–1198. [CrossRef] [PubMed]

5.  Baker, R.J. New reactivity of the uranyl ion. *Chem. Eur. J.* **2012**, *18*, 16258–16271. [CrossRef] [PubMed]

6.  Natrajan, L.S. Developments in the photophysics and photochemistry of actinide ions and their coordination compounds. *Coord. Chem. Rev.* **2012**, *256*, 1583–1603. [CrossRef]

7.  Drobot, B.; Steudtner, R.; Raff, J.; Geipel, G.; Brendler, V.; Tsushima, S. Combining luminescence spectroscopy, parallel factor analysis and quantum chemistry to reveal metal speciation—A case study of uranyl(VI) hydrolysis. *Chem. Sci.* **2015**, *6*, 964–972. [CrossRef]

8.  Hashem, E.; McCabe, T.; Schulzke, C.; Baker, R.J. Synthesis, structure and photophysical properties of [UO2X2(O=PPh3)2] (X = Cl, Br, I). *Dalton Trans.* **2014**, *43*, 1125–1131. [CrossRef] [PubMed]

9.  Redmond, M.P.; Cornet, S.M.; Woodall, S.D.; Whittaker, D.; Collison, D.; Helliwell, M.; Natrajan, L.S. Probing the local coordination environment and nuclearity of uranyl(VI) complexes in non-aqueous media by emission spectroscopy. *Dalton Trans.* **2011**, *40*, 3914–3926. [CrossRef] [PubMed]

10. Aoyagi, N.; Shimojo, K.; Brooks, N.R.; Nagaishi, R.; Naganawa, H.; van Hecke, K.; van Meervelt, L.; Binnemans, K.; Kimura, T. Thermochromic properties of low-melting ionic uranyl isothiocyanate complexes. *Chem. Commun.* **2011**, *47*, 4490–4492. [CrossRef] [PubMed]

11. Hashem, E.; Platts, J.A.; Hartl, F.; Lorusso, G.; Evangelisti, M.; Schulzke, C.; Baker, R.J. Thiocyanate complexes of uranium in multiple oxidation states: A combined structural, magnetic, spectroscopic, spectroelectrochemical, and theoretical study. *Inorg. Chem.* **2014**, *53*, 8624–8637. [CrossRef] [PubMed]

12. Crawford, M.-J.; Karaghiosoff, K.; Mayer, P. The homoleptic $U(NCSe)_8{}^{4-}$ anion in $(Pr_4N)_4U(NCSe)_8 \cdot 2CFCl_3$ and $Th(NCSe)_4(OP(NMe_2)_3)_4 \cdot 0.5CH_3CN \cdot 0.5H_2O$: First structurally characterized actinide isoselenocyanates. *Z. Anorg. Allg. Chem.* **2010**, *636*, 1903–1906. [CrossRef]

13. Rowland, C.E.; Kanatzidis, M.G.; Soderholm, L. Tetraalkylammonium uranyl isothiocyanates. *Inorg. Chem.* **2012**, *51*, 11798–11804. [CrossRef] [PubMed]

14. Straka, M.; Patzschke, M.; Pyykkö, P. Why are hexavalent uranium cyanides rare while U–F and U–O bonds are common and short? *Theor. Chem. Acc.* **2003**, *109*, 332–340. [CrossRef]

15. Surbella, R.G.; Cahill, C.L. The exploration of supramolecular interactions stemming from the $[UO_2(NCS)_4(H_2O)]^{2-}$ tecton and substituted pyridinium cations. *CrystEngComm* **2014**, *16*, 2352–2364. [CrossRef]

16. Carter, T.J.; Wilson, R.E. Coordination chemistry of homoleptic actinide(IV)–thiocyanate complexes. *Chem. Eur. J.* **2015**, *21*, 15575–15582. [CrossRef] [PubMed]

17. Fortier, S.; Hayton, T.W. Oxo ligand functionalization in the uranyl ion ($UO_2{}^{2+}$). *Coord. Chem. Rev.* **2010**, *254*, 197–214. [CrossRef]

18. Arunan, E.; Desiraju, G.R.; Klein, R.A.; Sadlej, J.; Scheiner, S.; Alkorta, I.; Clary, D.C.; Crabtree, R.H.; Dannenberg, J.J.; Hobza, P.; *et al.* Definition of the hydrogen bond (IUPAC Recommendations 2011). *J. Pure Appl. Chem.* **2011**, *83*, 1637–1641. [CrossRef]

19. Alvarez, S. A cartography of the van der Waals territories. *Dalton Trans.* **2013**, *42*, 8617–8636. [CrossRef] [PubMed]

20. Michalczyk, R.; Schmidt, J.G.; Moody, E.; Li, Z.; Wu, R.; Dunlap, R.B.; Odom, J.D.; Silks, L.A., III. Unusual C–H··· Se=C interactions in aldols of chiral *N*-acyl selones detected by gradient-selected [1]H–[77]Se HMQC NMR spectroscopy and X-ray crystallography. *Angew. Chem. Int. Ed.* **2000**, *39*, 3067–3070. [CrossRef]

21. Uhl, W.; Wegener, P.; Layh, M.; Hepp, A.; Wuerthwein, E.-U. Chalcogen capture by an Al/P-based frustrated lewis pair: Formation of Al-E-P bridges and intermolecular tellurium–tellurium interactions. *Organometallics* **2015**, *34*, 2455–2462. [CrossRef]

22. Kobayashi, K.; Masu, H.; Shuto, A.; Yamaguchi, K. Control of face-to-face π–π stacked packing arrangement of anthracene rings via chalcogen–chalcogen interaction:  9,10-Bis(methylchalcogeno)anthracenes. *Chem. Mater.* **2005**, *17*, 6666–6673. [CrossRef]

23. Bleiholder, C.; Gleiter, R.; Werz, D.B.; Koeppel, H. Theoretical investigations on heteronuclear chalcogen–chalcogen interactions: On the nature of weak bonds between chalcogen centers. *Inorg. Chem.* **2007**, *46*, 2249–2260. [CrossRef] [PubMed]

24. Bleiholder, C.; Werz, D.B.; Koeppel, H.; Gleiter, R. Theoretical investigations on chalcogen–chalcogen interactions: What makes these nonbonded interactions bonding? *J. Am. Chem. Soc.* **2006**, *128*, 2666–2674. [CrossRef] [PubMed]

25. Fazekas, Z.; Yamamura, T.; Tomiyasu, H. Deactivation and luminescence lifetimes of excited uranyl ion and its fluoro complexes. *J. Alloys Compd.* **1998**, *271–273*, 756–759. [CrossRef]

26. Sornein, M.-O.; Cannes, C.; le Naour, C.; Lagarde, G.; Simoni, E.; Berthet, J.-C. Uranyl complexation by chloride ions. Formation of a tetrachlorouranium(VI) complex in room temperature ionic liquids [Bmim][Tf$_2$N] and [MeBu$_3$N][Tf$_2$N]. *Inorg. Chem.* **2006**, *45*, 10419–10421. [CrossRef] [PubMed]

27. Hardwick, H.C.; Royal, D.S.; Helliwell, M.; Pope, S.J.A.; Ashton, L.; Goodacre, R.; Sharrad, C.A. Structural, spectroscopic and redox properties of uranyl complexes with a maleonitrile containing ligand. *Dalton Trans.* **2011**, *40*, 5939–5952. [CrossRef] [PubMed]

28. Clark, D.L.; Conradson, S.D.; Donohoe, R.J.; Keogh, D.W.; Morris, D.E.; Palmer, P.D.; Rogers, R.D.; Tait, C.D. Chemical speciation of the uranyl ion under highly alkaline conditions. Synthesis, structures, and oxo ligand exchange dynamics. *Inorg. Chem.* **1999**, *38*, 1456–1466. [CrossRef]

29. Yaprak, D.; Spielberg, E.T.; Bäcker, T.; Richter, M.; Mallick, B.; Klein, A.; Mudring, A.-V. A roadmap to uranium ionic liquids: Anti-crystal engineering. *Chem. Eur. J.* **2014**, *20*, 6482–6493. [CrossRef] [PubMed]

30. Ogura, T.; Takao, K.; Sasaki, K.; Arai, T.; Ikeda, Y. Spectroelectrochemical identification of a pentavalent uranyl tetrachloro complex in room-temperature ionic liquid. *Inorg. Chem.* **2011**, *50*, 10525–10527. [CrossRef] [PubMed]

31. Ikeda, Y.; Hiroe, K.; Asanuma, N.; Shirai, A. Electrochemical studies on uranyl(VI) chloride complexes in ionic liquid, 1-butyl-3-methylimidazolium chloride. *J. Nucl. Sci. Technol.* **2009**, *46*, 158–162. [CrossRef]

32. Takao, K.; Tsushima, S.; Ogura, T.; Tsubomura, T.; Ikeda, Y. Experimental and theoretical approaches to redox innocence of ligands in uranyl complexes: What is formal oxidation state of uranium in reductant of uranyl(VI)? *Inorg. Chem.* **2014**, *53*, 5772–5780. [CrossRef] [PubMed]

33. Wilkerson, M.P.; Burns, C.J.; Paine, R.T.; Scott, B.L. Synthesis and crystal structure of UO$_2$Cl$_2$(THF)$_3$: A simple preparation of an anhydrous uranyl reagent. *Inorg. Chem.* **1999**, *38*, 4156–4158. [CrossRef]

# 2

# Formation of Micro and Mesoporous Amorphous Silica-Based Materials from Single Source Precursors

**Mohd Nazri Mohd Sokri** [1,2], **Yusuke Daiko** [1,3], **Zineb Mouline** [1], **Sawao Honda** [1,3] and **Yuji Iwamoto** [1,3,*]

[1]   Department of Frontier Materials, Graduate School of Engineering, Nagoya Institute of Technology, Gokiso-cho, Showa-ku, Nagoya 466-8555, Japan; nazrisokri@utm.my (M.N.M.S.); daiko.yusuke@nitech.ac.jp (Y.D.); zinebmln@gmail.com (Z.M.); honda@nitech.ac.jp (S.H.)

[2]   Department of Energy Engineering and Advanced Membrane Technology Research Centre, Faculty of Chemical and Energy Engineering, Universiti Teknologi Malaysia (UTM), Johor Bahru 81310, Malaysia

[3]   Core Research for Evolutional Science and Technology (CREST), Japan Science and Technology Agency, Gokiso-cho, Showa-ku, Nagoya 466-8555, Japan

*   Correspondence: iwamoto.yuji@nitech.ac.jp

Academic Editors: Samuel Bernard, André Ayral and Philippe Miele

**Abstract:** Polysilazanes functionalized with alkoxy groups were designed and synthesized as single source precursors for fabrication of micro and mesoporous amorphous silica-based materials. The pyrolytic behaviors during the polymer to ceramic conversion were studied by the simultaneous thermogravimetry-mass spectrometry (TG-MS) analysis. The porosity of the resulting ceramics was characterized by the $N_2$ adsorption/desorption isotherm measurements. The Fourier transform infrared spectroscopy (FT-IR) and Raman spectroscopic analyses as well as elemental composition analysis were performed on the polymer-derived amorphous silica-based materials, and the role of the alkoxy group as a sacrificial template for the micro and mesopore formations was discussed from a viewpoint to establish novel micro and mesoporous structure controlling technologies through the polymer-derived ceramics (PDCs) route.

**Keywords:** perhydropolysilazane; alkyl alcohol; single source precursor; polymer-derived ceramics; micro and mesoporous silica

## 1. Introduction

The organometallic precursor route has received increased attention as an attractive ceramic processing method since it has inherent advantages over conventional powder processing methods such as purity control, compositional homogeneity in the final ceramic product and lower processing temperatures in the ceramics fabrication [1–3]. Moreover, this route provides alternatives towards the synthesis of advanced silicon-based non-oxide ceramics such as silicon nitride ($Si_3N_4$)-based ceramics, particularly starting from silicon-based polymers such as polysilazanes [4–6].

Recently, micro and mesoporous structure formations have been often discussed for the polymer-derived amorphous silicon nitride [7,8], silicon carbide [9–14], silicon carbonitride (Si–C–N) [15], silicon oxycarbide (Si–O–C) [16–18] and quaternary Si–M–C–N (M = B [19,20], Ni [21]). During the crosslinking and subsequent high-temperature pyrolysis under an inert atmosphere of polymer precursors, by-product gases such as $CH_4$, $NH_3$ and $H_2$ were detected [5,22], and the microporosity in the polymer-derived non-oxide amorphous ceramics could be assigned to the release of the small gaseous species formed *in-situ* [8,22].

Iwamoto *et al.* [23] reported an approach in synthesizing micro and mesoporous amorphous silica using organo-substituted polysilazane precursors, in which the organic moieties acted as a "sacrificial template" during polymer to ceramic conversion by heat treatment in air, thus allowing the micro and mesoporous structure formation.

In our recent studies, commercially available perhydropolysilazanes (PHPS) was used as a starting polymer. The PHPS contains many reactive Si–H and N–H groups which can react with chemical modifiers to yield novel ternary or quaternary Si-based non-oxide ceramics [24–27]. In addition, compared with polysiloxanes often used as a precursor for silica, PHPS can be easily oxidized at low temperatures to yield pure silica in high ceramic yield due to the weight gain as shown in the following equation [28–30].

$$[SiH_2 - NH]_n + nO_2 \rightarrow nSiO_2 + nNH_3 \uparrow \qquad (1)$$

In our previous studies, alkoxy groups were introduced to PHPS, and the resulting chemically modified PHPS was converted to microporous amorphous silica by oxidative crosslinking at 270 °C followed by heat treatment at 600 °C in air (route R1 in Figure 1) [31]. Microporous amorphous silica was also synthesized through the two step conversion route, (i) room temperature oxidation of the chemically modified PHPS to afford an air-stable alkoxy group-functionalized amorphous silica-based inorganic-organic hybrid; and (ii) subsequent heat treatment at 600 °C in air (route R2 in Figure 1) [32]. In continuation of our ongoing studies, we herein report the effect of the heat treatment atmosphere on porous structure development of the polymer-derived amorphous silica-based materials. The polymer/ceramics thermal conversion under an argon flow was performed on the polymer precursors and the polymer-derived hybrid silica (route R3 and R4 in Figure 1). The thermal conversion behaviors were *in-situ* analyzed by the simultaneous thermogravimetry-mass spectrometry (TG-MS) analysis. The relationships between the heat treatment atmosphere, gaseous species formed *in-situ* during the thermal conversion and the micro and mesoporosity in the resulting amorphous silica-based materials are discussed and compared with our previous results [32] in order to achieve a better understanding of the micro and mesoporous structure development in the polymer-derived amorphous-silica based materials.

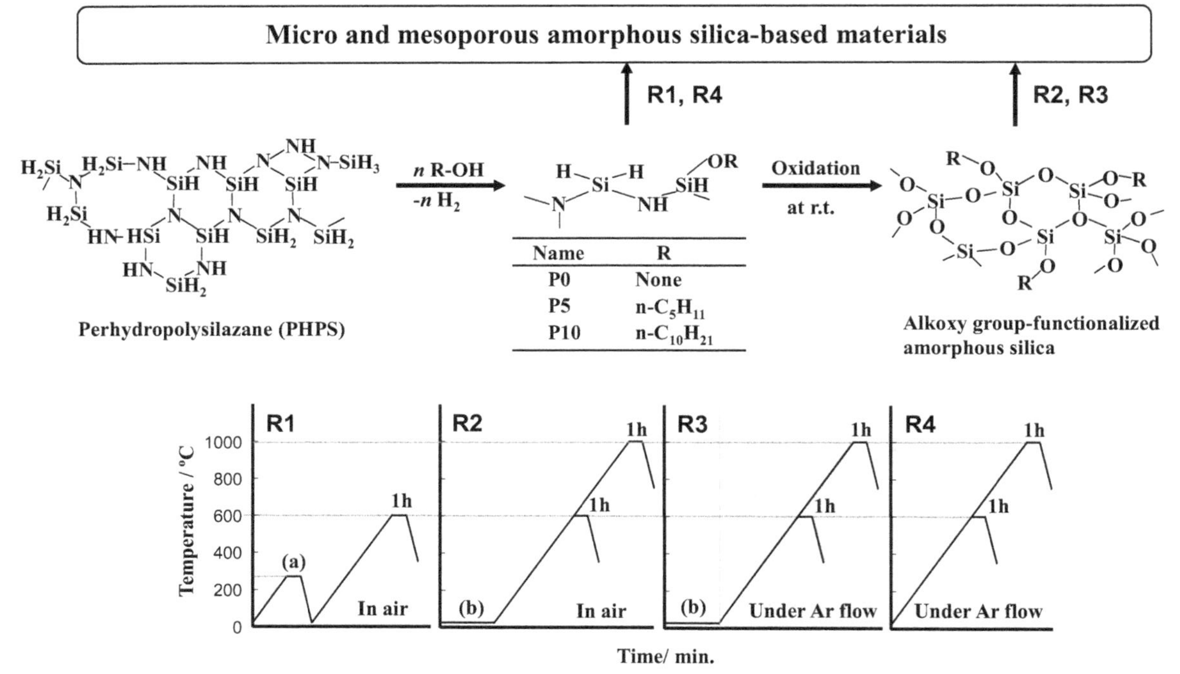

**Figure 1.** Synthesis of micro- and mesoporous amorphous silica-based materials through polymer derived ceramics (PDCs) routes investigated in this study.

## 2. Results and Discussion

### 2.1. Polymer Precursors and Alkoxy Group-Functionalized Amorphous Silica

The chemical structures of the polymer precursors and those of the alkoxy group-functionalized inorganic-organic hybrid silica were assessed based on their FT-IR spectra. As shown in Figure 2a, the as-received PHPS exhibited absorption bands at 3400 ($\nu$N–H), 2150 ($\nu$Si–H), 1180 ($\delta$N–H) and 840–1020 cm$^{-1}$ ($\delta$Si–N–Si) [4,31–33]. The PHPSs modified with the alcohols (Figure 2b,c) presented additional absorption bands at 2950–2850 ($\nu$C–H), 1450 ($\delta$CH$_3$) and 1090 cm$^{-1}$ ($\nu$Si–OR) [31,32,34]. As shown in Figure 2a, after the room temperature oxidation, the absorption bands corresponding to the as-received PHPS completely disappeared and new absorption bands appeared at 3400 (Si–OH) and 1090 cm$^{-1}$ (Si–O–Si) [32,34]. In addition to these bands, the chemically modified PHPSs exhibited C–H absorption bands in the vicinity of 2950 to 2850 cm$^{-1}$ (Figure 2b,c).

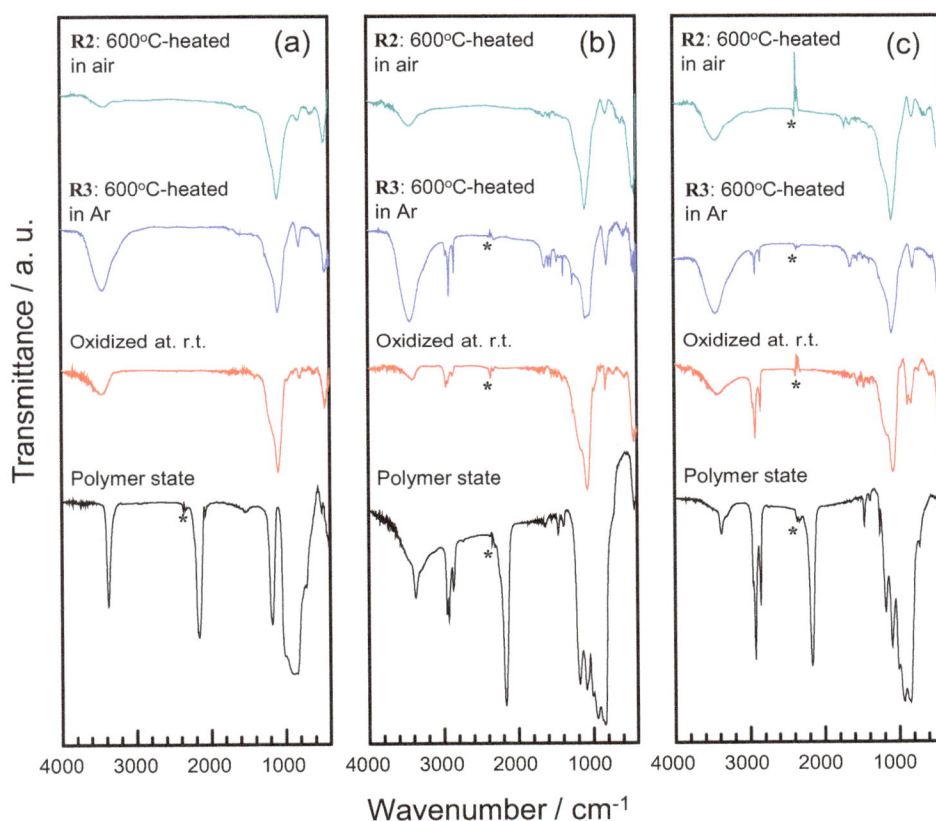

**Figure 2.** FT-IR spectra for polymer precursors and the precursor-derived samples. Polymer precursor: (a) **P0**; (b) **P5** and (c) **P10** (* indicating background, absorption due to CO$_2$).

The chemical structure of the oxidized samples was further studied by the solid state $^{29}$Si magic angle spinning (MAS) NMR spectroscopy. As a typical result, the spectrum for the sample derived from PHPS chemically modified with $n$-C$_{10}$H$_{21}$OH (**P10**) is shown in Figure 3. The characteristic peaks due to the structural units of as-received PHPS at around −35 ppm (SiHN$_3$/SiH$_2$N$_2$) and −50 ppm (SiH$_3$N) [35] completely disappeared, and the spectrum presented new broad peaks assigned to the units which composed amorphous silica at −78, −89, −85 to −94 and −112 ppm assigned to Q1($\equiv$SiO–SiX$_3$), Q2 (($\equiv$SiO)$_2$SiX$_2$), Q3 (($\equiv$SiO)$_3$SiX) and Q4 (($\equiv$SiO)$_4$Si)) (X = OH, OC$_{10}$H$_{21}$), respectively. These results indicate that the as-received PHPS and the chemically modified PHPSs were successfully converted to amorphous silica, and alkoxy group-functionalized inorganic-organic hybrid silica, respectively.

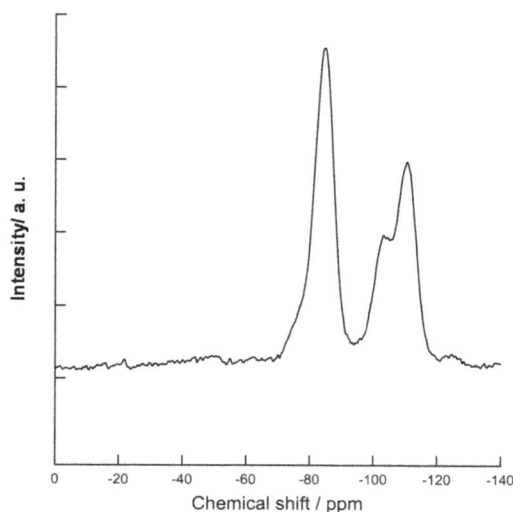

**Figure 3.** $^{29}$Si MAS NMR spectrum for alkoxy group-functionalized amorphous silica synthesized by room temperature oxidation of **P10** (PHPS chemically modified with $n$-$C_{10}H_{21}OH$).

## 2.2. Thermal Conversion to Amorphous Silica-Based Materials

The oxidative cross-linking at 270 °C and subsequent heat treatment at 600 °C in air of the polymer precursors (route R1) yielded carbon-free amorphous silica (Table 1, route R1, 600 °C). The alkoxy group-functionalized inorganic-organic hybrid silica could be also converted to carbon-free amorphous silica by the 600 °C-heat treatment in air (Table 1, route R2, 600 °C), and the C–H absorption bands completely disappeared in their FT-IR spectra (R2 in Figure 2b,c). On the other hand, the 600 °C-heat treatment in argon (Ar) resulted in the detection of C–H absorption bands (R3 in Figure 2b,c). The amount of carbon remained in the **P5** and **P10**-derived samples were 6.2 and 9.5 wt %, respectively (Table 1, route R3, 600 °C).

**Table 1.** Chemical composition of polymer-derived amorphous silica-based materials synthesized in this study.

| Polymer | Conversion Route | Flow Gas | Temp./°C | Composition/wt % | | | | | Empirical Ratio |
|---|---|---|---|---|---|---|---|---|---|
| | | | | Si | C | O | N | H | |
| **P0** | R1 | air | 600 | 68.1 | 0.0 | 30.2 | 0.7 | 0.0 | $Si_{1.0}C_{0.00}O_{0.78}N_{0.02}H_{0.00}$ |
| | R2 | air | 600 | 59.1 | 0.0 | 39.3 | 0.6 | 0.0 | $Si_{1.0}C_{0.00}O_{1.17}N_{0.02}H_{0.00}$ |
| | R3 | Ar | 600 | 54.5 | 0.0 | 42.4 | 1.9 | 0.2 | $Si_{1.0}C_{0.00}O_{1.37}N_{0.07}H_{0.10}$ |
| | R4 | Ar | 600 | 65.8 | 0.9 | 4.8 | 26.9 | 0.6 | $Si_{1.0}C_{0.03}O_{0.13}N_{0.82}H_{0.25}$ |
| | R3 | Ar | 1000 | 58.1 | 0.0 | 40.2 | 0.5 | 0.2 | $Si_{1.0}C_{0.00}O_{1.21}N_{0.02}H_{0.10}$ |
| | R4 | Ar | 1000 | 63.4 | 0.9 | 5.0 | 29.1 | 0.6 | $Si_{1.0}C_{0.03}O_{0.14}N_{0.92}H_{0.26}$ |
| **P5** | R1 | air | 600 | 64.7 | 0.0 | 34.2 | 0.1 | 0.0 | $Si_{1.0}C_{0.00}O_{0.93}N_{0.003}H_{0.00}$ |
| | R2 | air | 600 | 60.7 | 0.0 | 38.2 | 0.1 | 0.0 | $Si_{1.0}C_{0.00}O_{1.10}N_{0.003}H_{0.00}$ |
| | R3 | Ar | 600 | 55.6 | 6.2 | 36.2 | 0.6 | 0.4 | $Si_{1.0}C_{0.26}O_{1.13}N_{0.02}H_{0.20}$ |
| | R3 | Ar | 1000 | 52.7 | 4.1 | 41.8 | 0.3 | 0.1 | $Si_{1.0}C_{0.18}O_{1.39}N_{0.01}H_{0.05}$ |
| **P10** | R1 | air | 600 | 64.7 | 0.0 | 34.3 | 0.0 | 0.0 | $Si_{1.0}C_{0.00}O_{0.93}N_{0.00}H_{0.00}$ |
| | R2 | air | 600 | 63.1 | 0.0 | 35.9 | 0.0 | 0.0 | $Si_{1.0}C_{0.00}O_{1.00}N_{0.00}H_{0.00}$ |
| | R3 | Ar | 600 | 51.9 | 9.5 | 36.6 | 0.1 | 0.9 | $Si_{1.0}C_{0.43}O_{1.24}N_{0.004}H_{0.48}$ |
| | R4 | Ar | 600 | 57.7 | 6.7 | 30.2 | 3.0 | 1.4 | $Si_{1.0}C_{0.27}O_{0.92}N_{0.10}H_{0.68}$ |
| | R2 | air | 1000 | 63.6 | 0.0 | 35.4 | 0.0 | 0.0 | $Si_{1.0}C_{0.00}O_{0.98}N_{0.00}H_{0.00}$ |
| | R3 | Ar | 1000 | 49.4 | 6.1 | 42.3 | 0.0 | 1.2 | $Si_{1.0}C_{0.29}O_{1.50}N_{0.00}H_{0.68}$ |
| | R4 | Ar | 1000 | 59.6 | 8.4 | 14.9 | 14.4 | 1.7 | $Si_{1.0}C_{0.33}O_{0.44}N_{0.48}H_{0.79}$ |

## 2.3. TG-MS Analysis

To study the effect of atmosphere on the pyrolytic behaviors more extensively, TG-MS analysis was performed on the selected samples. The results were summarized and shown in Figures 4 and 5. The PHPS-derived amorphous silica synthesized by the room temperature oxidation showed a slight weight loss at 150 to 400 °C (Figure 4(Aa)). The MS spectrum measured at 200 °C was composed of $m/z$ ratios at 106, 91, 77, 63, 51 and 39, which was identical to that reported for xylene (Figure 4(Ab)) [36,37]. Since the molecular ion ($m/z = 106$) and the tropylium ion ($m/z = 91$) were detected at 400 °C and below (Figure 4(Ac)), the observed weight loss is mainly due to the residual xylene [32]. When the PHPS was directly heat-treated under the inert atmosphere, the weight loss up to 400 °C increased to approximately 20 wt % (Figure 4(Ba)). A typical MS spectrum measured at 200 to 250 °C is shown in Figure 4(Bb). The $m/z$ ratios at 76 and 30 were assigned to $SiH_2-NH-SiH_3^+$ and $SiH_2^+$, respectively. The smaller species, $m/z$ ratios at 28 and 17 were assigned to $N_2^+$ and $NH_3^+$, respectively. These fragment ions could be formed in-situ by the thermal decomposition of the silazane rings. The evolutions of these gaseous species were found to be completed at around 400 °C (Figure 4(Bc)).

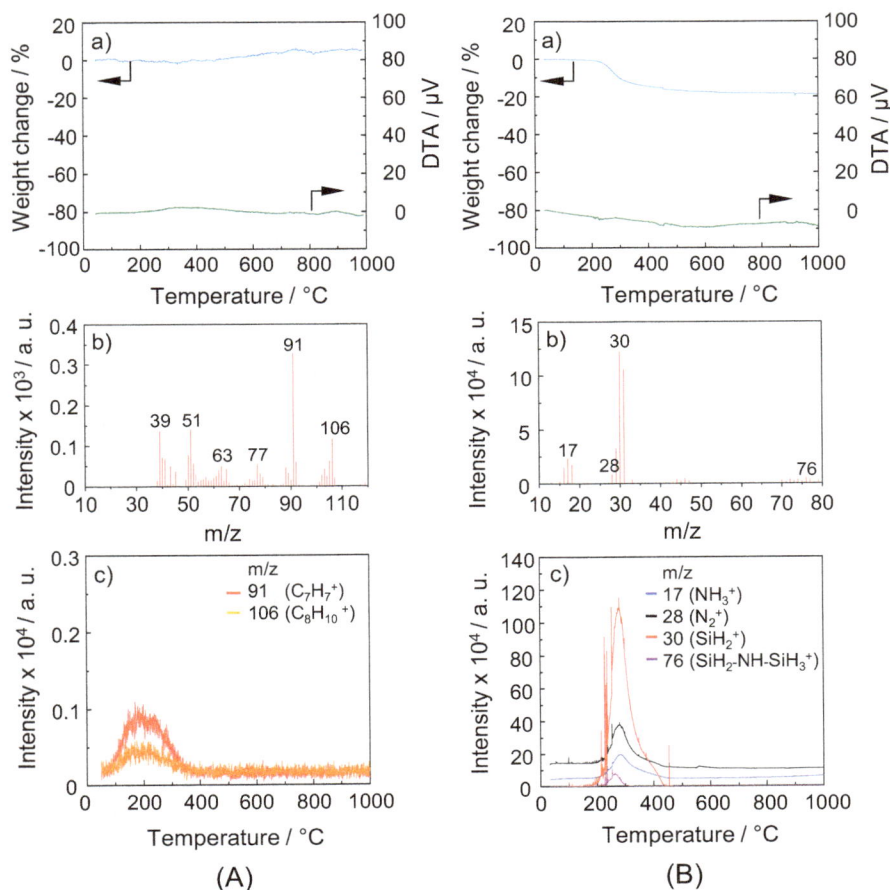

**Figure 4.** Thermal behavior of (**A**) PHPS (**P0**)-derived amorphous silica under oxidative atmosphere (route R2); and (**B**) as-received PHPS under an inert atmosphere (route R4). (**a**) TG-DTA; (**b**) monitoring of gaseous species by mass spectrometry at 200 °C and (**c**) continuous in-situ monitoring of gaseous species by mass spectrometry.

As shown in Figure 5(Aa), under the oxidative atmosphere (route R2), the **P10**-derived hybrid silica having $n$-$C_{10}H_{21}O$ groups exhibited a weight loss of approximately 50% at 150 to 600 °C, and the differential thermal analysis (DTA) resulted in the detection of a dominant exothermic peak centered at 320 °C. The MS spectrum measured at 320 °C is shown in Figure 5(Ab). The sequential peaks 14 mass units apart at $m/z = 71$, 57 and 43 can be assigned to the fragment ions derived from hydrocarbons formed *in situ* by the typical $\alpha$-cleavage in the $n$-$C_{10}H_{21}O$ group, followed by the sequential C–C bond cleavage to release methylene units [36,38]. The $m/z$ ratios at $m/z = 70$ ($C_5H_{10}^{\bullet+}$) and 55 ($C_4H_7^+$) could be derived from traces of unreacted $n$-$C_{10}H_{21}OH$ [36,39]. The smaller species, $m/z$ ratios at 44 and 18 were assigned to $CO_2^+$ and $H_2O^+$, respectively. By monitoring the fragment ions with $m/z$ ratios at 85, 71, 57 and 43, it was confirmed that the decomposition of $n$-$C_{10}H_{21}O$ group mainly occurred at 250 to 500 °C (Figure 5(Ac)). Simultaneously, combustion of the hydrocarbon species started at around 250 °C, and the evolution of $CO_2$ and $H_2O$ continued to approximately 650 °C (Figure 5(Ad)).

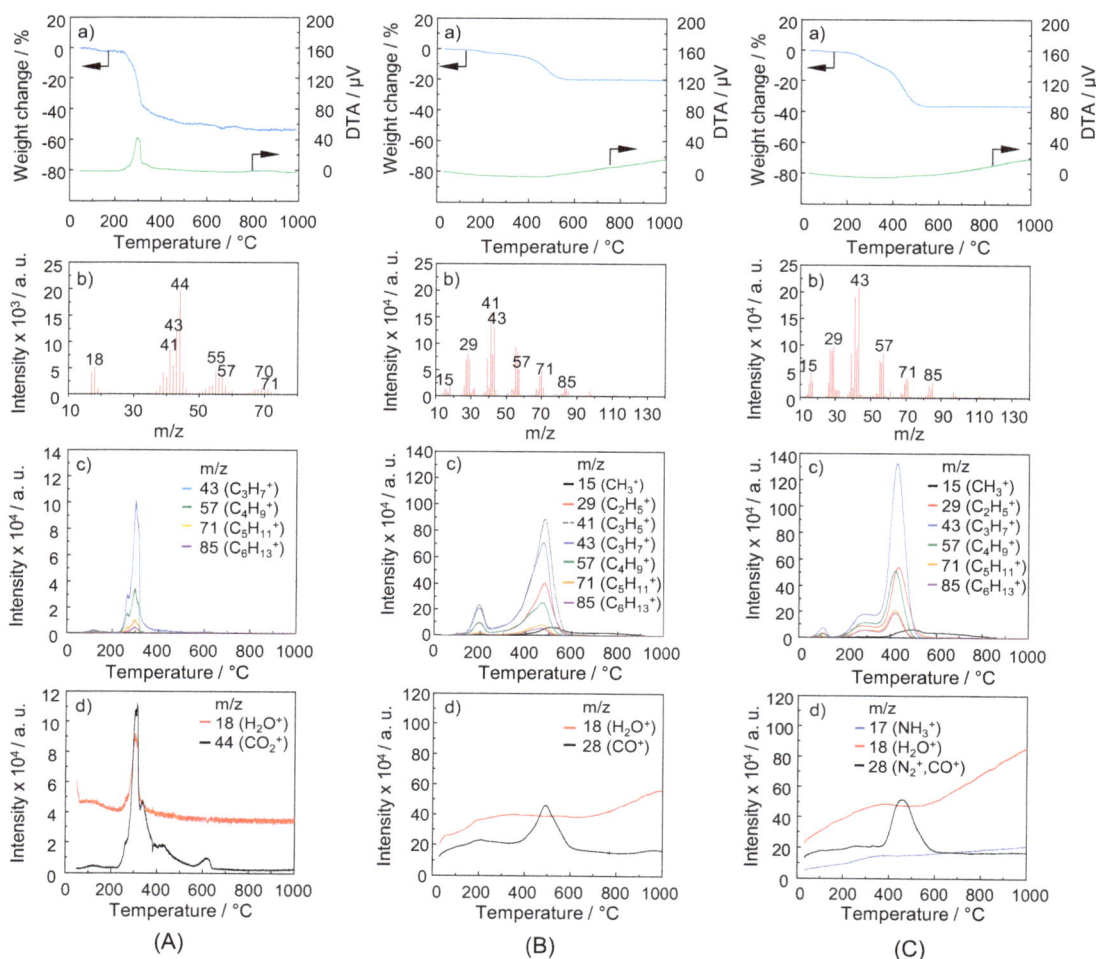

**Figure 5.** Thermal behavior of $n$-$C_{10}H_{21}O$ group-functionalized hybrid silica under (**A**) oxidative atmosphere (route R2); (**B**) inert atmosphere (route R3) and (**C**) that of PHPS functionalized with $n$-$C_{10}H_{21}O$ groups (**P10**) under inert atmosphere (route R4). (a) TG-DTA; (b) monitoring of gaseous species by mass spectrometry at 320 °C; (c,d) continuous *in situ* monitoring of gaseous species by mass spectrometry.

Under the inert atmosphere (route R3) (Figure 5(Ba)), the weight loss up to 600 °C of the **P10**-derived hybrid silica was decreased to 20%. The MS spectrum measured at 320 °C was composed of the above mentioned species derived from $n$-$C_{10}H_{21}O$ groups and those with the $m/z$ ratios of 41 ($C_3H_5^+$), 29 ($C_2H_5^+$) and 15 ($CH_3^+$) (Figure 5(Bb)). The gaseous species formed *in situ* were detected mainly at 100 to 600 °C (Figure 5(Bc)). When the **P10** was directly heat-treated under an inert atmosphere (route R4), the weight loss up to 600 °C was rather suppressed and measured to be 38%. The thermal decomposition behavior of the **P10** was similar to that characterized for the

$n$-$C_{10}H_{21}O$-functionalized hybrid silica, and it was found that the relative volume fraction of the *in situ* formed $C_3H_7^+$ ($m/z = 43$) remarkably increased (Figure 5C).

## 2.4. $N_2$ Adsorption/Desorption Isotherm and Porosities Evaluated for the Polymer-Derived Amorphous Silica-Based Materials

The textural properties of the polymer-derived amorphous silica-based materials were studied by $N_2$ physisorption at −196 °C. The adsorption/desorption isotherms are presented in Figure 6. Each sample exhibited an $N_2$ uptake above $p/p_o = 0.9$, which was thought to be related to the porosity generated by agglomeration of the powdered sample [31].

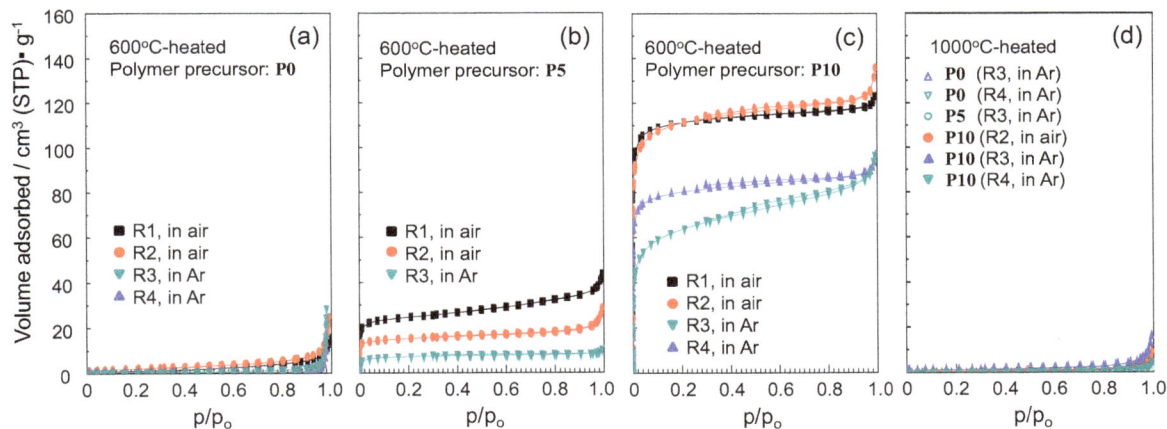

**Figure 6.** $N_2$ sorption isotherms of polymer-derived samples.

Regardless of the thermal conversion condition investigated in this study, the PHPS (**P0**)-derived amorphous silica generated a type III isotherm according to the International Union of Pure and Applied Chemistry (IUPAC) classification method [40,41] (Figure 6a), and the **P0**-derived samples were characterized as non-porous (Figure 7a). Consequently, it was clarified that the evolution below 400 °C of the relatively large gaseous species (shown in Figure 4) did not contribute to the porous structure formation.

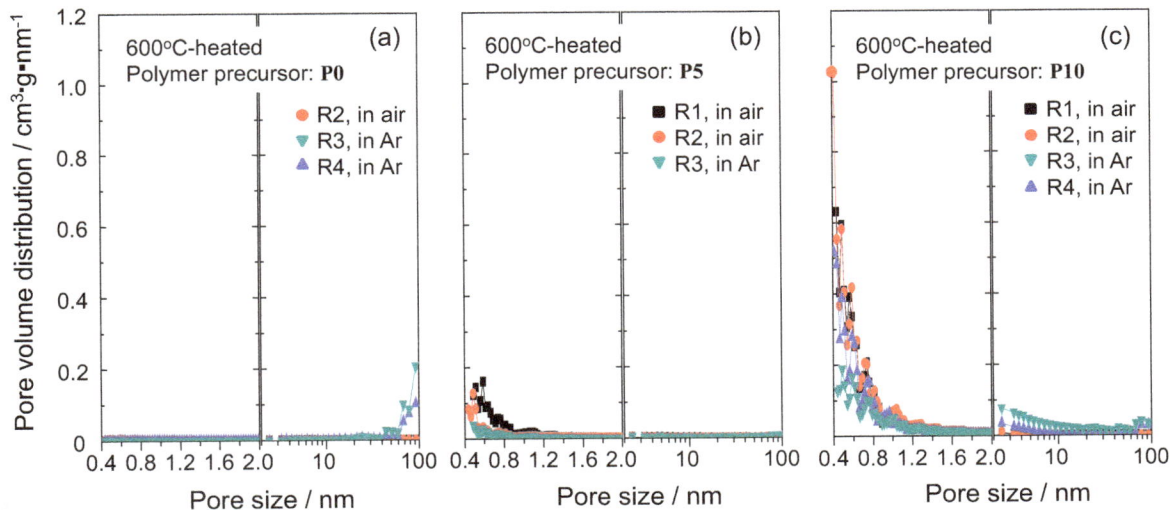

**Figure 7.** Pore size distribution of the polymer-derived samples synthesized by heat treatment at 600 °C for 1 h.

The isotherm of the **P5**-derived sample synthesized through the route R1 exhibited a type I without distinct hysteresis loops. The slight $N_2$ uptake at the relative pressure lower than $p/p_o = 0.2$ was related to the micropore filling (Figure 6b), and the micropore sizes evaluated by using $N_2$ as probe molecule were in the range of 0.43 to 1.2 nm [31] (Figure 7b).

The isotherm of the **P10**-derived sample synthesized through the route R1 also presented a similar type (Figure 6c). The $N_2$ uptake below $p/p_o = 0.2$ was apparently increased by increasing the carbon number in the alkoxy group from 5 to 10, and the resulting micropore size distribution plot peaked at 0.43 nm and extended to approximately 1.6 nm in size [31] (Figure 7c).

On the other hand, when the heat treatment of the **P5**-derived hybrid silica was performed under flowing Ar, the $N_2$ uptake below $p/p_o = 0.2$ apparently decreased (R3 in Figure 6b). The micropore sizes evaluated for the resulting **P5**-derived sample were below 1.0 nm, and thus almost unchanged, while the micropore volume decreased (R3 in Figure 7b). In addition to the similar microporosity change (R3 in Figures 6c and 7c, the **P10**-derived sample exhibited an increase in the volume of mesopores having a size range of approximately 2 to 20 nm (R3 in Figure 7c).

The heat treatment under an Ar flow of the **P10** also resulted in the micro and mesoporosity changes similar to those observed for the **P10**-derived hybrid silica (R4 in Figures 6c and 7c. However, regardless of the atmospheric condition, the porosity in the **P5** and **P10**-derived samples was lost during further heat treatment from 600 to 1000 °C (Figure 6d).

Figure 8a presents the relations between the number of the carbon in the alkoxy group introduced to PHPS, the conversion route and the resulting micropore volume of the heat-treated samples. In our previous studies on the alkoxy group-functionalized PHPS [31] and hybrid silica [32], the alkoxy group with C3 hydrocarbon did not contribute to the porous structure development [31], while those having a carbon chain longer than or equal to C5 hydrocarbon were found to be effective as a sacrificial template for generating microporosity under the thermal conversion in air [32]. Moreover, the micropore volume increased with increasing the total volume of the released gaseous species having a kinetic diameter below 0.45 nm. As a result, the micropore volume of the **P10**-derived sample reached 0.17 $cm^3$/g (R2 in Figure 8a), and this value was compatible with that of the amorphous silica fabricated without room temperature oxidation (route R1 in Figure 8a, 0.173 $cm^3$/g). It should also be noted that the specific surface area (SSA) of this sample was 321.2 $m^2$/g, and almost the same as that achieved by the sample fabricated through the route R1 (328 $m^2$/g) [32].

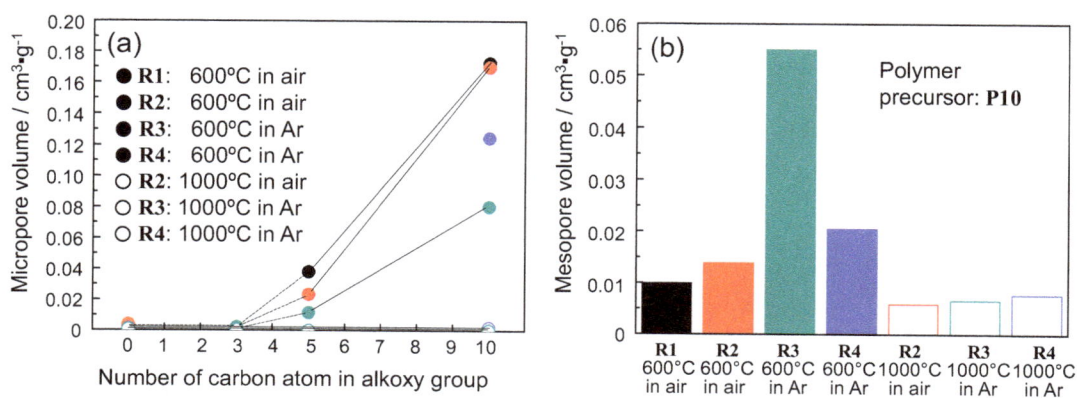

**Figure 8.** Porosity of polymer-derived amorphous silica-based materials. (**a**) micropore volume of the synthesized samples and (**b**) mesopore volume of **P10**-derived samples.

Under the present thermal conversion in Ar, the gaseous species formed *in situ* were C4–C1 hydrocarbons and CO with a kinetic diameter below 0.45 nm [42], which contributed to the microporosity formation in the **P5** and **P10**-derived samples. However, the total volume of the released gaseous species essentially decreased due to the lack of $CO_2$ evolution, which led to a decrease in the micropore volume and a considerable amount of carbon remaining in the resulting amorphous silica-based material. Moreover, it was found that the decrease in the micropore volume became much more pronounced when the thermal conversion of **P10** was performed after the room temperature oxidation (R3 at the number of carbon = 10 in Figure 8a). Simultaneously, the mesopore volume in this sample turned out to be much higher (0.055 $cm^3$/g) than those in other samples (approximately 0.01–0.02 $cm^3$/g) (Figure 8b). One possible reason for the decrease in the micropore volume associated with the increase in mesopore volume may be attributed to the pore growth from micro- to mesopore size range within the amorphous silica matrix immediately densified at the early stage of the heat treatment approximately below 400 °C.

Figure 9 presents the Raman spectra for the **P10**-derived amorphous silica synthesized by heat treatment in air (route R2), and their assignments are listed in Table 2. The spectra for the 600 °C heat-treated sample exhibited broad bands related to the O–Si–O and Si–O–Si asymmetrical vibrations at 800 and 978 $cm^{-1}$, respectively [43]. The weak broad bands above 1000 $cm^{-1}$ were thought to be attributed to the $SiO_4$ asymmetrical vibration [44–46]. On the other hand, the lower wavenumber bands below 700 $cm^{-1}$ were attributed to the Si–O–Si symmetric vibrations originated from structural defects such as 3-, 4- and 6-membered Si–O–Si rings within the amorphous silica matrix [47–49].

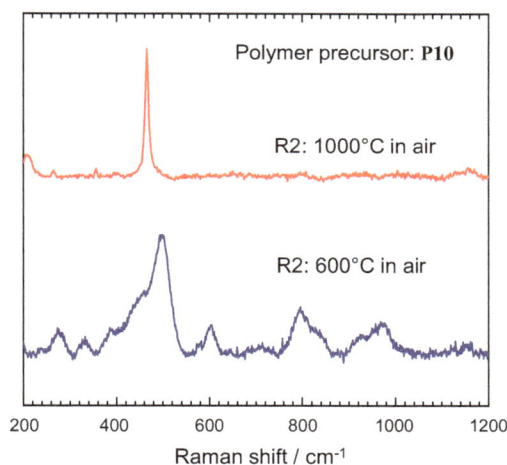

**Figure 9.** Raman spectra for **P10**-derived amorphous silica synthesized by heat treatment in air (route R2).

**Table 2.** Assignment for the Raman spectra shown in Figure 9.

| ν, (Si–O–Si)/$cm^{-1}$ | Number of $SiO_4$ in Rings | Notes | Reference |
|---|---|---|---|
| 381 ($ν_s$) | 6 | Observe in α-Carnegieite | Matson et al. [48] |
| 463 ($ν_s$) | 6 | Observe in moganite | Kingma and Hemley [49] |
| 465 ($ν_s$) | 6 | Observe in α-quartz | Kingma and Hemley [49] |
| 498 ($ν_s$) | 4 + 6 | Observe in Leucite | Matson et al. [48] |
| 606 ($ν_s$) | 3 | - | Uchino et al. [48] |
| 800 ($ν_{as}$) | - | O–Si–O vibration (asymmetric stretching mode) | Li et al. [43] |
| 978 ($ν_{as}$) | - | Si–O–Si bond due to defect by surface silanol group (asymmetric stretching mode) | Li et al. [43] |
| 1000–1200 ($ν_{as}$) | - | Asymmetric stretching $SiO_4$ vibrations | Apopei et al. [44] Makreski et al. [45] Ventura et al. [46] |

On the other hand, after the high-temperature heat treatment up to 1000 °C, the spectrum presented one distinct band at 465 $cm^{-1}$ due to the six-membered Si–O–Si ring [49]. Since the intensity of the characteristic bands related to the Si–O–Si and O–Si–O vibrations remarkably decreased, the 1000 °C heat-treated sample was thought to be composed of a highly rigid silica network.

It has been suggested that PHPS contains 2-, 3- and 4-membered Si–N–Si ring clusters as microporous units [5,22]. Upon oxidation at room temperature, the ring clusters were readily oxidized to form Si–O–Si linkages, which could serve as nucleation sites for subsequent growth of micropores. Finally, after the 1000 °C-heat treatment, the six-membered Si–O–Si rings remained to form the intrinsic microporosity in amorphous silica. The estimated diameter of the six-membered ring is approximately 0.3 nm [50,51], which is too small for the $N_2$ probe molecule to access (0.364 nm) [42]. Consequently, the resulting 1000 °C heat-treated sample was characterized as non-porous. On the other hand, the walls of micro- and mesopores formed by the thermal decomposition of the alkoxy groups are thought to be composed of disordered silica having considerable amount of dangling bonds, and the thermal stability of the polymer-derived micro- and mesoporous amorphous silica-based materials synthesized in this study is estimated to be around 600 °C.

## 3. Experimental Procedures

### 3.1. Precursor Synthesis

Commercially available perhydropolysilazane (PHPS, Type NN110, 20% xylene solution, AZ Electronic Materials, Tokyo, Japan) was used as the starting polymer. n-Pentanol ($n$-$C_5H_{11}OH$) and n-decanol ($n$-$C_{10}H_{21}OH$) were used for the chemical modification of the PHPS. According to the published procedures [32], the reaction between the as-received PHPS and each alcohol was carried out under a dry Ar atmosphere using Schlenk techniques. A PHPS (Si basis)/ROH molar ratio of 4:1 was applied in all cases. In each synthesis, the alcohol was added dropwise to a xylene solution of as-received PHPS with magnetic stirring at room temperature, followed by the addition of toluene to decrease the PHPS concentration from 20 to 1 wt %. After the addition was complete, the mixture was refluxed for 1 h under an Ar flow and cooled to room temperature. The xylene and toluene were removed from the reaction mixture under vacuum to afford the alcohol adduct as a viscous liquid.

### 3.2. Conversion of Polymer Precursor to Amorphous Silica-Based Materials

In this study, the polymer precursor was converted to amorphous silica-based material through the following four different routes shown in Figure 1.

R1: Oxidative crosslinking and subsequent heat treatment in air [31]

Polymer precursor was cured by heating at 270 °C in air to promote oxidative crosslinking with a heating rate of 100 °C/h and dwell time of 1 h in an alumina tube furnace. After cooling down to room temperature, the crosslinked polymer was obtained as white solid. The crosslinked polymer precursor was ground to a fine powder using a mortar and pestle, then heat-treated in an alumina tube furnace under an air flow by heating from room temperature to 600 °C in 6 h, maintaining the temperature at 600 °C for an additional 1 h, and finally cooling down to room temperature to give amorphous silica as white powders.

R2: Room temperature oxidation followed by heat treatment in air [32]

Polymer precursor was converted to amorphous silica-based hybrid by exposing the polymer precursor to vapour from aqueous ammonia ($NH_3$) according to the procedure reported by Kubo et al. [28,29]. In this process, 15 mL of aqueous $NH_3$ was placed in a 200 mL beaker with a slightly open lid and the polymer precursor (ca. 600 mg) was suspended over the aqueous $NH_3$ until the silica-based hybrid formed as white powder. The resulting amorphous silica-based hybrid was ground to a fine powder using a mortar and pestle, then heat-treated in an alumina tube furnace under an air flow up to 600 °C for a period of time of 6 h. Then the temperature was maintained at 600 °C for an additional hour, and finally cooled down to room temperature to afford amorphous silica as white powders.

R3: Room temperature oxidation followed by heat treatment in Ar

The powdered sample of polymer-derived amorphous silica-based hybrid was heat-treated in an alumina tube furnace under an Ar flow up to 600 °C at a heating rate of 10 °C/min. Then the temperature was maintained at 600 °C for an additional hour, and finally cooled down to room temperature.

R4: Heat treatment in Ar

Polymer precursor was heat-treated in an alumina tube furnace under Ar flow up to 600 °C for a period of time of 1 h (heating rate of 10 °C/min). Then the temperature was maintained at 600 °C for an additional 1 h, and finally cooled down to room temperature.

As shown in Figure 1, 1000 °C-heat treatment was also performed for the polymer/ceramics conversion (R2, R3 and R4).

### 3.3. Characterization

FT-IR spectra of the polymers and the polymer-derived amorphous silica-based materials were recorded using KBr pellets over the range of 4000 to 400 cm$^{-1}$ (FT/IR-4200 IF, Jasco, Tokyo, Japan). Solid-state $^{29}$Si magic angle spinning (MAS) nuclear magnetic resonance (NMR) spectra of the amorphous silica-based hybrids were measured at 79.45 MHz, 6.0 μs of 90° pulse length, 120 s of decay time between pulses, and spinning rate of 3500 Hz (Unity 400 plus, Varian, Palo Alto, CA, USA). Raman spectra of the amorphous silica-based materials were measured using 532 or 785 nm solid laser (Renishaw, Wotton-under-Edge, UK). The laser power of 5% was applied for the measurements.

The thermal behaviors up to 1000 °C were studied by simultaneous TG and MS analyses (Model TG/DTA6300, Hitachi High Technologies Ltd., Tokyo, Japan/Model JMS-Q1050GC, JEOL, Tokyo, Japan). The measurements were performed under flowing mixed gas of helium (He) and oxygen ($O_2$) (He:$O_2$ = 4:1, 100 mL/min) or He with a heating rate of 20 °C/min. To avoid the presence of dominant fragment peaks at around $m/z$ = 32 ($O_2^+$) related to the presence of $O_2$, the $m/z$ ratios in the range of 20 to 35 were excluded from the MS spectra recorded under the flowing mixed gas of He and $O_2$.

Elemental analyses were performed on the heat-treated samples for oxygen, nitrogen and hydrogen (inert-gas fusion method, Model EMGA-930, HORIBA, Ltd., Kyoto, Japan), and carbon (non-dispersive infrared method, Model CS744, LECO Co., St Joseph, MI, USA). The silicon content in the samples was calculated as the difference of the sum of the measured C, N, O and H content to 100 wt % [35].

The pore size distribution for the heat-treated samples was determined using a nitrogen ($N_2$) sorption technique with the relative pressure of the $N_2$ gas ranging from 0 to 0.99 (Belsorp Max, BEL Japan Inc., Osaka, Japan). The micropores ($r_{pore}$ < 2.0 nm) and mesopores (2.0 nm ≤ $r_{pore}$ < 50 nm) of the polymer-derived amorphous silica-based materials were characterized by the SF [52] and BJH [53] methods, respectively.

## 4. Conclusions

In this study, micro and mesoporous amorphous silica-based materials were successfully synthesized from single source precursors, PHPS functionalized with alkoxy groups (OR, R = $n$-$C_5H_{11}$, $n$-$C_{10}H_{21}$).

The textural properties evaluated for the polymer-derived amorphous silica-based materials suggested the potential of the polymer precursors employed in this study to develop a novel microporous structure controlling technology by manipulating the carbon chain in the alkoxy groups introduced to PHPS. It is also interesting to note that the PDC routes investigated in this study show some potential to synthesize mesoporous amorphous silica-based materials by the two-step conversion process, the room temperature oxidation followed by the thermal treatment under an inert atmosphere. These porous structure control technologies are expected to be useful to synthesize amorphous silica-based materials for energy application such as a highly efficient catalyst support and a gas separation membrane having a hyper-organized porous structure both in the micro- and mesopore size ranges.

**Acknowledgments:** This work was supported by CREST (Core Research for Evolutional Science and Technology) of Japan Science and Technology Corporation (JST).

**Author Contributions:** Mohd Nazri Mohd Sokri performed experiments and wrote the paper; Yusuke Daiko contributed NMR and FT-IR spectroscopic analyses; Sawao Honda contributed TG-MS analysis; Zineb Mouline contributed nitrogen adsorption/desorption measurements; Yuji Iwamoto conceived and designed experiments and wrote the paper.

## References

1.   Rice, R.W. Ceramics from polymer pyrolysis, opportunities and needs. A materials perspective. *Am. Ceram. Soc. Bull.* **1983**, *62*, 889–892.

2.   Wynne, K.J.; Rice, R.W. Ceramics via polymer pyrolysis. *Annu. Rev. Mater. Sci.* **1984**, *14*, 297–334. [CrossRef]

3.   Riedel, R.; Dressler, W. Chemical formation of ceramics. *Ceram. Int.* **1996**, *22*, 233–239. [CrossRef]

4.   Seyferth, D.; Strohmann, C.; Dando, N.R.; Perrotta, A. Poly(ureidosilazanes): Preceramic polymeric precursors for silicon carbonitride and silicon nitride. Synthesis, characterization and pyrolytic conversion to $Si_3N_4$/SiC ceramics. *J. Chem. Mater.* **1995**, *7*, 2058–2066. [CrossRef]

5.   Kroke, E.; Li, Y-L.; Konetschny, C.; Lecomte, E.; Fasel, C.; Riedel, R. Silazane derived ceramics and related materials. *Mater. Sci. Eng. R* **2000**, *26*, 97–199. [CrossRef]

6.   Funayama, O.; Arai, M.; Tashiro, Y.; Aoki, H.; Suzuki, T.; Tamura, K.; Kaya, H.; Nishii, H.; Isoda, T. Tensile strength of silicon nitride fibers produced from perhydropolysilazane. *J. Ceram. Soc. Jpn.* **1990**, *98*, 104–107. [CrossRef]

7.   Miyajima, K.; Eda, T.; Ohta, H.; Ando, Y.; Nagaya, S.; Ohba, T.; Iwamoto, Y. Development of Si–N based hydrogen separation membrane. *Ceram. Trans.* **2010**, *213*, 87–94.

8.   Schitco, C.; Bazarjani, M.S.; Riedel, R.; Gurlo, A. $NH_3$-assisted synthesis of microporous silicon oxycarbonitride ceramics from preceramic polymers: A combined $N_2$ and $CO_2$ adsorption and small angle X-ray scattering study. *J. Mater. Chem. A* **2015**, *3*, 805–818. [CrossRef]

9.   Kusakabe, K.; Li, Z.Y.; Maeda, H.; Morooka, S. Preparation of supported composite membrane by pyrolysis of polycarbosilane for gas separation at high temperature. *J. Membr. Sci.* **1995**, *103*, 175–180. [CrossRef]

10.  Li, Z.Y.; Kusakabe, K.; Morooka, S. Preparation of thermostable amorphous Si–C–O membrane and its application to gas separation at elevated temperature. *J. Membr. Sci.* **1996**, *118*, 159–168.

11.  Nagano, T.; Sato, K.; Saito, T.; Iwamoto, Y. Gas permeation properties of amorphous SiC membranes synthesized from polycarbosilane without oxygen-curing process. *J. Ceram. Soc. Jpn.* **2006**, *114*, 533–538. [CrossRef]

12.  Suda, H.; Yamauchi, H.; Uchimaru, Y.; Fujiwara, I.; Haraya, K. Preparation and gas permeation properties of silicon carbide-based inorganic membranes for hydrogen separation. *Desalination* **2006**, *193*, 252–255. [CrossRef]

13.  Mourhatch, R.; Tsotsis, T.T.; Sahimi, M. Network model for the evolution of the pore structure of silicon-carbide membranes during their fabrication. *J. Membr. Sci.* **2010**, *356*, 138–146. [CrossRef]

14.  Takeyama, A.; Sugimoto, M.; Yoshikawa, M. Gas permeation property of SiC membrane using curing of polymer precursor film by electron beam irradiation in helium atmosphere. *Mater. Trans.* **2011**, *52*, 1276–1280. [CrossRef]

15.  Völger, K.W.; Hauser, R.; Kroke, E.; Riedel, R.; Ikuhara, Y.H.; Iwamoto, Y. Synthesis and characterization of novel non-oxide sol-gel derived mesoporous amorphous Si–C–N membranes. *J. Ceram. Soc. Jpn.* **2006**, *114*, 567–570. [CrossRef]

16.  Soraru, G.D.; Liu, Q.; Interrante, L.V.; Apple, T. Role of precursor molecular structure on the microstructure and high temperature stability of silicon oxycarbide glasses derived from methylene-bridged polycarbosilanes. *Chem. Mater.* **1998**, *10*, 4047–4054. [CrossRef]

17.  Liu, Q.; Shi, W.; Babonneau, F.; Interrante, L.V. Synthesis of polycarbosilane/siloxane hybrid polymers and their pyrolytic conversion to silicon oxycarbide ceramics. *Chem. Mater.* **1997**, *9*, 2434–2441. [CrossRef]

18.  Lee, L.; Tsai, D.S. A hydrogen-permselective silicon oxycarbide membrane derived from polydimethylsilane. *J. Am. Ceram. Soc.* **1999**, *82*, 2796–2800. [CrossRef]

19.  Hauser, R.; Nahar-Borchard, S.; Riedel, R.; Ikuhara, Y.H.; Iwamoto, Y. Polymer-derived SiBCN ceramic and their potential application for high temperature membranes. *J. Ceram. Soc. Jpn.* **2006**, *114*, 524–528. [CrossRef]

20. Prasad, R.M.; Iwamoto, Y.; Riedel, R.; Gurlo, A. Multilayer amorphous Si–B–C–N/$\gamma$-Al$_2$O$_3$/$\alpha$-Al$_2$O$_3$ membranes for hydrogen purification. *Adv. Eng. Mater.* **2010**, *12*, 522–528. [CrossRef]

21. Bazarjani, M.S.; Müller, M.M.; Kleebe, H.-J.; Jüttke, Y.; Voigt, I.; Yazdi, M.B.; Alff, L.; Riedel, R.; Gurlo, A. High-temperature stability and saturation magnetization of superparamagnetic nickel nanoparticles in microporous polysilazane-derived ceramics and their gas permeation properties. *Appl. Mater. Interfaces* **2014**, *6*, 12270–12278. [CrossRef] [PubMed]

22. Weinmann, M. Chapter 7 Polysilazanes. In *Inorganic Materials*; De Jaeger, R., Gleria, M., Eds.; Nova Science Publishers Inc.: New York, NY, USA, 2007; pp. 371–413.

23. Iwamoto, Y.; Sato, K.; Kato, T.; Inada, T.; Kubo, Y. A hydrogen-permselective amorphous silica membrane derived from polysilazane. *J. Eur. Ceram. Soc.* **2005**, *25*, 257–264. [CrossRef]

24. Funayama, O.; Kato, T.; Tashiro, Y.; Isoda, T. Synthesis of a polyborosilazane and its conversion into inorganic compounds. *J. Am. Ceram. Soc.* **1993**, *76*, 717–723. [CrossRef]

25. Funayama, O.; Tashiro, Y.; Aoki, T.; Isoda, T. Synthesis and pyrolysis of polyaluminosilazane. *J. Ceram. Soc. Jpn.* **1994**, *102*, 908–912. [CrossRef]

26. Iwamoto, Y.; Kikuta, K.; Hirano, S. Microstructural development of Si$_3$N$_4$–SiC–Y$_2$O$_3$ ceramics derived from polymeric precursors. *J. Mater. Res.* **1998**, *13*, 353–361. [CrossRef]

27. Iwamoto, Y.; Kikuta, K.; Hirano, S. Synthesis of poly-titanosilazanes and conversion into Si$_3$N$_4$–TiN ceramics. *J. Ceram. Soc. Jpn.* **2000**, *108*, 350–356. [CrossRef]

28. Kubo, T.; Tadaoka, E.; Kozuka, H. Formation of silica coating films from spin-on polysilazane at room temperature and their stability in hot water. *J. Mater. Res.* **2004**, *19*, 635–642. [CrossRef]

29. Kubo, T.; Kozuka, H. Conversion of perhydropolysilazane-to-silica thin films by exposure to vapor from aqueous ammonia at room temperature. *J. Ceram. Soc. Jpn.* **2006**, *114*, 517–523. [CrossRef]

30. Miyajima, K.; Eda, T.; Nair, B.N.; Iwamoto, Y. Organic-inorganic layered membrane for selective hydrogen permeation together with dehydration. *J. Membr. Sci.* **2012**, *421–422*, 124–130. [CrossRef]

31. Sokri, M.N.M.; Daiko, Y.; Honda, S.; Iwamoto, Y. Synthesis of microporous amorphous silica from perhydropolysilazane chemically modified with alcohol derivatives. *J. Ceram. Soc. Jpn.* **2015**, *123*, 292–297. [CrossRef]

32. Sokri, M.N.M.; Onishi, T.; Mouline, Z.; Daiko, Y.; Honda, S.; Iwamoto, Y. Polymer-derived amorphous silica-based inorganic-organic hybrids having alkoxy groups: Intermediates for synthesizing microporous amorphous silica materials. *J. Ceram. Soc. Jpn.* **2015**, *123*, 732–738. [CrossRef]

33. Seyferth, D.; Wiseman, G.; Prud'homme, C. A liquid silazane precursor to silicon nitride. *J. Am. Ceram. Soc.* **1983**, *66*, C-13–C-14. [CrossRef]

34. Silverstein, R.M.; Bassler, G.C.; Morrill, T.C. *Spectrometric Identification of Organic Compounds*, 5th ed.; John Wiley and Sons, Inc.: New York, NY, USA, 1991.

35. Iwamoto, Y.; Matsunaga, K.; Saito, T.; Völger, W.; Kroke, E.; Riedel, R. Crystallization behaviors of amorphous Si–C–N ceramics derived from organometallic precursors. *J. Am. Ceram. Soc.* **2001**, *84*, 2170–2178. [CrossRef]

36. Stein, S.E. Mass Spectra in NIST Chemistry WebBook, NIST Standard Reference Database Number 69. Available online: http://webbook.nist.gov (accessed on 2 April 2015).

37. Cruz, J.M.D.; Lozovoy, V.V.; Dantus, M. Quantitative mass spectrometric identification of isomers applying coherent laser control. *J. Phys. Chem. Lett. A* **2005**, *109*, 8447–8450.

38. Johnstone, R.A.W. *Mass Spectrometry for Organic Chemists*; Cambridge Chemistry Texts; Cambridge University Press: Cambridge, UK, 1972; pp. 64–67.

39. Watson, J.T.; Sparkman, D. *Introduction to Mass Spectrometry: Instrumentation, Applications, and Strategies for Data Interpretation*, 4th ed.; John Wiley and Sons, Inc.: New York, NY, USA, 2013; pp. 368–376.

40. Sing, K.S.W.; Everett, D.H.; Haul, R.A.W.; Moscou, L.; Pierotti, R.A.; Rouquerol, J.; Siemieniewska, T. Reporting physisorption data for gas/solid systems with special reference to the determination of surface area and porosity. *Pure Appl. Chem.* **1985**, *57*, 603–619. [CrossRef]

41. Lowell, S.; Shields, J.E.; Thomas, M.A.; Thommes, M. *Characterization of Porous Solid and Powders*; Springer: Dordrecht, The Netherlands, 2004.

42. Breck, D.W. *Zeolite Molecular Sieves*; John Wiley and Sons, Inc.: New York, NY, USA, 1974; p. 636.

43. Li, Y.; Feng, Z.; Lian, Y.; Sun, K.; Zhang, L.; Jia, G.; Yang, Q.; Li, C. Direct synthesis of highly ordered Fe-SBA-15 mesoporous materials under weak acidic conditions. *Microporous Mesoporous Mater.* **2005**, *84*, 41–49. [CrossRef]

44.    Apopei, A.I.; Buzgar, N.; Buzatu, A. Raman and infrared spectroscopy of kaersutite and certain common amphiboles. *AUI Geol.* **2011**, *57*, 35–58.

45.    Makreski, P.; Jovanovski, G.; Gajovic, A. Minerals from Macedonia XVII. Vibrational spectra of some common appearing amphiboles. *Vib. Spectrosc.* **2006**, *40*, 98–109. [CrossRef]

46.    Ventura, G.D.; Robert, J.-L.; Beny, J.-M. Tetrahedrally coordinated $Ti^{4+}$ in synthetic Ti-rich potassic richterite: Evidence from XRD, FTIR, and Raman studies. *Am. Miner.* **1991**, *76*, 1134–1140.

47.    Uchino, T.; Kitagawa, Y.; Yoko, T. Structure, energies, and vibrational properties of silica rings in $SiO_2$ glass. *Phys. Rev. B* **2000**, *61*, 234–240. [CrossRef]

48.    Matson, D.W.; Sharma, S.K.; Philpotts, J.A. Raman spectra of some tectosilicates and of glasses along the orthoclase-anorthite and nepheline-anorthite joins. *Am. Miner.* **1986**, *71*, 694–704.

49.    Kingma, K.J.; Hemley, R.J. Raman spectroscopic study of microcrystalline silica. *Am. Miner.* **1994**, *79*, 269–273.

50.    Oyama, S.; Lee, D.; Hacarlioglu, P.; Saraf, R.F. Theory of hydrogen permeability in nonporous silica membranes. *J. Membr. Sci.* **2004**, *244*, 45–53. [CrossRef]

51.    Hacarlioglu, P.; Lee, D.; Gibbs, G.V.; Oyama, S. Activation energies for permeation of He and $H_2$ through silica membranes: An *ab initio* calculation study. *J. Membr. Sci.* **2008**, *313*, 277–283. [CrossRef]

52.    Saito, A.; Foley, H. Curvature and parametric sensitivity in models for adsorption in micropores. *AIChE J.* **1991**, *37*, 429–436. [CrossRef]

53.    Barrett, P.; Joyner, G.; Halenda, P.H. The determination of pore volume and area distributions in porous substances. I. Computations from nitrogen isotherms. *J. Am. Chem. Soc.* **1951**, *73*, 373–380. [CrossRef]

# Direct Control of Spin Distribution and Anisotropy in Cu-Dithiolene Complex Anions by Light

Hiroki Noma [1], Keishi Ohara [1] and Toshio Naito [1,2,*]

1   Graduate School of Science and Engineering, Ehime University, 2-5, Bunkyo-cho, Matsuyama 790-8577, Japan; b851015m@mails.cc.ehime-u.ac.jp (H.N.); ohara.keishi.mg@ehime-u.ac.jp (K.O.)
2   Division of Material Science, Advanced Research Support Center (ADRES), Ehime University, 2-5, Bunkyo-cho, Matsuyama 790-8577, Japan
*   Correspondence: tnaito@ehime-u.ac.jp

Academic Editor: Duncan H. Gregory

**Abstract:** Electrical and magnetic properties are dominated by the (de)localization and the anisotropy in the distribution of unpaired electrons in solids. In molecular materials, these properties have been indirectly controlled through crystal structures using various chemical modifications to affect molecular structures and arrangements. In the molecular crystals, since the energy band structures can be semi-quantitatively known using band calculations and solid state spectra, one can anticipate the (de)localization of unpaired electrons in particular bands/levels, as well as interactions with other electrons. Thus, direct control of anisotropy and localization of unpaired electrons by locating them in selected energy bands/levels would realize more efficient control of electrical and magnetic properties. In this work, it has been found that the unpaired electrons on Cu(II)-complex anions can be optically controlled to behave as anisotropically-delocalized electrons (under dark) or isotropically-localized electrons like free electrons (under UV), the latter of which has hardly been observed in the ground states of Cu(II)-complexes by any chemical modifications. Although the compounds examined in this work did not switch between conductors and magnets, these findings indicate that optical excitation in the $[Cu(dmit)_2]^{2-}$ salts should be an effective method to control spin distribution and anisotropy.

**Keywords:** Cu(II)-dithiolene complex; electron spin resonance; $\pi$–d interaction; quantum chemical calculation; molecular crystal

## 1. Introduction

Electrical and magnetic properties are based on unpaired electrons in solids. Delocalized unpaired electrons can exhibit electrical conduction, while localized unpaired electrons can exhibit magnetism. This trend is true for any solid, including molecular materials. In addition, the (an)isotropy in the interactions among unpaired electrons also matters in these properties. Thus, the control of conduction and/or magnetism requires the control of the (de)localization and anisotropy of the unpaired electrons. If one can switch unpaired electrons between localized and delocalized states, then one may switch the solids between conductive and magnetic materials. This point of view is different from that of light-induced excited spin-state trapping (LIESST), where the spin states of transition-metal complexes are switched between high- and low-spins by photoexcitations at low temperature [1–9]. Additionally, the abovementioned point of view is also different from that of photo-induced phase transitions (PIPTs), where phase transitions, such as metal-insulator and ionic-neutral transitions at certain temperatures ($T_C$s), are brought about by photo-irradiation near the $T_C$s [10,11]. The purpose of the present work is the control of spin distribution and anisotropy independent of thermodynamic conditions, unlike PIPTs. Thus far, most of the studies to control the electrical and magnetic properties have utilized chemical modifications to affect molecular structures and arrangements. However, after tens of years of worldwide extensive research, this strategy has turned out to be challenging for the fine and precise control of conduction and magnetism [12–14]. In both organic and inorganic materials, the slightest chemical modifications often result in unexpectedly marked differences in crystal structures [15,16]. In addition, the slightest differences in crystal structures would sometimes make qualitative differences in the physical properties. In particular, in molecular crystals, there are many kinds of weak intermolecular interactions with keeping a subtle balance to form complicated electronic and crystal structures. Because of this characteristic situation, molecular crystals containing unpaired electrons often have several metastable states, as well as unique excited states with peculiar relaxation mechanisms [17] and are sensitive to perturbation, such as photo-irradiation [18]. All of these properties are interesting in the development of electrical and magnetic materials. However, a precise control of atomic/molecular arrangements in crystals remains impossible. Accordingly, it is interesting, as well as important to develop new methods to control the conductive and magnetic properties of the molecular crystals.

Paying attention to such unique features of molecular crystals, the control of conduction [19,20], or magnetism [21,22], or other physical properties [23–28] was carried out using photochemical redox reactions in solid states. In the reactions, photo-induced electron transfer occurs between two different kinds of molecular species in the crystal, e.g., the cations and the anions; one species is responsible for conduction and/or magnetism, and the other serves as the charge reservoir in high-$T_C$ cuprates [29]. The latter species is not involved in either conductive or magnetic properties. As photochemical reactions generally have high spatial resolutions, this method is utilized to make junction structures of a molecular single crystal for electronic devices when the electron transfer is irreversible [19,30]. Furthermore, reversible and simultaneous control of conduction and magnetism has been recently realized using charge transfer (CT) transition between two different molecular species in salts [31,32], instead of chemical reactions. In this case, one of the molecular species accommodates localized spins for magnetism, in addition to serving as the charge reservoir. The remaining species is optically doped to have excited carriers, like photoconductors. "Giant photoconductivity", which possibly originates from reversible melting of charge order, has been observed by UV irradiation upon a related molecular crystalline salt [33]. The advantage of this photochemical or optical method lies in the fact that control of the crystal structure is not required at any level. Now that a crystal with an appropriate structure and physical properties, which are known under dark conditions, is selected, nothing but photo-irradiation is required to excite and produce localized and/or delocalized unpaired electrons in the crystal. As long as the electronic band structure and the solid state spectra are known, selected photoexcitation with appropriate wavelength and intensity can take place with accidental or uncontrollable factors being minimized in the control of conduction and magnetism. In addition to the abovementioned sensitivity to light and unique photo-physics of molecular crystals, this kind of photo-control is based on the fact that molecular crystals often contain different kinds of molecules for different roles and also the fact that they have well-defined crystal and band structures, which are clarified by standard experimental and calculation techniques. This idea will make a step forward when a single kind of molecule is responsible for both magnetism and conduction. This will be realized when the distribution of unpaired electrons on the molecule can be controlled by irradiation. In other words, the desired molecular orbitals for unpaired electrons should be selected by appropriate photoexcitation. If the unpaired electrons are excited to be localized, the magnetic properties dominate, and *vice versa*. Based on this idea, a unique Cu(II)-dithiolene complex anion has been found to be suitable for this purpose, after a number of transition-metal complexes were examined.

Metal-dithiolene complex molecules, $M(dmit)_2$ (M = Ni, Pd, Pt, Au, *etc.*, $dmit^{2-}$ = 1,3-dithiole-2-thione-4,5-dithiolate, Figure 1), have attracted attention for a long time as building blocks for conducting, magnetic and optical materials [34–48]. Among a great number of molecular species, $M(dmit)_2$ and their derivatives have always shown us new possibilities of states of matter, such as single-component metals/superconductors [49–59] and spin liquids [60]. This is because $M(dmit)_2$ have uniquely many degrees of freedom, *i.e.*, spin, charge and orbital degrees of freedom, originating from d-orbitals at the metal centers and π-orbitals of the dmit-ligands. The $M(dmit)_2$ species often become stable radical anions with fractional formal charges in solid states. In the solid states, they closely interact with each other through the overlap of the molecular orbitals (MOs) accommodating the unpaired electrons. Among the $M(dmit)_2$ and related complexes, the $Cu(dmit)_2$ anions [61–67] are unique in that the number and distribution of the unpaired electrons may be reversibly controlled with UV radiation [68]. In this work, the electronic and spin densities of $[Cu(dmit)_2]^{2-}$ anions are compared in detail in salts with various cations. Among them, three salts, $(nBu_4N)_2[Cu(dmit)_2]$ (**1**), $[(DABCO)H]_2[Cu(dmit)_2]CH_3CN$ (DABCO = 1,4-diazabicyclo[2.2.2]octane) (**2**) and $BP_2DBF[Cu(dmit)_2]$ ($BP_2DBF^{2+}$ = dibenzofuran-2,2'-bis($N$-methylene-4,4'-bipyridinium)) (**3**) (Figure 1), exhibited systematic differences in structural and physical properties. The following are discussed in this paper for **1–3**: the crystal and molecular structures, the temperature-dependent electrical resistivities, the calculated charge and spin densities, the electron spin resonance (ESR) spectra and the temperature-dependent magnetic susceptibilities.

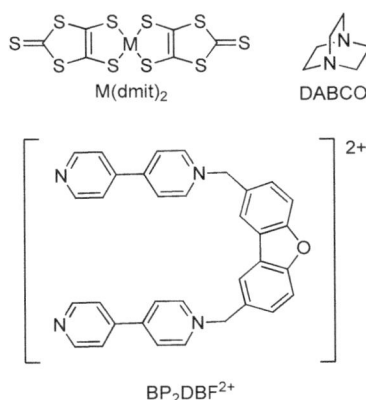

**Figure 1.** Molecular structures of abbreviated chemical species. $dmit^{2-}$ = 1,3-dithiole-2-thione-4,5-dithiolate; DABCO, 1,4-diazabicyclo[2.2.2]octane; $BP_2DBF^{2+}$ = dibenzofuran-2,2'-bis($N$-methylene-4,4'-bipyridinium).

## 2. Results and Discussion

### 2.1. Crystal and Molecular Structures

The crystal and molecular structures of **1–3** are shown in Figure 2, and the crystal data are summarized in Table S1 in the Supplementary Materials. The crystal and molecular structures of **1–3** remain unchanged from room to low temperature, except for a few percentage of thermal shrinkage in the unit cell volumes $V$ ($V(100\text{ K})/V(296\text{ K})$ = 96.6%, 97.5% and 97.9% for **1–3**, respectively). The electrical and magnetic properties are cooperative phenomena and are mainly based on intermolecular interactions through sulfur–sulfur (S–S) interatomic contacts in **1–3**. With respect to the arrangement of the $[Cu(dmit)_2]^{2-}$ anions, the shortest intermolecular S–S distances at 296/100 K in **1–3** are 6.138(1)/5.8082(7) Å in **1**, 3.499(1)/3.429(1) Å in **2** and 3.921(3)/3.848(5) Å in **3**. Those in **1** and **3** are larger than twice the van der Waals radius of a sulfur atom (1.85 × 2 = 3.70 Å; simply called the "van der Waals distance" below). Thus, there should hardly be any interaction through the overlap of MOs between adjacent $[Cu(dmit)_2]^{2-}$ anions in **1** and **3**. This suggests that both salts should be insulators, which is supported by electrical measurements below. A tight-binding band calculation (TBBC) was carried out for **2** (Figure 3). The calculated band is slightly dispersive along such lines as $\Gamma(0,0,0)$–X(0.5,0,0), Y(0,0.5,0)–Z(0,0,0.5) and Z(0,0,0.5)–R(0.5,0.5,0.5). There are no partially-filled bands, and there is a band gap of ~0.07 eV (~800 K) at the Fermi level $E_F$. These results mean that **2** should be an insulator in its ground state, but around room temperature (RT), it can become semiconductive. As for the remaining salts, a TBBC suggests that **1** and **3** should also be either semiconductors or insulators, since the calculated bands are too narrow to exhibit metallic conduction (**1**, Figure S1c in the

Supplementary Materials) or do not include partially-filled bands (**3**, Figure S1o in the Supplementary Materials). These results are consistent with the observed electrical properties (see below).

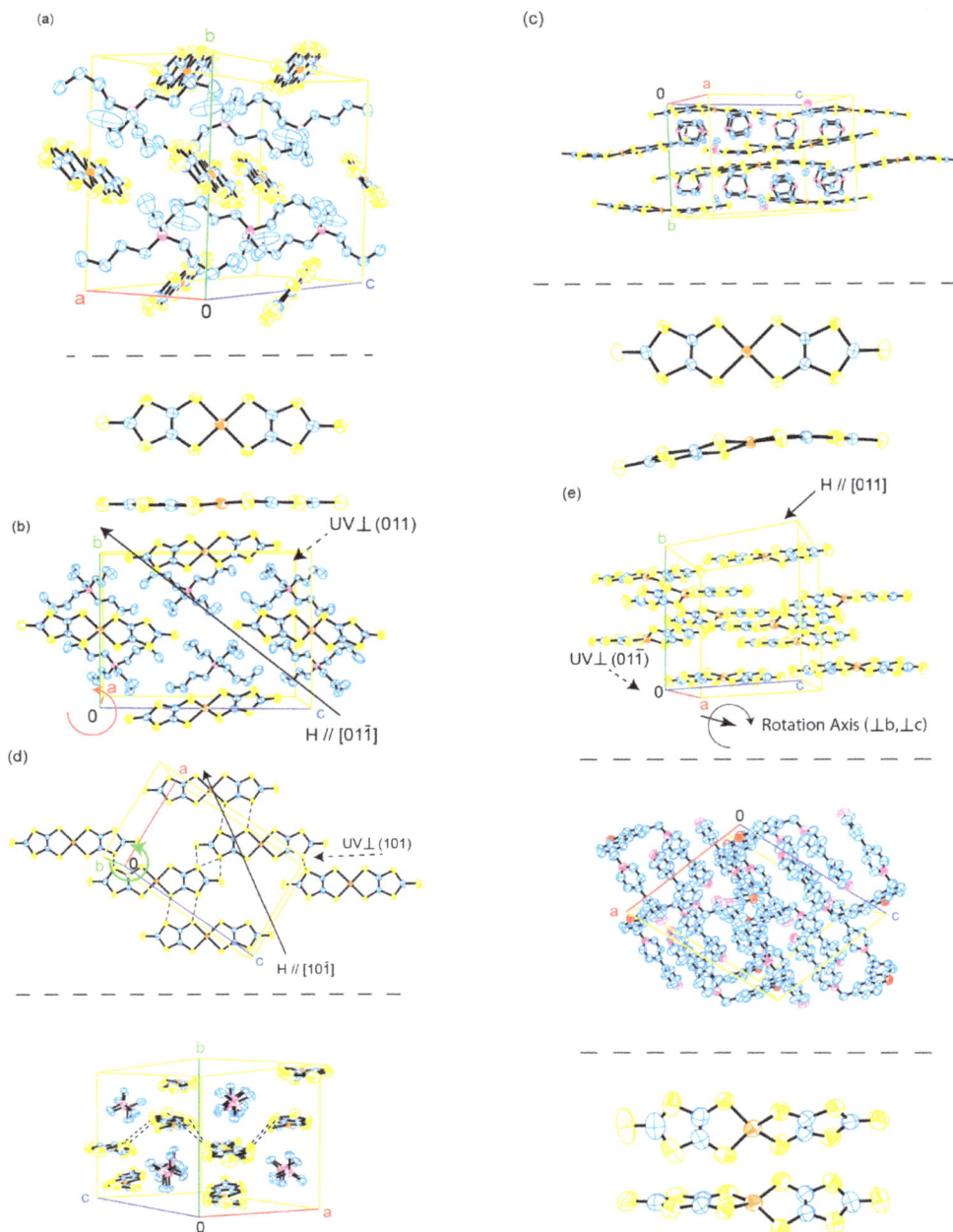

**Figure 2.** Crystal and molecular structures of **1**–**3** at 297 K: (**a**,**b**) **1**; (**c**,**d**) **2**; and (**e**) **3**. Yellow lines designate the edges of the unit cells. Hydrogen atoms are omitted for clarity. Yellow, blue, brown, pink and red octants designate S, C, Cu, N and O atoms, respectively. In (**d**), broken lines designate S–S short contacts, and only [Cu(dmit)$_2$]$^{2-}$ anions are drawn in order to show the conduction pathways in the *ac*-planes (**upper**). In (**e**), [Cu(dmit)$_2$]$^{2-}$ and BP$_2$DBF$^{2+}$ are drawn separately in the upper and middle parts, respectively, for clarity. In (**b**,**d**,**e**), the directions of the magnetic fields (*H*), those of the UV irradiation and the rotation axes of the single crystals relative to *H* (circular arrows around the *a*-axis in **1**, around the *b*-axis in **2** and around the axis perpendicular to both the *b*- and *c*-axes in **3**) in angle-dependent ESR measurements are shown.

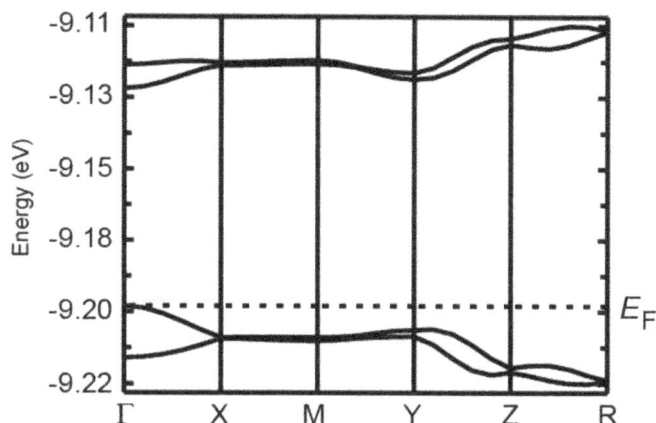

**Figure 3.** Calculated band structure of **2** using a tight-binding approximation. The broken line ($E_F$) designates the Fermi level. $\Gamma$, X, M, Y, Z and R designate points in reciprocal space located at (0,0,0), (0.5,0,0), (0.5,0.5,0), (0,0.5,0), (0,0,0.5) and (0.5,0.5,0.5), respectively. Band structures of other salts and those of wider energy ranges are shown in Figure S1 in the Supplementary Materials.

*2.2. Molecular Structures and Spin Delocalization*

Unlike other kinds of metal-dithiolene-complex anions, not only the crystal structures, but also the molecular structures of the $[Cu(dmit)_2]^{2-}$ anions vary with the counter-ion (Figure 2). The geometries at the metal center ($CuS_4$) are almost square-planar, slightly distorted square-planar and distorted-tetrahedral in **1–3**, respectively. Such flexibility of the $CuS_4$ geometry has been known for long [56–59,61–68], yet the reason for their thermodynamical stabilities has never been discussed by considering both crystal and molecular structures. The structure of the $[Cu(dmit)_2]^{2-}$ anions in **2** is unique in that it is also distorted along the long molecular axis, which is disadvantageous for delocalization energy of the dmit ligands. In order to elucidate whether the crystal structures affect the coordination geometries or not, the structural optimization of the $[Cu(dmit)_2]^{2-}$ anions as isolated molecules was carried out using Gaussian 09 [69]. The results indicate that both the planar and non-planar coordination geometries are stable and that the stable structure depends on the initial structure (Figure S2). Assuming the initial structures to exist as observed in **1** and **3**, the optimized structures do not significantly differ from the initial structures, respectively. This means that the molecular structures of the $[Cu(dmit)_2]^{2-}$ anions are governed by their own stability and are not governed by intermolecular interactions in the solid state. The calculation results indicate that the differences in stability between the square-planar (like that in **1**) and distorted-tetrahedral (like that in **3**) coordination geometries are small (~0.1 eV) and are subject to subtle conditions of crystallization. This situation realizes the flexibility and variety of the coordination geometries, *i.e.*, the degrees of freedom in the molecular structures of the $[Cu(dmit)_2]^{2-}$ anions. However, the same calculation for **2** gave a rather different molecular structure from the initial, *i.e.*, the observed structure in the crystal (Figure S2). The optimized structure is close to that for **3**, a distorted-tetrahedral geometry with planar ligands. Thus, the molecular structure in **2** is affected by surrounding chemical species in the solid state, resulting in a uniquely-distorted structure. Such a "frustrated" molecular structure can be related to unusually small *g*-values observed for **2** discussed below. Figure 4 shows the SOMOs (SOMO = the singly-occupied MO) and spin densities calculated for **1–3** using Gaussian 09. The electron densities are distributed with high symmetry over the entire molecule in all three salts **1–3**. Similarly, the spin distributions do not differ from each other in **1–3**. The extent of delocalization of the unpaired electrons at the $Cu^{2+}$ ions was experimentally examined by measurement of the electrical resistivity, the magnetic susceptibility and the electron spin resonance (ESR), as discussed in the next section.

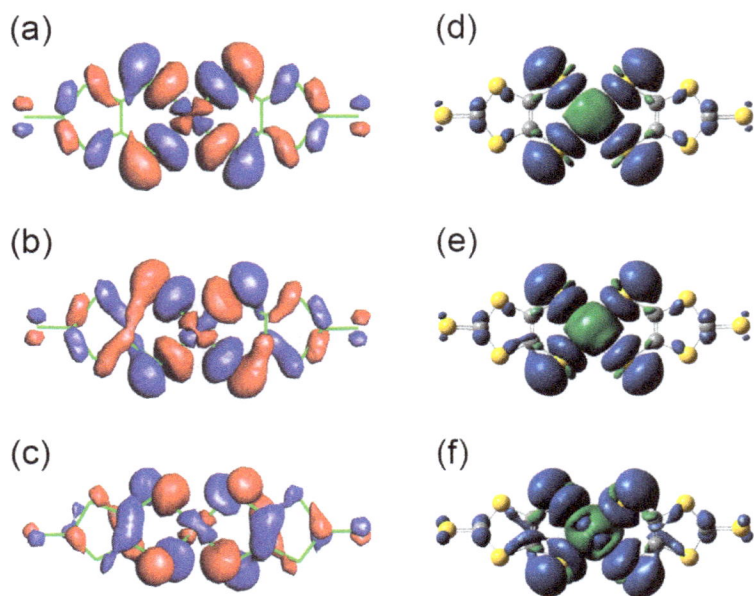

**Figure 4.** Calculated singly-occupied molecular orbitals (SOMO) ((a) **1**; (b) **2**; and (c) **3**) via extended Hückel calculation and spin densities ((d) **1**; (e) **2**; and (f) **3**) for $[Cu(dmit)_2]^{2-}$ anions via Gaussian 09 (B3LYP/6-31+G(3d)). Red and blue in (a)–(c) and green and blue in (d)–(f) designate different orbital lobes. Green lines in (a)–(c) and grey lines in (d)–(f) designate chemical bonds. In (d)–(f) yellow and grey spheres designate S and C atoms, respectively. For each molecular structure, see Figure 2. MOs of excited states are shown in Figure S9 in the Supplementary Materials.

## 2.3. Electrical Properties under Dark Conditions

Irrespective of the relative direction of the applied electric field, salts **1** and **3** are insulators ($>10^7$ ohm·cm at 300 K). On the other hand, resistivity measurements along the $b$-axis indicate that **2** is a semiconductor at room temperature (298 K), with a resistivity $\rho_{RT}$ of = $(6.5 \pm 3.5) \times 10^5$ ohm·cm and an activation energy $E_a$ of = $0.15 \pm 0.02$ eV (Figure 5a). These observations are qualitatively consistent with the crystal structures described above. However, the calculated energy gap (~0.06 eV) is significantly smaller than that estimated from the observed $E_a$, *i.e.*, the gap is estimated to be $2E_a \sim 0.30$ eV. This inconsistency suggests that electron correlation effects are strong in **2**, since the tight-binding approximation does not take them into consideration. The correlation effects originate from electron–electron Coulombic repulsion and often open a band gap at $E_F$ [70].

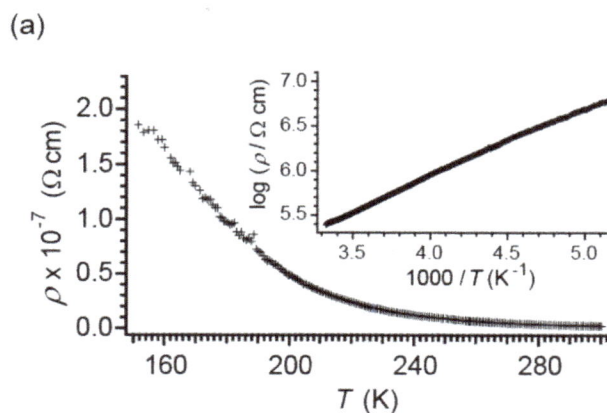

**Figure 5.** *Cont.*

**(b)**

**(c)**

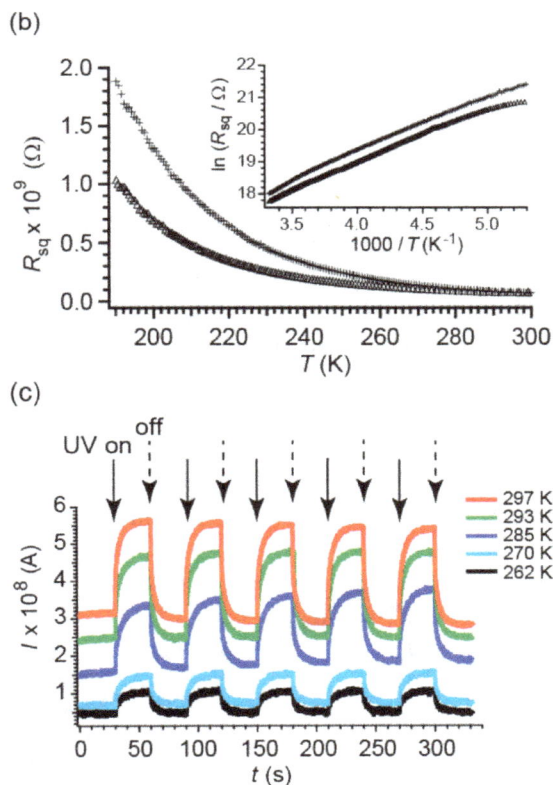

**Figure 5.** Temperature dependence of electrical resistivity along the $b$-axis of **2**. (**a**) Resistivity $\rho$ in the dark; (inset) Arrhenius plot of the same data; (**b**) surface resistivity $R_{sq}$ in the dark (crosses) and under UV radiation (triangles); (inset) Arrhenius plot of the same data; (**c**) response of photoconduction to switching the light source on/off.

## 2.4. Electrical Properties under UV Irradiation

Upon UV irradiation (250–450 nm; 3 Wcm$^{-2}$), only **2** exhibited a response; the resistivity clearly decreased. The temperature dependence of the resistivity under continuous UV irradiation is shown in Figure 5b. As shown in Figure 5c, the change in the resistivity was quick and reversible upon the commencement/cessation of UV irradiation; the resistivity instantly increased/decreased. The observed current ratio between dark and UV irradiated conditions $I_{UV}/I_{dark}$ gradually increased with decreasing temperature: ~1.7 at 297 K, ~1.8 at 293 K, ~1.9 at 285 K, ~2.0 at 270 K and ~2.1 at 262 K. This is quantitatively consistent with the temperature dependency of the resistivity under dark and UV irradiated conditions (Figure 5b). Since the activation energy remains fundamentally unchanged during irradiation and since the resistivity change upon irradiation is small, the thermally-excited carriers and the photoexcited carriers are considered to coexist, and the former should be dominant in **2**. This observation suggests that the molecular and crystal structures in the photoexcited states remain practically unchanged, since both activation energy and conductivity remain practically unchanged under irradiation. This interpretation is experimentally corroborated by the ESR measurements discussed below.

## 2.5. Magnetic Susceptibilities under Dark Conditions

The temperature ($T$) dependence of the magnetic susceptibilities ($\chi$) of **1**–**3** clearly shows that all are diamagnetic (Figure 6). Both the zero-field-cooling and field-cooling processes were examined and were confirmed to yield identical results within experimental error. The increases in $\chi$ at low temperature ($T \leqslant 50$–60 K), as well as small jumps in $\chi$ at ~50 K suggest that **1** and **2** contained impurities, such as oxygen adsorbed on the sample. Curve-fitting analyses using Equations (1) (for $\chi$ vs. $T$) and (2) (for $\chi T$ vs. $T$) consistently and unambiguously distinguished the intrinsic behavior from those of impurities:

$$\chi_{obs} = \frac{C}{T} + \chi_{dia} \qquad (1)$$

$$\chi_{obs}T = \chi_{dia}T + C \qquad (2)$$

where $\chi_{obs}$, $\chi_{dia}$ and $C$ designate the observed $\chi$, the diamagnetic contribution from the samples and the paramagnetic contribution from impurities, respectively. The diamagnetic susceptibilities obtained ($\chi_{dia} \times 10^2$ (emu mol$^{-1}$)) for **1–3** are $-1.06 \pm 0.05$, $-1.24 \pm 0.03$ and $-0.80 \pm 0.08$, respectively. The observed diamagnetism of **1–3** means that the unpaired electrons on the [Cu(dmit)$_2$]$^{2-}$ anions interact with each other rather strongly in antiferromagnetic ways in all three of the salts and that there are no phase transitions at 2–300 K in any of them. The diamagnetism is a rather unexpected result, since all of the intermolecular distances in the [Cu(dmit)$_2$]$^{2-}$ anions of **1–3** are long. Such long-range spin–spin interactions are known as "through-space" interactions [71]. The observed diamagnetism for the three salts is consistent with their insulating properties at the ground states.

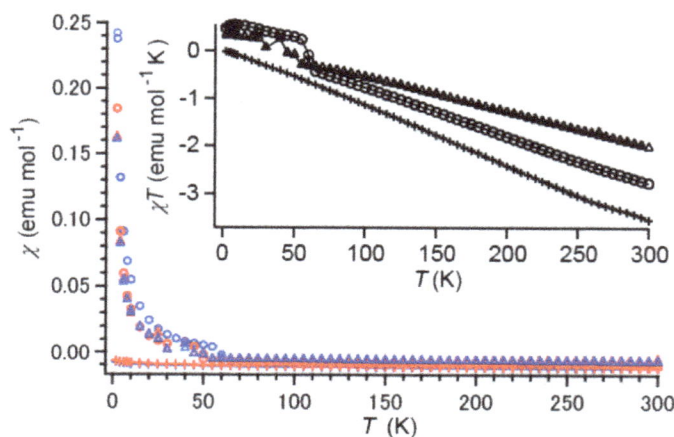

**Figure 6.** Temperature ($T$) dependence of the magnetic susceptibility ($\chi$) of **1** (circles), **2** (crosses) and **3** (triangles). Both field-cooling (blue) and zero-field-cooling (red) processes are shown. (Inset) $\chi T$ vs. $T$ of **1–3** for the same field-cooling data as those in the main panel; the small jumps around 50 K in **1** and **3** are extrinsic, possibly due to residual adsorbed oxygen.

## 2.6. Electron Spin Resonance under Dark Conditions

Next, we should discuss the ESR spectra under dark conditions. In the following discussion, $g$-values and their anisotropies in this work mean those averaged in an entire crystal instead of those for isolated molecules. The latter properties of [Cu(dmit)$_2$]$^{2-}$ are reported using solution- [63] or magnetically-diluted single-crystalline samples [62]. The angle ($\theta$)-dependence of the $g$-values for **1–3** is shown in Figure 7a–c. Since all of the spectra under dark conditions were symmetric without fine structures, the $g$-values were approximately estimated from the maxima of the integrated ESR spectra. The $\theta$-dependence indicates that the $g$-values are in the range of ~2.01–2.06, depending on the relative angle of the applied field and the salts. For comparison, the calculated $g$-values, i.e., the principal values of $g$-tensors for the isolated [Cu(dmit)$_2$]$^{2-}$ anions, using Gaussian 09 based on the optimized molecular structures are $g_{xx} = 2.0403$, $2.0314$, $g_{yy} = 2.0644$, $2.0407$ and $g_{zz} = 2.0919$, $2.1210$ for **1** and **3**, respectively. The observed $g$-values for **1** (2.024–2.028) and **3** (~2.02–2.06) are smaller than the calculated values and those generally observed for Cu(II)-complexes (for the spectra measured with internal standards, see Figure S7 in the Supplementary Materials). Larger $g$-values were observed in other directions under dark conditions (2.02–2.04 for **1**, 2.02–2.09 for **2** and 2.02–2.05 for **3**; Figure S8 in the Supplementary Materials). Thus, the observed $g$-values indicate rather large anisotropy of the three salts under the dark conditions. Both observed $g$-values and their $\theta$-dependence are smaller in **2** than in **1** and **3**, because the single crystals were rotated in such a way that the magnetic field was always approximately in the molecular plane of the $[Cu(dmit)_2]^{2-}$ anions in **2**, while the magnetic field was rotated between nearly parallel and nearly perpendicular to the molecular planes in **1** and **3**. The minima and maxima of the $g$-values are consistent with the crystal structures. For **1**, $\theta \sim 45°$ (minimum) and $\theta \sim 135°$ (maximum) coincide with where the long molecular axis of all of the $[Cu(dmit)_2]^{2-}$ anions becomes nearly parallel with and perpendicular to $H$, respectively. For **2**, $\theta \sim 25°$ (minimum) and $\theta \sim 115°$ (maximum) also coincide with where the long molecular axis of $[Cu(dmit)_2]^{2-}$ becomes nearly parallel with and perpendicular to $H$, respectively. Similarly, for **3**, $\theta \sim 50°$ (minimum) coincides with where the long molecular axis of $[Cu(dmit)_2]^{2-}$ becomes nearly parallel with $H$. ESR signals were observed only for $\sim 15° \leqslant \theta \leqslant \sim 90°$ in **3**. The signal almost disappeared in other directions (Figure S3 in the Supplementary Materials). This is consistent with the following explanation. In **3**, the observed S–S distances (<4.0 Å) (Figure S6 in the Supplementary Materials) suggest that every pair of $[Cu(dmit)_2]^{2-}$ anions should closely interact with each other to form a weak dimer. The overlap and the transfer integrals within the dimer are calculated to be $1.5 \times 10^{-3}$ and $-0.03$ eV via an extended Hückel method. The calculated values of these integrals suggest that the interactions should be moderately strong and antiferromagnetic, which is supported by the observed diamagnetism (Figure 6). The best fitting parameters for a simulation of the observed ESR spectra (Tables S4, S6 and S8 in the Supplementary Materials) are comparable or consistent with those of related compounds [35,37]. The observed and simulated ESR spectra under dark conditions for **1–3** well agree with each other (Figure S4 in the Supplementary Materials). The ESR spectra under dark conditions can be interpreted as follows. The signals observed for **1–3** in the dark originate from the spin delocalized over the $[Cu(dmit)_2]^{2-}$ anions. The anisotropy mainly originates from that of the molecular orbital geometries. The hyperfine structures characteristic of the spin on Cu(II) generally observed in various Cu(II)-complexes were not observed in **1–3**. The reason is considered to be a broadening effect caused by the following two kinds of interaction; intermolecular antiferromagnetic interactions and spin-orbit coupling on the sulfur atoms in the ligands (dmit).

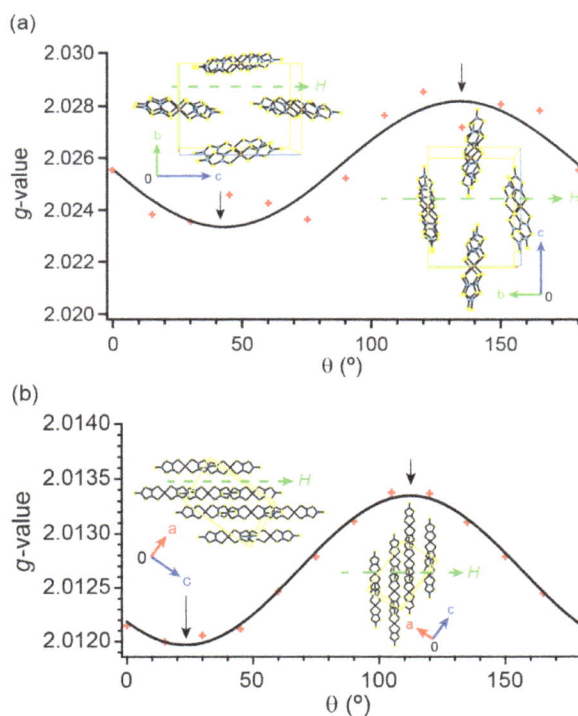

**Figure 7.** *Cont.*

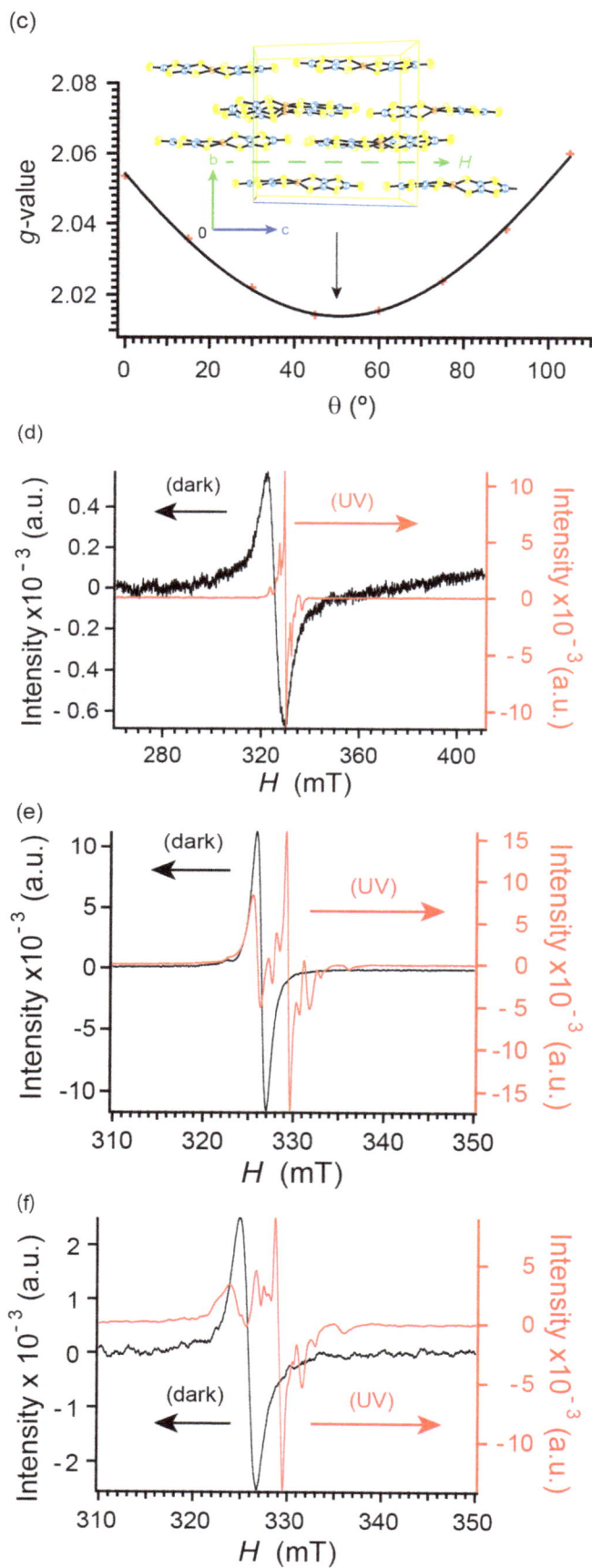

**Figure 7.** *Cont.*

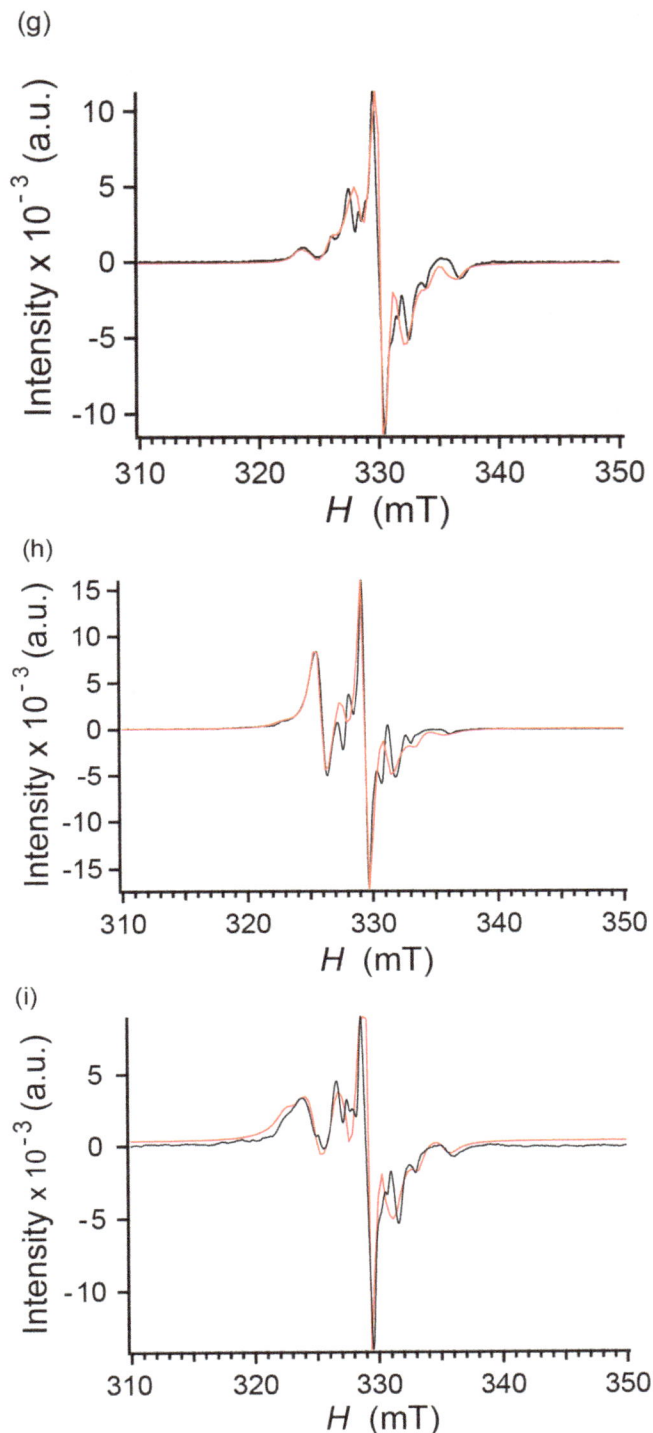

**Figure 7.** Electron spin resonance (ESR) spectra for **1–3**, measured on single crystals. Anisotropy of the *g*-values at RT in: (**a**) **1**; (**b**) **2**; and (**c**) **3**; observed (red crosses) and best fit (black sine curves). Black arrows designate the angle $\theta$ corresponding to the orientation of the crystal shown by the arrows. Comparison of the ESR spectra under dark (black) and UV irradiated (red) conditions at 123 K for: (**d**) **1** ($\theta = 0$); (**e**) **2** ($\theta = 0$); and (**f**) **3** ($\theta = 45°$). Comparison of the observed (black) and simulated (red) ESR spectra under UV irradiated conditions for: (**g**) **1** (from [68]); (**h**) **2**; and (**i**) **3**. The UV irradiated spectra are the same as those in (**d–f**). The comparison between the observed and simulated ESR spectra in the dark for **1–3** and the parameters for all spectral simulation are shown in the Supplementary Materials (Figures S4 and S5 and Tables S4 and S9, respectively).

## 2.7. *Electron Spin Resonance under UV Irradiation*

Under UV irradiation (250–450 nm), clearly different ESR spectra were obtained for all three salts from those obtained in the dark (Figure 7d–f); the intensities obviously increased, and hyperfine structures appeared for **1–3**. This means that the UV excitation of the unpaired electrons alters their distribution from delocalized to localized pictures. After ceasing irradiation, the spectra recovered their original shapes, and the reversibility was confirmed via repeated measurements on different single crystals. In **1–3**, an energy of 250–450 nm (2.75–5.00 eV) brings about excitation from an orbital $\varphi_b$ to a higher orbital $\varphi_a$, comprised of $Cu^{2+}$ d-orbitals and $dmit^{2-}$ $\pi$-orbitals from a molecular orbital perspective (Figure S9 in the Supplementary Materials). Firstly, we should consider the origin of ESR signals under UV irradiation. As for **1** and **2**, the possible final states, *i.e.*, $\varphi_a$, by the excitation of 250–450 nm are limited to several groups of degenerate bands. They are those around −5.02 eV (#311–312), −5.94 eV (#309–310) and −5.96 eV (#307–308) in **1** (Figure S1a,b in the Supplementary Materials) and those around −4.83 eV (#429–432) and −5.56–6.14 eV (#421–428) in **2** (Figure S1e,f) in the Supplementary Materials, respectively. Their energies from the Fermi levels correspond to the wavelengths of 299, 384 and 386 nm for **1**, while 284, 342 and 407 nm for **2**, respectively. These excitation energies are consistent with the solution spectra containing the $[Cu(dmit)_2]^{2-}$ anions (Figure 8a). All of these excited states are almost purely (>99.9%) comprised of the MOs of the $[Cu(dmit)_2]^{2-}$ anions, and the cations' contribution is negligible. For **1** and **2**, all of the ESR spectral features under UV irradiation must originate from the excited $[Cu(dmit)_2]^{2-}$ anions, because the cations do not absorb light in 250–450 nm (Figure 8). On the other hand, in **3**, the TBBC results suggest that there should be anion-cation mixed bands around −5.9 eV (Figure S1k in the Supplementary Materials), *i.e.*, 3.5 eV ≈ 355 nm above the $E_F$ (~−9.4 eV). This means that CT transitions from the anion to the cation can occur under UV radiation of 250–450 nm for **3**. In addition, the solution spectra (Figure 8a) show that the cation in **3** strongly absorbs UV light shorter than ~300 nm (⩾4.1 eV). Thus, UV radiation of 250–450 nm can cause local excitations in the $BP_2DBF^{2+}$ cations. The possibilities of the CT transitions and the local excitations consistently suggest the existence of unpaired electrons on the $BP_2DBF^{2+}$ cations during UV irradiation upon **3**. On the other hand, the orbitals, located immediately below $E_F$, are primarily comprised of a $dmit^{2-}$ contribution (70%, 71% and 78% for **1**, **2** and **3**, respectively, based on a TBBC). There are several degenerate bands at $E_F$ or immediately below $E_F$ in **1–3** originating from the fact that there are several $[Cu(dmit)_2]^{2-}$ in their unit cells. However, their dmit contributions are almost equal in the respective materials; the differences are less than 0.1% among the degenerate bands. All spectra were analyzed on this basis and were reproduced by simulations considering anisotropy, nuclear magnetic moments and natural abundances of isotopes of the magnetic nuclei in the $[Cu(dmit)_2]^{2-}$ anions (for **1** and **2**) or in both the anions and the cations (for **3**) (Figure 7g,i). The following factors are considered; $^{63}Cu$:$^{65}Cu$ = 69.09:30.91; nuclear magnetic moment $\beta_n$ = 2.2206 ($^{63}Cu$) and 2.3790 ($^{65}Cu$); $^{14}N$ ($I$ = 1, 99.635%). In the meantime, $^{33}S$ ($I$ = 3/2, 0.74%), $^{13}S$ ($I$ = 1/2, 1.108%) and $^{15}N$ ($I$ = 1/2, 0.365%) are ignored. Next, we shall discuss the spin distribution on the $[Cu(dmit)_2]^{2-}$ anions based on the simulation results from a molecular orbital perspective. UV (250–450 nm) radiation reversibly increases the contribution from the spins on the copper atoms in all of the ESR spectra for **1–3**; during UV irradiation, the Cu contribution increased from $I_{rel}$ (Cu) = 9%–38% for **1**, 17%–31% for **2**, and 13%–33% for **3**, respectively (Tables S4–S9 in the Supplementary Materials). This observation is qualitatively consistent with the calculated band structures in that the copper atomic d-orbitals have larger contributions in the excited states $\varphi_a$ than in the ground (dark) states $\varphi_b$; $\psi(Cu)$ = 91% *vs.* 30% (the bands #307–312 at −5.97~−5.02 eV for **1**), 89% *vs.* 29% (the bands #421–432 at −6.14~−4.83 eV for **2**) and 82% *vs.* 22% (the bands #589–624 at −6.56~−4.96 eV for **3**), respectively. Here is a point worthy of note. Now, the approximate g-values under UV irradiation are tentatively defined by the centers of the newly-emerged peaks during UV irradiation. This is confirmed to be a valid approximation by detailed analyses using spectral simulation (Figure S5 in the Supplementary Materials). Then, the observed g-values under UV irradiation are even smaller than those under dark conditions. They are comparable with that of free electrons (~2.0023) irrespective of the relative direction of $H$ (Figure S7 in the Supplementary Materials). Such small $g$-values with almost no anisotropy have never been reported either for Cu(II)-complexes or transition-metal-dithiolene complexes. This can be explained by the characteristics of SOMOs of various [M(dmit)$_2$] radical anions in excited and ground states in a unified picture. Unpaired electrons either delocalized in the extended dithiolene π-orbitals or localized in the metal (except for Cu) d-orbitals generally exhibit larger $g$-values due to spin-orbit coupling with many sulfur atoms or with anisotropic d-orbitals. In both cases, the spin-orbit coupling enhances the anisotropy of the $g$-values at the same time. This trend is slightly observed in the dark states of **1–3** and is more evident in the M(dmit)$_2$ radical anions (M = Ni, Pd, Pt, *etc.*) under dark conditions, where the hybridization between ligands' π- and metals' d-orbitals (π–d mixing) is less important than that in the [Cu(dmit)$_2$]$^{2-}$ anions (Table S14 in the Supplementary Materials) [68,72–79]. In contrast, whether under dark or UV irradiated conditions, the SOMOs of [Cu(dmit)$_2$]$^{2-}$ anions have a much higher degree of π–d mixing than that in other metal-dithiolene complex anions. As shown in Figure S9 in the Supplementary Materials, some of the excited MOs (e.g., MO#52 of all three salts) are localized around the Cu atoms with rather spherical geometries because of the π–d mixing. Thus, the orbital angular momentum is practically quenched, which makes spin-orbit coupling negligible. The findings in this work mean that UV excitation in [Cu(dmit)$_2$]$^{2-}$ can realize an unusual spin state, which cannot be realized by thermal excitation, the variation of counter cations or chemical modification of the metal center and/or ligands.

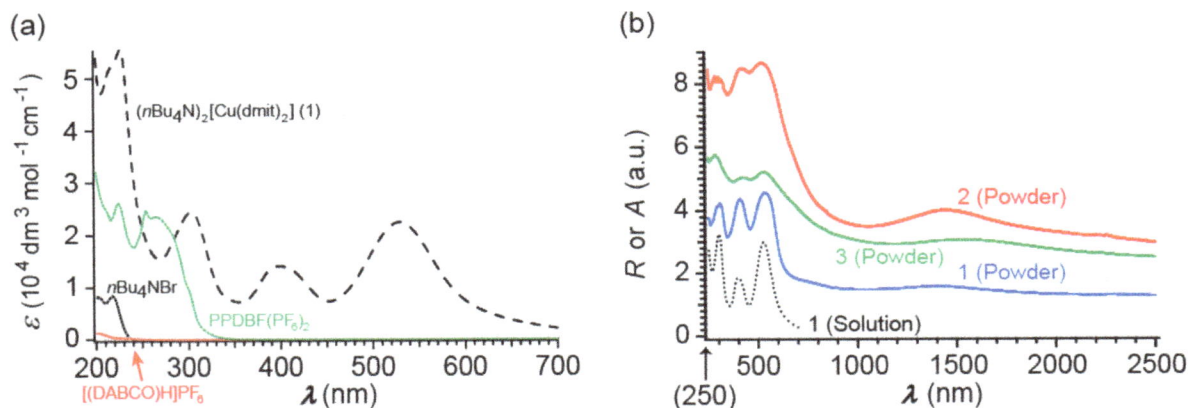

**Figure 8.** Solution and powder spectra for **1–3** (20 °C). (**a**) UV-VIS absorption ($\varepsilon$) spectra in CH$_3$CN for **1–3**; (**b**) UV-VIS-NIR diffuse reflectance ($R$) spectra in KBr pellets for **1–3**. The solution spectrum ($A$) for **1** (same spectrum as that in (**a**)) is also shown for comparison.

## 3. Experimental Section

### 3.1. Materials and Methods

All chemicals were purchased as the highest grade and used as received. Compound **1** was synthesized according to the literature [80]. [(DABCO)H]PF$_6$ and BP$_2$DBF(PF$_6$)$_2$ were provided by Solid-State Chemistry Laboratory, Department of Chemistry, Faculty of Science, Hokkaido University, Sapporo, Japan. Single crystals of **1** suitable for X-ray structural analysis and physical property measurements were obtained during the recrystallization of the crude product from CH$_3$CN. Single crystals of **2** and **3** were obtained from slow, double decompositions of saturated solutions. DABCO·HPF$_6$ (3 mg) in CH$_3$CN (Wako, Osaka, Japan; 10 mL) (for **2**) or (BP$_2$DBF)(PF$_6$)$_2$ (3 mg) in CH$_3$CN (20 mL) (for **3**) and **1** (3 mg) in CH$_3$CN (10 mL for both **2** and **3**) were dissolved using a supersonic wave machine (AS ONE, Osaka, Japan, Ultrasonic Cleaner US-5A) for 1 min. After filtration, one of the solutions was added to the other. The mixed solution was sealed and left to stand in the dark for 14–17 days at −30 °C (**2**) or at 24 °C (**3**) to yield shiny black blocks ~0.5 mm on the longest side. Single crystal X-ray structural analysis was carried out with a Rigaku R-AXIS RAPID

Cu-K$\alpha$ radiation ($\lambda$ = 1.54187 Å) or Rigaku Saturn724 Mo-K$\alpha$ radiation ($\lambda$ = 0.71075 Å) diffractometer at (296 $\pm$ 1) and (100 $\pm$ 1) K. Cambridge Crystallographic Data Centre (CCDC) numbers 991696 (**1**, 296 K), 994604 (**1**, 100 K), 1037127 (**2**, 296 K), 1037128 (**2**, 100 K), 1037129 (**3**, 296 K) and 991697 (**3**, 100 K) contain supplementary crystallographic data for this paper. These can be obtained free of charge from the CCDC via the internet (www.ccdc.cam.ac.uk/data_request/cif). Details of the molecular orbital and band structure calculations are described in the Supplementary Materials.

### 3.2. UV Irradiation

UV irradiation of the samples was carried out in a He atmosphere (ESR) or in liquid $N_2$ (resistivity measurements at low temperatures). A Hg/Xe lamp (SAN-EI Electric, Osaka, Japan, SUPERCURE-203S; 200 W, 220–1100 nm) equipped with a band pass filter (transparent only at 250–450 nm) and a quartz light guide of a multimode optical fiber (5 mm in the core diameter) were employed. The actual power at a particular wavelength was estimated from the spectrum of the light source ($I$ (mW cm$^{-2}$ nm$^{-1}$) vs. $\lambda$ (nm)) provided by the supplier. The specifications of the light source used in this work have been described in previous papers [19,25]. In **2** and **3**, the most developed crystal surfaces were selected for irradiation. For **3**, the radiation was incident on the $(01\,\bar{1})$-planes, which do not belong to any crystal surface.

### 3.3. Physical Property Measurements: General

Except for the magnetic susceptibility measurements, all of the single crystals used in the measurements were briefly checked using X-ray oscillation photographs to identify the crystal quality and the directions of the crystallographic axes. For the magnetic susceptibility measurements, several of the single crystals were subjected to X-ray oscillation photographs to identify the crystal phase. The data are shown as averaged values, and the errors were estimated based on scattering of the data of independent measurements. After the measurements under UV radiation, the sample crystals were again checked with X-ray photographs and/or by the same measurements in the dark to confirm that there was no crystallographic deterioration upon irradiation. To check reproducibility and sample dependence, the magnetic susceptibility and resistivity measurements were examined repeatedly using the same samples, as well as different samples independently prepared/irradiated.

### 3.4. Magnetic Susceptibility Measurements

The temperature dependence of the magnetic susceptibility was measured on polycrystalline samples using a superconducting quantum interference device (SQUID) MPMS-XL7minBXR3 (Quantum Design, San Diego, CA, USA) using the DC method at 300–2 K. The measurements were carried out under a low pressure He atmosphere in the dark. The polycrystalline sample (typical amount, ~1 mg; 1 $\times$ 10$^{-6}$ mol) was set in the field. The sample was placed in a gelatin capsule (5 mm in inner diameter, 13.9 mm in height, EM Japan, Tokyo, Japan No. G7330, No.4), which was fixed in the middle of a polyethylene straw (Quantum Design). The data correction for the sample holder was negligibly small (ca. 1 $\times$ 10$^{-8}$ cm$^3\cdot$mol$^{-1}$) across the temperature range, as proven by measuring its susceptibility from room temperature to 4 K. The temperature was varied by 0.5–2 K/min. The applied fields were 0.8–1 T. Linearity of the magnetization up to 1.2 T was checked at 300 K by measuring the magnetization curves.

### 3.5. Electrical Resistivity Measurements

The electrical resistivity measurements were carried out using a personal computer-controlled home-made cryostat system. The system is composed of an insertion module with a high-vacuum jacket and a pumping/helium gas inlet valve, a Keithley 2400 SourceMeter, a Keithley 2182A Nanovoltmeter, a Keithley 6487 Picoammeter/Voltage Source, a Keithley 7001 Switch System (Keithley, Cleveland, Ohio, USA) and a LakeShore 331 Temperature Controller (Lake Shore Cryotronics, Inc., Westerville, Ohio, USA) with a silicon diode thermometer DT-470-SD. Linearity between the applied current and the observed voltage drop in a sample (the ohmic contact) was checked at the beginning of every set of measurements. The electrical resistivity was measured via a two-probe method along the longitudinal direction of the needle (1); elongated hexagonal plate (2); or irregularly-shaped block (3) single crystal. These directions approximately coincide with the crystallographic $a$-axis for **1**, the $b$-axis for **2** and nearly the $a$-axis for **3** (the $a$-axis makes an angle of $20°$ with the long axis of the crystal in **3**). The resultant currents were measured under a constant applied voltage. The resistivity of **2** was also measured along the $a$-axis at RT and gave a similar value of resistivity to that along the $b$-axis, both in the dark and under UV radiation. Gold wires (25 μm in diameter) and gold paste (No.8560, Tokuriki Chemical Research Co., Ltd., Tokyo, Japan) were used as electrical contacts.

### 3.6. Electron Spin Resonance

The electron spin resonance spectra of the X-band (9.3 GHz) were measured on a single crystal in the temperature range 120–300 K using a JEOL JES FA100 equipped with a continuous flow-type liquid $N_2$ cryostat with a digital temperature controller (JEOL, Tokyo, Japan). The temperature was controlled so as not to allow the temperature variation to exceed $\pm0.5$ K during the field sweep. A single crystal of **1–3** was mounted on a Teflon piece settled with a minimal amount of Apiezon N grease, sealed in a 5 mm diameter quartz sample tube in a low-pressure (~20 mmHg) helium atmosphere. In some measurements, a single crystal of DPPH (1,1-diphenyl-2-picrylhydrazyl; $g = 2.00366 \pm 0.00004$) [81–83] was also mounted on the Teflon piece beside **1–3** as an internal standard of $g$-values. In the measurements under UV irradiation, the standard sample was mounted on the opposite face of the Teflon piece in order to avoid UV irradiation on the standard sample. The magnetic field was applied perpendicular to the directions where the closest intermolecular interactions were expected based on the TBBC. They were parallel to the $[01\bar{1}]$-direction for **1**, parallel to the $[10\bar{1}]$-direction for **2** and parallel to the $[011]$-direction for **3**. The magnetic field was corrected by a gauss meter (JEOL NMR Field Meter ES-FC5) at the end of every measurement. The UV light was incident on the (011)-planes for **1**, on the (101)-planes for **2** and on the $(01\bar{1})$-planes for **3**. In the measurements of anisotropy, the angle $\theta$ defines the rotation angle around the $a$-axis in **1**, the $b$-axis in **2** and the axis normal to the $bc$-plane in **3**. With $\theta$ increasing, the single crystals were rotated around the fixed $H$ in the direction indicated by the circular arrows in Figure 2b,d,e, respectively. Magnetic fields ($H$) are initially ($\theta = 0$) applied parallel to $[01\bar{1}]$ for **1**, $[10\bar{1}]$ for **2** and $[011]$ for **3**. The molecular orientations relative to $H$ at the maximum and minimum values of $g$-values are shown in Figure 7. The time constant, sweep time and the modulation were common for **1–3**: 0.01 s, 30 s and 100 kHz, respectively. Both dark and UV irradiated spectra were measured under identical amplitudes, modulations and time constants for all three salts. The spectra simulation was carried out using Anisotropic Simulation software AniSim/FA ver. 2.2.0 by JEOL. In the simulation of the ESR spectra containing both signals from DPPH and the Cu(dmit)$_2$ salts, the parameters for the DPPH-signal of $g$-value, hyperfine coupling constant $A$ (mT) of $^{14}$N, linewidth $\Gamma$ (mT) and the ratio between Lorentzian and Gaussian were unified among different spectra, which was obtained from the independent measurement and simulation analysis on the single crystal of DPPH alone (Figure S7n), and only the relative intensity $I_{rel}$ (%) of the DPPH-signal was varied to reproduce the observed spectra. Under UV irradiation, the ESR signals generally contain both contributions from dark and irradiated parts of each sample, as UV does not always penetrate the single crystals. Thus, it was required in the simulation to assume many oscillators to reproduce the observed spectra under UV irradiation. However, the number of oscillators thus required was always beyond the limitation of the software. Therefore, we tentatively carried out the simulation assuming a few (a maximum of four) oscillators to reproduce a part of the spectrum and then carried out the simulation again for the same observed spectrum assuming different parameters in order to reproduce a different part of the spectrum. This procedure was repeated until all spectral features were covered. Finally, all of the obtained simulated spectra were overlapped to compare to the observed spectrum, and the parameters were modified to reproduce the observed spectrum better. These procedures were repeated until the simulated spectra best agreed with the observed spectrum. As a result, the spectral features were sensitive to the parameters concerning Cu(II), and reproduction of the observed spectra required two independent parameter sets for Cu(II) corresponding to the dark and UV irradiated states. However, the separation of contribution from the dmit ligands into dark and UV irradiated states had arbitrariness, and one could not determine the unique parameter sets of the dmit ligands corresponding to the dark and UV irradiated states in each spectrum. Such a difference between Cu(II) and the dmit ligands probably originates from the differences in the linewidths, the hyperfine structures and spectral differences between dark and UV irradiated conditions of their ESR signals. Therefore, in the simulation of the ESR spectra under UV irradiation, two sets of parameters were obtained for Cu(II), while only one set of parameters was obtained for the dmit ligands (Tables S5, S9 and S10–S12). The former corresponds to the contributions from dark and UV irradiated parts, respectively, while the latter contains both contributions from the dark and UV irradiated parts.

### 3.7. Calculations

The tight-binding band calculation was carried out via an extended Hückel method using Caesar 1.0 and 2.0 (PrimeColor Software, Inc., Raleigh, North Carolina, USA). The quantum chemistry calculations of the molecular orbitals, structural optimization and Mulliken analyses for spin and electron densities were carried out using Gaussian 09, and the results were illustrated using GaussView 5.0 [84]. The parameter- and calculation-level dependencies were examined, and it was confirmed that there were no significant differences among the calculation results with different parameters, basis sets or levels of calculation. For example, Mulliken analyses of the $[Cu(dmit)_2]^{2-}$ using Gaussian 09 gave nearly identical results to those from an extended Hückel method using Caesar 2.0. A series of quantum chemistry calculations for isolated $[Cu(dmit)_2]^{2-}$ anions, B3LYP/6-31G(3d), B3LYP/6-31G+(3d), B3LYP/6-311G+(3d) and B3LYP/6-311G+(3df), were utilized as basis sets and gave nearly identical results to each other. Thus, the discussion and conclusion above are not considered to be affected or altered in any qualitative way by different parameters/basis sets or different levels of calculations within the range we studied. In the calculation of $g$-values for **1** and **2**, B3LYP/6-31+G(3d) was utilized as the basis set, and the optimized molecular structures were assumed, which were nearly identical to the observed structures in the crystals. The contributions of dmit $\psi$(dmit) and Cu $\psi$(Cu) to the MOs $\varphi_a$ and $\varphi_b$ were respectively estimated based on the summation of the squares of the coefficients ($a_i$ and $b_i$ in Equations (3) and (4), respectively) of the atomic orbitals ($\varphi_i$ in Equations (3) and (4)) involved in the extended Hückel MOs of $[Cu(dmit)_2]^{2-}$:

$$\varphi_a = \sum_i a_i \phi_i \tag{3}$$

$$\varphi_b = \sum_i b_i \phi_i \tag{4}$$

where summation in Equations (3) and (4) is carried out on the atomic orbitals Cu $4s$, $4p$, $3d$; S $3s$, $3p$; C $2s$, $2p$.

$$\psi(\mathbf{A}) = \frac{\sum_{j \in \mathbf{A}} c_j^2}{\sum_i c_i^2} \tag{5}$$

where A designates dmit or Cu, the summation on $j$ is carried out on the atomic orbitals belonging to the molecular fragment A in regard to $\varphi_a$ or $\varphi_b$ orbitals and the summation on $i$ is carried out on all of the atomic orbitals involved in $\varphi_a$ or $\varphi_b$. In **3**, there are mixed bands of $[Cu(dmit)_2]^{2-}$ and $BP_2BDF^{2+}$ in the possible final states under UV radiation of 250–450 nm. For simplicity, the corresponding mixing was ignored in the estimation of $\psi$(Cu), since such approximation does not significantly alter the contribution ratio between Cu and dmit in any band.

## 4. Conclusions

The $[Cu(dmit)_2]^{2-}$ anions in **1–3** adapt the flexible coordination geometries in concert with different counter-ions. Calculations show that this is not because of intermolecular interactions, but because of the thermodynamic stability of the anion for **1** and **3**, while this is because of the intermolecular interactions for **2**. Although such a change of the counter cations alters the molecular structures and the electrical properties, the charge and spin distributions on $[Cu(dmit)_2]^{2-}$ are not different from each other at their ground states. This is because $[Cu(dmit)_2]^{2-}$ flexibly changes its molecular structure in the given solid states using the many degrees of freedom to have the lowest energy possible at ambient temperature ($T \leqslant 300$ K). As the charge and spin distributions on $[Cu(dmit)_2]^{2-}$ dominate the anion's energy, the same thermodynamic condition gives similar charge and spin distributions. On the other hand, UV irradiation alters the spin distribution and anisotropy in a qualitative and reversible way, because the excited states have rather different molecular orbitals from that of the ground state. Such photo-response in spin degrees of freedom is characteristic of $[Cu(dmit)_2]^{2-}$ anions. Although Compounds **1–3** did not switch between conductors and magnets, these findings indicate that optical excitation in the $[Cu(dmit)_2]^{2-}$ anions should be an effective method to control spin distribution and anisotropy.

**Acknowledgments:** The authors are grateful for Tamotsu Inabe at Hokkaido University for providing us DABCO· $HPF_6$ and $BP_2DBF(PF_6)_2$. The authors acknowledge financial support from Grant-in-Aid for Research Promotion, Ehime University, from The Dean's Research Grant, Faculty of Science, Ehime University, from Collaborative Research Grant, Graduate School of Science and Engineering, Ehime University, and also technical support from the Integrated Center for Sciences, Ehime University. The magnetic susceptibility measurements were carried out using the SQUID at the Department of Physics, Faculty of Science, Ehime University.

**Author Contributions:** Toshio Naito conceived of and designed the experiments; Hiroki Noma performed the experiments. Toshio Naito and Hiroki Noma analyzed the data; Keishi Ohara contributed with the ESR measurements and their analysis tools. Toshio Naito wrote the paper.

## References

1. Tissot, A. Photoswitchable spin crossover nanoparticles. *New J. Chem.* **2014**, *38*, 1840–1845. [CrossRef]
2. Hatcher, L.E.; Lauren, E.; Raithby, P.R. Dynamic single-crystal diffraction studies using synchrotron radiation. *Coord. Chem. Rev.* **2014**, *277–278*, 69–79. [CrossRef]
3. Gütlich, P.; Gasper, A.B.; Garcia, Y. Spin states switching in iron coordination compounds. *Beilstein J. Org. Chem.* **2013**, *9*, 342–391. [CrossRef] [PubMed]
4. Letard, J.-F. Photomagnetism of iron(II) spin crossover complexes—The *T*(LIESST) approach. *J. Mater. Chem.* **2006**, *16*, 2550–2559. [CrossRef]
5. Varret, F.; Boukheddaden, K.; Codjovi, E.; Enachescu, C.; Linares, J. On the competition between relaxation and photoexcitations in spin crossover solids under continuous irradiation. In *Spin Crossover in Transition Metal Compounds II, Topics in Current Chemistry*; Springer-Verlag Berlin: Berlin, Germany, 2004; Volume 234, pp. 199–229.
6. Gütlich, P.; Garcia, Y.; Spering, H. Spin transition phenomena. In *Magnetism: Molecules to Materials IV*; Miller, J.S., Drillon, M., Eds.; Wiley-VCH Verlag GmbH & Co.: Weinheim, Germany, 2003; pp. 271–344.
7. Spiering, H.; Kohlhaas, T.; Romstedt, H.; Hauser, A.; Bruns-Yilmaz, C.; Kusz, J.; Gütlich, P. Correlations of the distribution of spin states in spin crossover compounds. *Coord. Chem. Rev.* **1999**, *190–192*, 629–647. [CrossRef]
8. Hauser, A.; Jeftic, J.; Romstedt, H.; Hinek, R.; Spiering, H. Cooperative phenomena and light-induced bistability in iron(II) spin-crossover compounds. *Coord. Chem. Rev.* **1999**, *190–192*, 471–491. [CrossRef]
9. Gütlich, P.; Ensling, J.; Tuczek, F. Metastable electronic states induced by nuclear decay and light. *Hyperfine Interact.* **1994**, *84*, 447–469. [CrossRef]
10. Koshihara, S. Photo-induced phase transitions in organic and inorganic semiconductors. *J. Lumin.* **2000**, *87–89*, 77–81. [CrossRef]

11.  Först, M.; Hoffmann, M.C.; Dienst, A.; Kaiser, S.; Rini, M.; Tobey, R.I.; Gensch, M.; Manzoni, C.; Cavalleri, A. THz control in correlated electron solids: Sources and applications. In *Terahertz Spectroscopy and Imaging*; Peiponen, K.-E., Zeitler, A., Kuwata-Gonokami, M., Eds.; Springer Berlin Heidelberg: Berlin, Germany, 2013; pp. 611–631.

12.  Mori, T. Structural genealogy of BEDT-TTF-based organic conductors I. Parallel molecules: β and β″ phases. *Bull. Chem. Soc. Jpn.* **1998**, *71*, 2509–2526. [CrossRef]

13.  Mori, T.; Mori, H.; Tanaka, S. Structural genealogy of BEDT-TTF-based organic conductors II. Inclined molecules: θ, α and κ phases. *Bull. Chem. Soc. Jpn.* **1999**, *72*, 179–197. [CrossRef]

14.  Mori, T. Structural genealogy of BEDT-TTF-based organic conductors III. Twisted molecules: δ and α′ phases. *Bull. Chem. Soc. Jpn.* **1999**, *72*, 2011–2027. [CrossRef]

15.  West, A.R. Electrical properties. In *Basic Solid State Chemistry*; John Wiley & Sons: Chichester, UK, 1994; pp. 281–375.

16.  Cox, P.A. Electronic structure of solids. In *Solid State Chemistry: Compounds*; Cheetham, A.K., Day, P., Eds.; Clarendon Press: Oxford, UK, 1992; pp. 1–30.

17.  Ishikawa, T.; Hayes, S.A.; Keskin, S.; Corthey, G.; Hada, M.; Pichugin, K.; Marx, A.; Hirscht, J.; Shionuma, K.; Onda, K.; *et al.* Direct observation of collective modes coupled to molecular orbital-driven charge transfer. *Science* **2015**, *350*, 1501–1505. [CrossRef] [PubMed]

18.  Naito, T., Ed.; *Molecular Electronic and Related Materials: Control and Probe with Light*; Transworld Research Network: Kerala, India, 2010; pp. 1–320.

19.  Naito, T.; Inabe, T.; Niimi, H.; Asakura, K. Light-induced transformation of molecular materials into devices. *Adv. Mater.* **2004**, *16*, 1786–1790. [CrossRef]

20.  Miyamoto, T.; Niimi, H.; Chun, W.-J.; Kitajima, Y.; Sugawara, H.; Inabe, T.; Naito, T.; Asakura, K. Chemical states of Ag in Ag(DMe-DCNQI)$_2$ photoproducts and a proposal for its photoinduced conductivity change mechanism. *Chem. Lett.* **2007**, *36*, 1008–1009. [CrossRef]

21.  Naito, T.; Kakizaki, A.; Wakeshima, M.; Hinatsu, Y.; Inabe, T. Photochemical modification of magnetic properties in organic low-dimensional conductors. *J. Solid State Chem.* **2009**, *182*, 2733–2742. [CrossRef]

22.  Naito, T.; Kakizaki, A.; Inabe, T.; Sakai, R.; Nishibori, E.; Sawa, H. Growth of nanocrystals in a single crystal of different materials: A way of giving function to molecular crystals. *Cryst. Growth Des.* **2011**, *11*, 501–506. [CrossRef]

23.  Miyamoto, T.; Kitajima, Y.; Sugawara, H.; Naito, T.; Inabe, T.; Asakura, K. Origin of photochemical modification of resistivity of Ag(DMe-DCNQ)$_2$ studied by X-Ray absorption fine structure. *J. Phys. Chem. C* **2009**, *113*, 20476–20480. [CrossRef]

24.  Miyamoto, T.; Niimi, H.; Kitajima, Y.; Naito, T.; Asakura, K. Ag L$_3$-edge X-Ray absorption near-edge structure of 4d$^{10}$ (Ag$^+$) compounds: Origin of the edge peak and its chemical relevance. *J. Phys. Chem. A* **2010**, *114*, 4093–4098. [CrossRef] [PubMed]

25.  Naito, T.; Sugawara, H.; Inabe, T.; Kitajima, Y.; Miyamoto, T.; Niimi, H.; Asakura, K. UV-VIS induced vitrification of a molecular crystal. *Adv. Funct. Mater.* **2007**, *17*, 1663–1670. [CrossRef]

26.  Naito, T.; Sugawara, H.; Inabe, T. Mechanism of spatially resolved photochemical control of resistivity of a molecular crystalline solid. *Nanotechnology* **2007**, *18*, 424008. [CrossRef] [PubMed]

27.  Tsutsumi, T.; Miyamoto, T.; Niimi, H.; Kitajima, Y.; Sakai, Y.; Kato, M.; Naito, T.; Asakura, K. Energy-filtered X-Ray photoemission electron microscopy and its applications to surface and organic materials. *Solid State Electron.* **2007**, *51*, 1360–1366. [CrossRef]

28.  Saiki, T.; Mori, S.; Ohara, K.; Naito, T. Capacitor-like behavior of molecular crystal β-DiCC[Ni(dmit)$_2$]. *Chem. Lett.* **2014**, *43*, 1119–1121. [CrossRef]

29.  Burdett, J.K. Structural and compositional basis of high-temperature superconductivity: Properties of the magic electronic state. *Inorg. Chem.* **1993**, *32*, 3915–3922. [CrossRef]

30.  Yamamoto, H.M.; Ito, H.; Shigeto, K.; Tsukagoshi, K.; Kato, R. Direct formation of micro-/nanocrystalline 2,5-dimethyl-*N,N*′-dicyanoquinonediimine complexes on SiO$_2$/Si substrates and multiprobe measurement of conduction properties. *J. Am. Chem. Soc.* **2006**, *128*, 700–701. [CrossRef] [PubMed]

31.  Naito, T.; Karasudani, T.; Mori, S.; Ohara, K.; Konishi, K.; Takano, T.; Takahashi, Y.; Inabe, T.; Nishihara, S.; Inoue, K. Molecular photoconductor with simultaneously photocontrollable localized spins. *J. Am. Chem. Soc.* **2012**, *134*, 18656–18666. [CrossRef] [PubMed]

32. Naito, T.; Karasudani, T.; Ohara, K.; Takano, T.; Takahashi, Y.; Inabe, T.; Furukawa, K.; Nakamura, T. Simultaneous control of carriers and localized spins with light in organic materials. *Adv. Mater.* **2012**, *24*, 6153–6157. [CrossRef] [PubMed]

33. Naito, T.; Karasudani, T.; Nagayama, N.; Ohara, K.; Konishi, K.; Mori, S.; Takano, T.; Takahashi, Y.; Inabe, T.; Kinose, S.; *et al.* Giant photoconductivity in NMQ[Ni(dmit)$_2$]. *Eur. J. Inorg. Chem.* **2014**, *2014*, 4000–4009. [CrossRef]

34. Alvarez, S.; Vicente, R.; Hoffmann, R. Dimerization and stacking in transition-metal bisdithiolenes and tetrathiolates. *J. Am. Chem. Soc.* **1985**, *107*, 6253–6277. [CrossRef]

35. Cassoux, P.; Valade, L.; Kobayashi, H.; Kobayashi, A.; Clark, R.A.; Underhill, A.E. Molecular metals and superconductors derived from metal complexes of 1,3-dithiole-2-thione-4,5-dithiolate (dmit). *Coord. Chem. Rev.* **1991**, *110*, 115–160. [CrossRef]

36. Williams, J.M.; Schultz, A.J.; Geiser, U.; Carlson, K.D.; Kini, A.M.; Wang, H.H.; Kwok, W.-K.; Whangbo, M.-H.; Schirber, J.E. Organic superconductors—New benchmarks. *Science* **1991**, *252*, 1501–1508. [CrossRef] [PubMed]

37. Olk, R.-M.; Olk, B.; Dietzsch, W.; Kirmse, R.; Hoyer, E. The chemistry of 1,3-dithiole-2-thione-4,5-dithiolate (dmit). *Coord. Chem. Rev.* **1992**, *117*, 99–131. [CrossRef]

38. Svenstrup, N.; Becher, J. The organic chemistry of 1,3-dithiole-2-thione-4,5-dithiolate (DMIT). *Synthesis* **1995**, *3*, 215–235. [CrossRef]

39. Canadell, E. Electronic structure of two-band molecular conductors. *New J. Chem.* **1997**, *21*, 1147–1159.

40. Kato, R.; Liu, Y.-L.; Hosokoshi, Y.; Aonuma, S.; Sawa, H. Se-substitution and cation effects on the high-pressure molecular superconductor, β-Me$_4$N[Pd(dmit)$_2$]$_2$—A unique two-band system. *Mol. Cryst. Liq. Cryst.* **1997**, *296*, 217–244. [CrossRef]

41. Rosa, A.; Ricciardi, G.; Baerends, E.J. Structural properties of M(dmit)$_2$-based (M = Ni, Pd, Pt; dmit$^{2-}$ = 2-thioxo-1,3-dithiole-4,5-dithiolato) molecular metals. Insights from density functional calculations. *Inorg. Chem.* **1998**, *37*, 1368–1379. [CrossRef] [PubMed]

42. Pullen, A.E.; Olk, R.-M. The coordination chemistry of 1,3-dithiole-2-thione-4,5-dithiolate (dmit) and isologs. *Coord. Chem. Rev.* **1999**, *188*, 211–262. [CrossRef]

43. Kato, R. Conducting metal dithiolene complexes: Structural and electronic properties. *Chem. Rev.* **2004**, *104*, 5319–5346. [CrossRef] [PubMed]

44. Mori, H. Materials viewpoint of organic superconductors. *J. Phys. Soc. Jpn.* **2006**, *75*, 051003. [CrossRef]

45. Valade, L.; Tanaka, H. Molecular inorganic conductors and superconductors. In *Molecular Materials*; Bruce, D.W., O'Hare, D., Walton, R.I., Eds.; John Wiley & Sons Ltd: West Sussex, UK, 2010; pp. 211–280.

46. Mercuri, M.L.; Deplano, P.; Pilia, L.; Serpe, A.; Artizzu, F. Interaction modes and physical properties in transition metal chalcogenolene-based molecular materials. *Coord. Chem. Rev.* **2010**, *254*, 1419–1433. [CrossRef]

47. De Bonneval, B.G.; Ching, K.I.M.-C.; Alary, F.; Bui, T.-T.; Valade, L. Neutral d$^8$ metal bis-dithiolene complexes: Synthesis, electronic properties and applications. *Coord. Chem. Rev.* **2010**, *254*, 1457–1467. [CrossRef]

48. Papavassiliou, G.C.; Anyfantis, G.C.; Mousdis, G.A. Neutral metal 1,2-dithiolenes: Preparations, properties and possible applications of unsymmetrical in comparison to the symmetrical. *Crystals* **2012**, *2*, 762–811. [CrossRef]

49. Tanaka, H.; Okano, Y.; Kobayashi, H.; Suzuki, W.; Kobayashi, A. A Three-dimensional synthetic metallic crystal composed of single-component molecules. *Science* **2001**, *291*, 285–287. [CrossRef] [PubMed]

50. Kobayashi, A.; Tanaka, H.; Kobayashi, H. Molecular design and development of single-component molecular metals. *J. Mater. Chem.* **2001**, *11*, 2078–2088. [CrossRef]

51. Kobayashi, A.; Fujiwara, E.; Kobayashi, H. Single-component molecular metals with extended-TTF dithiolate ligands. *Chem. Rev.* **2004**, *104*, 5243–5264. [CrossRef] [PubMed]

52. Yamamoto, K.; Fujiwara, E.; Kobayashi, A.; Fujishiro, Y.; Nishibori, E.; Sakata, M.; Tanaka, M.; Tanaka, H.; Okano, Y.; Kobayashi, H. Single-component molecular conductor [Zn(tmdt)$_2$] and related Zn complexes. *Chem. Lett.* **2005**, *34*, 1090–1091. [CrossRef]

53. Kobayashi, A.; Okano, Y.; Kobayashi, H. Molecular design and physical properties of single-component molecular metals. *J. Phys. Soc. Jpn.* **2006**, *75*, 051002. [CrossRef]

54. Zhou, B.; Shimamura, M.; Fujiwara, E.; Kobayashi, A.; Higashi, T.; Nishibori, E.; Sakata, M.; Cui, H.B.; Takahashi, K.; Kobayashi, H. Magnetic transitions of single-component molecular metal [Au(tmdt)$_2$] and its alloy systems. *J. Am. Chem. Soc.* **2006**, *128*, 3872–3873. [CrossRef] [PubMed]

55. Zhou, B.; Kobayashi, A.; Okano, Y.; Nakashima, T.; Aoyagi, S.; Nishibori, E.; Sakata, M.; Tokumoto, M.; Kobayashi, H. Single-component molecular conductor [Pt(tmdt)$_2$] (tmdt = trimethylenetetrathiafulvalenedithiolate)—An advanced molecular metal exhibiting high metallicity. *Adv. Mater.* **2009**, *21*, 3596–3600. [CrossRef]

56. Zhou, B.; Yajima, H.; Kobayashi, A.; Okano, Y.; Tanaka, H.; Kumashiro, T.; Nishibori, E.; Sawa, H.; Kobayashi, H. Single-component molecular conductor [Cu(tmdt)$_2$] containing an antiferromagnetic heisenberg chain. *Inorg. Chem.* **2010**, *49*, 6740–6747. [CrossRef] [PubMed]

57. Zhou, B.; Idobata, Y.; Kobayashi, A.; Cui, H.-B.; Kato, R.; Takagi, R.; Miyagawa, K.; Kanoda, K.; Kobayashi, H. Single-component molecular conductor [Cu(dmdt)$_2$] with three-dimensionally arranged magnetic moments exhibiting a coupled electric and magnetic transition. *J. Am. Chem. Soc.* **2012**, *134*, 12724–12731. [CrossRef] [PubMed]

58. Idobata, Y.; Zhou, B.; Kobayashi, A.; Kobayashi, H. Molecular alloy with diluted magnetic moments—Molecular kondo system. *J. Am. Chem. Soc.* **2012**, *134*, 871–874. [CrossRef] [PubMed]

59. Cui, H.B.; Kobayashi, H.; Ishibashi, S.; Sasa, M.; Iwase, F.; Kato, R.; Kobayashi, A. A single-component molecular superconductor. *J. Am. Chem. Soc.* **2014**, *136*, 7619–7622. [CrossRef] [PubMed]

60. Yamashita, S.; Yamamoto, T.; Nakazawa, Y.; Tamura, M.; Kato, R. Gapless spin liquid of an organic triangular compound evidenced by thermodynamic measurements. *Nat. Commun.* **2011**, *2*, 275. [CrossRef] [PubMed]

61. Steimecke, G.; Kirmse, R.; Hoyer, E. Dimercaptoisotrithione. New, unsaturated 1,2-dithiolate ligand. *Z. Chem.* **1975**, *15*, 28–29. [CrossRef]

62. Stach, J.; Kirmse, R.; Dietzsch, W.; Olk, R.M.; Hoyer, E. Single-crystal EPR spectra of tetra-*n*-butylammonium bis(isotrithione-3,4-dithiolato)cuprate(II) (copper-63). *Inorg. Chem.* **1984**, *23*, 4779–4780. [CrossRef]

63. Matsubayashi, G.; Takahashi, K.; Tanaka, T. X-ray crystal structure of bis(*N*-ethylpyridinium) bis[4,5-dimercapto-1,3-dithiole-2-thionate(2-)]copper(II) and electrical properties of its oxidized salts. *J. Chem. Soc. Dalton Trans.* **1988**, 967–972. [CrossRef]

64. Guo, W.F.; Sun, X.B.; Sun, J.; Yu, W.T.; Wang, X.Q.; Zhang, G.H.; Xu, D. Preparation, single crystal growth and characterization of bis(tetrabutylammonium)bis(4,5-dithiolato-1,3-dithiole-2-thione)copper. *Cryst. Res. Technol.* **2007**, *42*, 349–335.

65. Guo, W.F.; Sun, X.B.; Sun, J.; Yu, W.T.; Wang, X.Q.; Zhang, G.H.; Xu, D. Synthesis, crystal structure and third order nonlinear optical properties of bis(tetra-*n*-propylammonium) bis(2-thioxo-1,3-dithiole-4,5-dithiolato)cuprate(II). *Cryst. Res. Technol.* **2007**, *42*, 522–528. [CrossRef]

66. Wang, X.-Q.; Yu, W.-T.; Xu, D.; Wang, Y.-L.; Li, T.-B.; Zhang, G.-H.; Sun, X.-B.; Ren, Q. Bis(tetraethylammonium) bis(2-thioxo-1,3-dithiole-4,5-dithiolato)cuprate(II). *Acta Crystallogr. E* **2005**, *61*, m717–m719. [CrossRef]

67. Li, T.; Hu, Y.; Ma, C.; He, G.; Zhao, R.; Li, J. Crystal structure and third-order nonlinear optical property study of a copper complex constructed by DMIT ligand. *Mater. Chem. Phys.* **2011**, *130*, 835–838. [CrossRef]

68. Noma, H.; Ohara, K.; Naito, T. [Cu(dmit)$_2$]$^{2-}$ Building block for molecular conductors and magnets with photocontrollable spin distribution. *Chem. Lett.* **2014**, *43*, 1230–1232. [CrossRef]

69. Frisch, M.J.; Trucks, G.W.; Schlegel, H.B.; Scuseria, G.E.; Robb, M.A.; Cheeseman, J.R.; Scalmani, G.; Barone, V.; Mennucci, B.; Petersson, G.A.; *et al. Gaussian 09, Revision C.01*; Gaussian, Inc.: Wallingford, CT, USA, 2009.

70. Mott, N.F. *Metal-Insulator Transitions*, 2nd ed.; Taylor & Francis: London, UK, 1990; pp. 123–144.

71. Jazwinski, J. Theoretical aspects of indirect spin-spin couplings. *Nucl. Magn. Res.* **2014**, *43*, 159–182.

72. Kirmse, R.; Stach, J.; Dietzsch, W.; Steimecke, G.; Hoyer, E. Single-crystal EPR studies on nickel(III), palladium(III), and platinum(III) dithiolene chelates containing the ligands isotrithionedithiolate, *o*-xylenedithiolate, and maleonitriledithiolate. *Inorg. Chem.* **1980**, *19*, 2679–2685. [CrossRef]

73. Teschmit, G.; Strauch, P.; Barthel, A.; Reinhold, J.; Kirmse, R. A single crystal EPR investigation on (*n*-Bu$_4$N)$_2$[Cu(dmit)$_2$] in the antiferro-magnetically coupled host lattice (*n*-Bu$_4$N)$_2$[(dmit)Cu(tto)Cu(dmit)]: A contribution to the nature of the so-called "paramagnetic impurities". *Z. Naturforsch. B* **1999**, *54*, 832–838. [CrossRef]

74. Nakamura, T.; Takahashi, T.; Aonuma, S.; Kato, R. EPR investigation of the electronic states in β'-type [Pd(dmit)$_2$]$_2$ compounds (where dmit is 2-thioxo-1,3-dithiole-4,5-dithiolate). *J. Mater. Chem.* **2001**, *11*, 2159–2162. [CrossRef]

75. Hoffmann, S.K.; Goslar, J.; Lijewski, S.; Zalewska, A. EPR and ESE of CuS$_4$ complex in Cu(dmit)$_2$: *g*-Factor and hyperfine splitting correlation in tetrahedral Cu–sulfur complexes. *J. Magn. Reson.* **2013**, *236*, 7–14. [CrossRef] [PubMed]

76. Hoffmann, S.K.; Goslar, J.; Lijewski, S.; Tadyszak, K.; Zalewska, A.; Jankowska, A.; Florczak, P.; Kowalak, S. EPR and UV-VIS study on solutions of Cu(II) *dmit* complexes and the complexes entrapped in zeolite A and ZIF-Cu(IM)$_2$. *Microporous Mesoporous Mater.* **2014**, *186*, 57–64. [CrossRef]

77. Schmitt, R.D.; Maki, A.H. Electronic ground state of bis(maleonitrile-dithiolene)nickel monoanion. Sulfur-33 hyperfine interaction. *J. Am. Chem. Soc.* **1968**, *90*, 2288–2292. [CrossRef]

78. Kirmse, R.; Dietzsch, W. A single crystal EPR study of the palladium(III)-bis(maleonitriledithiolate) monoanion. *J. Inorg. Nucl. Chem.* **1976**, *38*, 255–257. [CrossRef]

79. Kirmse, R.; Dietzsch, W.; Solovev, B.V. A single crystal EPR study of the platinum (III)-bis(maleonitriledithiolate)-monoanion. *J. Inorg. Nucl. Chem.* **1977**, *39*, 1157–1160. [CrossRef]

80. Steimecke, G.; Sieler, H.-J.; Kirmse, R.; Hoyer, E. 1,3-dithiole-2-thione-4,5-dithiolate from carbon disulfide and alkali metal. *Phosphorus Sulfur* **1979**, *7*, 49–55. [CrossRef]

81. Yordanov, N.D. Quantitative EPR spectrometry—"State of the art". *Appl. Magn. Reson.* **1994**, *6*, 241–257. [CrossRef]

82. Kai, A.; Miki, T. Electron spin resonance of sulfite radicals in irradiated calcite and aragonite. *Radiat. Phys. Chem.* **1992**, *40*, 469–476. [CrossRef]

83. Inokuchi, H.; Kinoshita, M. The oxygen effect on electronic properties of α,α'-diphenyl-β-picrylhydrazyl. *Bull. Chem. Soc. Jpn.* **1960**, *33*, 1627–1629. [CrossRef]

84. Dennington, R.; Keith, T.; Millam, J. *GaussView, Version 5*; Semichem Inc.: Shawnee Mission, KS, USA, 2009.

# β,β-Isomer of Open-Wells–Dawson Polyoxometalate Containing a Tetra-Iron(III) Hydroxide Cluster: [{Fe$_4$(H$_2$O)(OH)$_5$}(β,β-Si$_2$W$_{18}$O$_{66}$)]$^{9-}$

Satoshi Matsunaga, Eriko Miyamae, Yusuke Inoue and Kenji Nomiya *

Department of Chemistry, Faculty of Science, Kanagawa University, Hiratsuka, Kanagawa 259-1293, Japan; matsunaga@kanagawa-u.ac.jp (S.M.); miyamae.527@gmail.com (E.M.); r201470042gl@jindai.jp (Y.I.)
* Correspondence: nomiya@kanagawa-u.ac.jp

Academic Editor: Duncan H. Gregory

**Abstract:** The β,β-isomer of open-Wells–Dawson polyoxometalate (POM) containing a tetra-iron(III) cluster, K$_9$[{Fe$_4$(H$_2$O)(OH)$_5$}(β,β-Si$_2$W$_{18}$O$_{66}$)]·17H$_2$O (potassium salt of **β,β-Fe$_4$-open**), was synthesized by reacting Na$_9$H[A-β-SiW$_9$O$_{34}$]·23H$_2$O with FeCl$_3$·6H$_2$O at pH 3, and characterized by X-ray crystallography, FTIR, elemental analysis, TG/DTA, UV–Vis, and cyclic voltammetry. X-ray crystallography revealed that the {Fe$^{3+}_4$(H$_2$O)(OH)$_5$}$^{7+}$ cluster was included in the open pocket of the β,β-type open-Wells–Dawson polyanion [β,β-Si$_2$W$_{18}$O$_{66}$]$^{16-}$ formed by the fusion of two trilacunary β-Keggin POMs, [A-β-SiW$_9$O$_{34}$]$^{10-}$, via two W–O–W bonds. The β,β-open-Wells–Dawson polyanion corresponds to an open structure of the standard γ-Wells–Dawson POM. **β,β-Fe$_4$-open** is the first example of the compound containing a geometrical isomer of α,α-open-Wells–Dawson structural POM.

**Keywords:** polyoxometalates; open-Wells–Dawson structural POM; iron; geometrical isomer

## 1. Introduction

Polyoxometalates (POMs) are discrete metal oxide clusters that are of interest as soluble metal oxides, with applications in catalysis, medicine, and materials science [1–13]. Recently, the open-Wells–Dawson POMs have been an emerging class of POMs [14–28]. The standard Wells–Dawson structural POM is regarded as an assembly of two trilacunary Keggin POMs via six W–O–W bonds. In 2014, Mizuno *et al.* reported the synthesis of a standard Wells–Dawson structural POM with a highly charged guest SiO$_4^{4-}$, TBA$_8$[α-Si$_2$W$_{18}$O$_{62}$]·3H$_2$O, by dimerization of a trilacunary Keggin POM, [α-SiW$_9$O$_{34}$]$^{10-}$, in an organic solvent [29]. However, in aqueous media, the electrostatic repulsion induced by the highly charged guest XO$_4^{4-}$ (X = Si, Ge) inhibits the assembly of the standard Wells–Dawson structure [30]. Therefore, the two trilacunary Keggin units with XO$_4^{4-}$ (X = Si, Ge) are linked by two W–O–W bonds, forming open-Wells–Dawson structural POMs. The open pocket of these POMs can accommodate up to six metals. Thus, these compounds constitute a promising platform for the development of metal-substituted POM-based materials and catalysts. To date, many compounds with various metal ions in their open pocket have been reported. For example, V$^{5+}$ [19], Mn$^{2+}$ [16,21], Fe$^{3+}$ [19], Co$^{2+}$ [14,16,20,21,25,27], Ni$^{2+}$ [16,21,24,27], Cu$^{2+}$ [15–17], Zn$^{2+}$ [23], Al$^{3+}$ [28], Ga$^{3+}$ [28], and lanthanoid (Eu$^{3+}$, Gd$^{3+}$, Tb$^{3+}$, Dy$^{3+}$, Ho$^{3+}$) [22,26].

The standard Wells–Dawson structural POM is one of the most deeply studied POMs. Baker and Figgis predicted the existence of six possible structural isomers of the Wells–Dawson POM (α, β, γ, α*, β*, γ*) [31]. So far, only the α-, β-, γ-, and γ*- isomers have been experimentally confirmed [32–37]. The α-Wells–Dawson POM has $D_{3h}$ symmetry (Figure 1, upper left). The β- and γ-Wells–Dawson isomers exhibit $C_{3v}$ and $D_{3h}$ symmetries, respectively, and are derived from the α-Wells–Dawson isomer by a 60° rotation of one (β-isomer) or two (γ-isomer) {M₃O₁₃} caps about the three-fold axis of the α-Wells–Dawson isomer (Figure 1, upper middle and right). Likewise, the α*-, β*-, and γ*-Wells–Dawson isomers are derived from the α-, β-, and γ-isomer by a 60° rotation of half {XM₉} units.

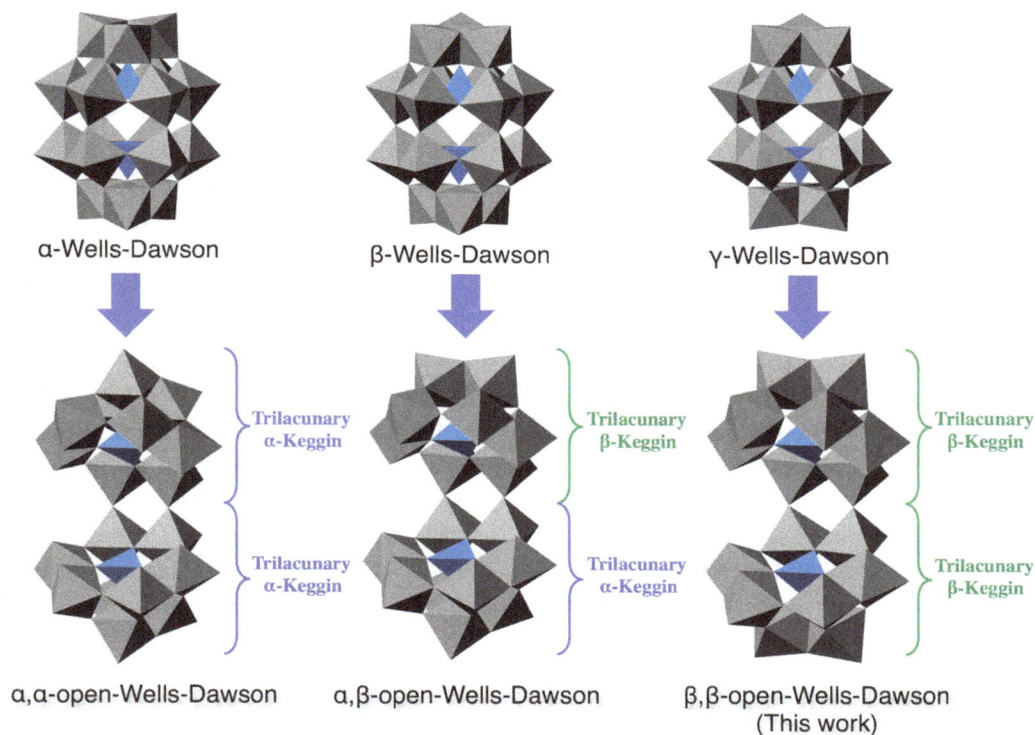

**Figure 1.** The isomers of the usual Wells–Dawson structural POMs and the open-Wells–Dawson POMs. Color code: $XO_4$, blue; $WO_6$, gray.

On the basis of the analogy with the standard Wells–Dawson isomers, the isomers of the open-Wells–Dawson POMs are possible. The well-known α,α-open-Wells–Dawson POM $[\alpha,\alpha\text{-}X_2W_{18}O_{66}]^{16-}$, built from two trilacunary α-Keggin units $[A\text{-}\alpha\text{-}XW_9O_{34}]^{10-}$ by two W–O–W bonds, corresponds to an open structure of the conventional α-Wells–Dawson POM (Figure 1, lower left). A similar assembly of two trilacunary β-Keggin units prompts the β,β-open-Wells–Dawson isomer, which is viewed as an open structure of the γ-Wells–Dawson POM (Figure 1, lower right). Likewise, an assembly of the trilacunary α- and β-Keggin POMs leads to the α,β-open-Wells–Dawson isomer corresponding to an open structure of the β-Wells–Dawson POM. However, only the α,α-isomer of the open-Wells–Dawson structural POM has been reported so far.

Herein, we report the synthesis of the β,β-isomer of the open-Wells–Dawson POM containing a tetra-iron(III) cluster, $K_9[\{Fe_4(H_2O)(OH)_5\}(\beta,\beta\text{-}Si_2W_{18}O_{66})]\cdot 17H_2O$ (potassium salt of **β,β-Fe₄-open**) and its characterization by X-ray crystallography, FTIR, elemental analysis, TG/DTA, UV–Vis, and cyclic voltammetry. The **β,β-Fe₄-open** is the first example of the compound containing a geometrical isomer of the α,α-open-Wells–Dawson structural POM.

## 2. Results and Discussion

### 2.1. Synthesis and Characterization

The potassium salts of **β,β-Fe₄-open**, $K_9[\{Fe_4(H_2O)(OH)_5\}(\beta,\beta\text{-}Si_2W_{18}O_{66})]\cdot 17H_2O$, were prepared by the reaction of the separately prepared $Na_9H[A\text{-}\beta\text{-}SiW_9O_{34}]\cdot 23H_2O$ with $FeCl_3\cdot 6H_2O$ in a 1:2 molar ratio at pH 3, followed by stirring the solution at 80 °C for 30 min. The β,β-type open-Wells–Dawson POM containing a tetra-iron(III) cluster was obtained as crystalline sample with 22.2% yield.

Equation (1) represents the formation of the polyoxoanion $[\{Fe_4(H_2O)(OH)_5\}(\beta,\beta\text{-}Si_2W_{18}O_{66})]^{9-}$.

$$2[\beta-SiW_9O_{34}]^{10-} + 4FeCl_3 + 4H_2O \rightarrow [\{Fe_4(H_2O)(OH)_5\}(\beta,\beta-Si_2W_{18}O_{66})]^{9-} + H^+ + 12Cl^- \qquad (1)$$

In the case of the $\alpha,\alpha$-analogue, i.e., $K_2Na_8[\{Fe_4(OH)_6\}(Si_2W_{18}O_{66})]\cdot 44H_2O$ ($\alpha,\alpha$-**Fe$_4$-open**) [19], the $K^+$ ions incorporating the open-Wells–Dawson POM, i.e., $K_{13}[\{K(H_2O)_3\}_2\{K(H_2O)_2\}(\alpha,\alpha\text{-}Si_2W_{18}O_{66})]\cdot 19H_2O$ (**K-open**), were used as a precursor, whereas the potassium salt of $\beta,\beta$-**Fe$_4$-open** was prepared from the trilacunary $\beta$-Keggin POM, i.e., $Na_9H[A\text{-}\beta\text{-}SiW_9O_{34}]\cdot 23H_2O$. Both $\alpha,\alpha$- and $\beta,\beta$-**Fe$_4$-open** were prepared at ca. pH 3. Excess $K^+$ ions are required for the formation of the $\alpha,\alpha$-open-Wells–Dawson POM [25,28], indicating that $K^+$ ions play an important role in the synthesis. In the case of the present $\beta,\beta$-open-Wells–Dawson POM, the addition of excess KCl was needed, indicating that $K^+$ ions also play an important role in the formation of the $\beta,\beta$-open-Wells–Dawson structural POMs.

The potassium salt of $\beta,\beta$-**Fe$_4$-open** was characterized via elemental analysis. Prior to analysis, the sample was dried overnight at room temperature under a vacuum of $10^{-3}$–$10^{-4}$ Torr. All elements (H, Fe, K, O, Si, and W) were analyzed for a total of 99.74%, indicating that the obtained compound was highly pure. The observed data was in accordance with the calculated values for the formula constituting four water molecules, i.e., $K_9[\{Fe_4(H_2O)(OH)_5\}(\beta,\beta\text{-}Si_2W_{18}O_{66})]\cdot 4H_2O$ (see Experimental Section). The weight loss observed during drying was 4.23% corresponding to ca. 13 crystallized water molecules (calcd. 4.33%), and therefore, the sample contained a total of 17 crystallized water molecules. On the other hand, during the TG/DTA measurements carried out under atmospheric conditions (Figure S2), a weight loss of 6.10%, observed at temperatures below 500 °C, corresponded to ca. 18 water molecules (calcd. 5.98%). Thus, the elemental analysis and TG/DTA displayed a presence of a total of 17–18 water molecules for the sample under atmospheric conditions. The formula for the potassium salt of $\beta,\beta$-**Fe$_4$-open** presented herein is decided as $K_9[\{Fe_4(H_2O)(OH)_5\}(\beta,\beta\text{-}Si_2W_{18}O_{66})]\cdot 17H_2O$ based on the results of the complete elemental analysis.

## 2.2. Molecular Structure

X-ray crystallography of $\beta,\beta$-**Fe$_4$-open** revealed that a $\beta,\beta$-type open-Wells–Dawson polyanion is formed by the fusion of two trilacunary $\beta$-Keggin POMs, $[A\text{-}\beta\text{-}SiW_9O_{34}]^{10-}$, via two W–O–W bonds, and the $\{Fe^{3+}_4(H_2O)(OH)_5\}^{7+}$ cluster is included in the open pocket of the $\beta,\beta$-type open-Wells–Dawson polyanion, $[\beta,\beta\text{-}Si_2W_{18}O_{66}]^{16-}$ (Figure 2). The structure of the $\beta,\beta$-open-Wells–Dawson polyanion moiety is derived by a 60° rotation of two $\{M_3O_{13}\}$ caps of the $\alpha,\alpha$-type isomer, corresponding to an open structure of the $\gamma$-Wells–Dawson POM; it is the first example of the isomer of the $\alpha,\alpha$-open-Wells–Dawson POM.

The four iron(III) centers in the open pocket were arranged in a rectangular array, and were connected to the neighboring iron atoms through edge-sharing oxygen atoms (O67, O68, O69, and O70) and corner-sharing oxygen atoms (O71 and O72). Bond valence sum (BVS) calculations [38] of the oxygen atoms connected to the iron atoms suggested that four of the oxygen atoms (O67, O69, O71, and O72) are protonated, i.e., they are ascribed to the hydroxide groups (the BVS values: O67, 1.106; O69, 1.157; O71, 1.170; O72, 1.255; Table S1). On the other hand, the BVS values of two inner edge-sharing oxygen atoms (O68, 0.796 and O70, 0.660; Table S1) were slightly lower than those of other oxygen atoms connected to the iron atoms. These results suggested that there is one proton between two oxygen atoms (O68 and O70), i.e., the iron(III) cluster in the open pocket can be represented as $\{Fe_4(H_2O)(OH)_5\}^{7+}$. All the iron atoms were in the +3 oxidation state (the BVS values: Fe1, 2.981; Fe2, 3.030; Fe3, 3.018; Fe4, 3.148; Table S1). These results were consistent with elemental analysis of the potassium salt of $\beta,\beta$-**Fe$_4$-open**. The Fe$\cdots$Fe distances connected through edge-sharing oxygen atoms were 3.103(4) (Fe1–Fe3) and 3.114(4) (Fe2–Fe4) Å, and through corner-sharing oxygen atoms were

3.648(5) (Fe1–Fe2) and 3.646(5) (Fe3–Fe4). This metal ion arrangement was similar to that of previously reported $\alpha,\alpha$-**Fe$_4$-open** [19], and [{M$_4$(OH)$_6$}($\alpha,\alpha$-Si$_2$W$_{18}$O$_{66}$)]$^{10-}$ (M = Al, Ga) [28].

The bite angle of β,β-**Fe$_4$-open** was 58.736°, similar to that of $\alpha,\alpha$-**Fe$_4$-open** (58.147°), and as for the structure around the open pocket, no clear difference was observed between the $\alpha,\alpha$- and β,β-**Fe$_4$-open**.

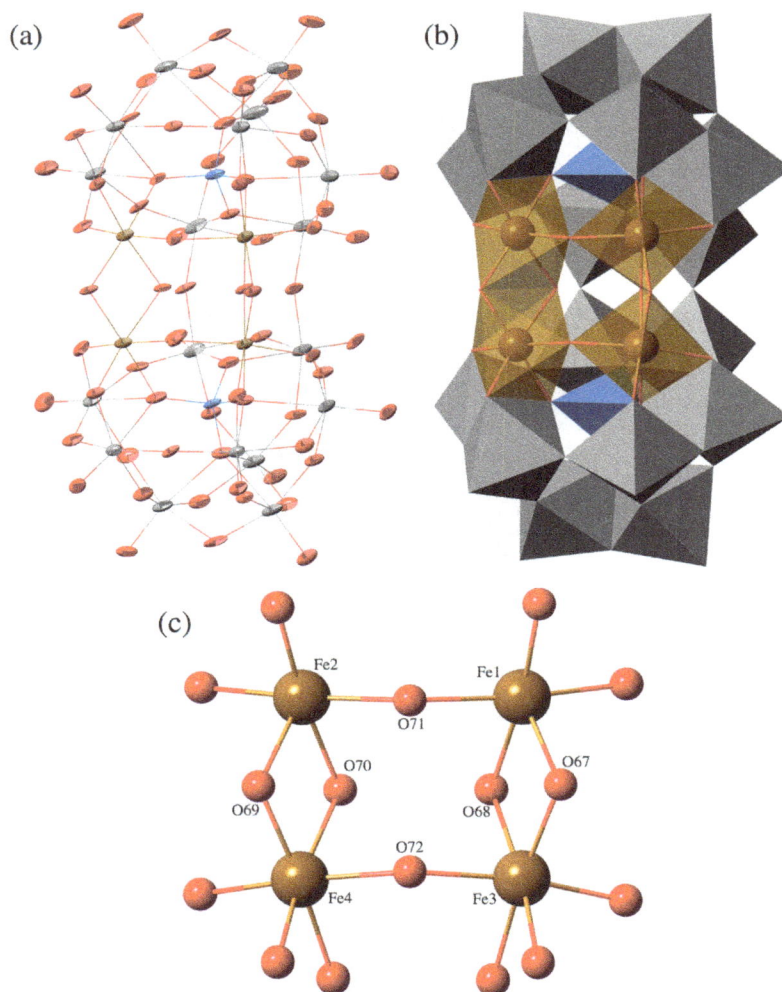

**Figure 2.** (**a**) Molecular structure of the polyoxoanion [{Fe$_4$(H$_2$O)(OH)$_5$}(β,β-Si$_2$W$_{18}$O$_{66}$)]$^{9-}$ of potassium salt of β,β-**Fe$_4$-open**; (**b**) its polyhedral representation; and (**c**) the partial structure around the Fe$_4$ center. Color code: Fe, brown; O, red; Si, blue; W, gray.

There are many interactions between the oxygen atoms of the polyanion moiety and K$^+$ in the crystal structure of β,β-**Fe$_4$-open**. In particular, the terminal (O15, O16, O37, O38) oxygen atoms of the WO$_6$ polyhedra serve as a hinge between the two trilacunary Keggin units interacting with K$^+$ cations (K1) (K1-O15, 2.800(14); K1-O16, 2.689(12); K1-O37, 2.685(14); K1-O38, 2.819(14) Å: Figure 3). These interactions play an important role in the formation of the β,β-type open-Wells–Dawson structural POMs and have been previously reported for the $\alpha,\alpha$-open-Wells–Dawson POMs [28].

## 2.3. Absorption Spectrum

The absorption spectrum of β,β-**Fe$_4$-open** in H$_2$O is shown in Figure 4. The shoulder band due to the O to Fe$^{3+}$ charge transfer [39] was observed around 450 nm ($\varepsilon$ = 98 M$^{-1}\cdot$cm$^{-1}$). This spectrum was similar to that of $\alpha,\alpha$-**Fe$_4$-open** (Figure S3).

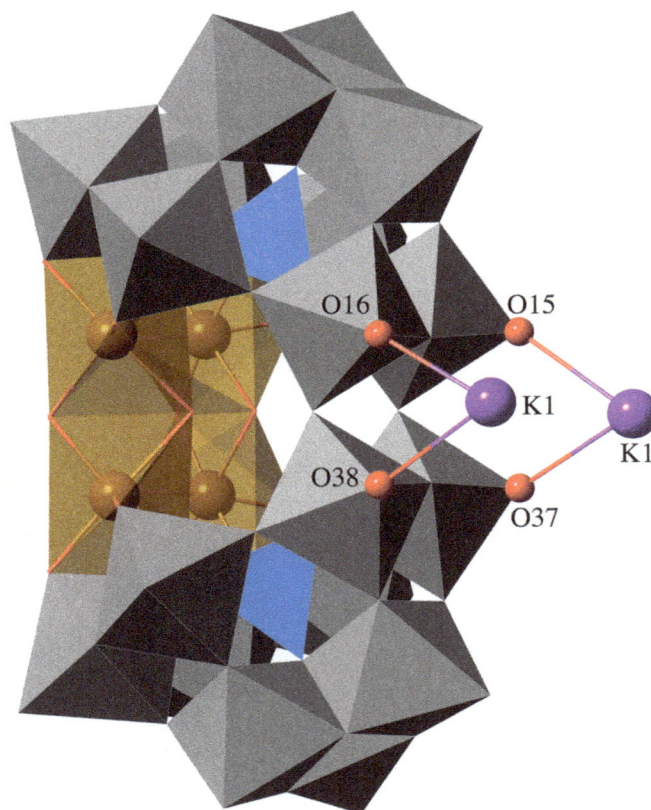

**Figure 3.** Interactions with $K^+$ ions of the terminal oxygen atoms of the $WO_6$ polyhedra serving as a hinge between the two trilacunary β-Keggin units of **β,β-Fe₄-open**.

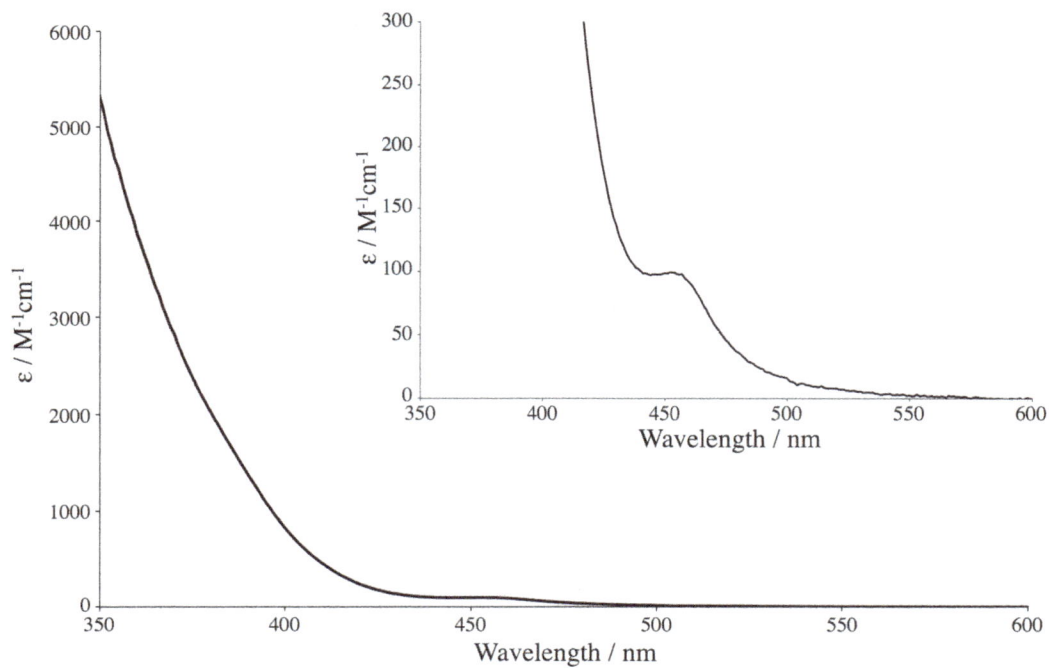

**Figure 4.** UV/Vis absorption spectrum of potassium salt of **β,β-Fe₄-open** in $H_2O$. Inset shows enlarged view.

## 2.4. Electrochemistry

The cyclic voltammogram of 0.5 mM **β,β-Fe₄-open** conducted in 0.5 M KOAc/HOAc buffer (pH 4.8) solution at a scan rate of 25 mV·s$^{-1}$ showed three characteristic peaks at −0.665, −0.794, and −0.939 V (*vs.* Ag/AgCl), respectively. These peaks were associated with $W^{6+}$ centered reduction processes (Figure 5). Hill *et al.* have reported similar redox processes based on Zn-containing open-Wells–Dawson POMs [23], and noted that the second reduction wave associated with $W^{6+}$ is a two-electron process. The cyclic voltammogram of 0.5 mM **α,α-Fe₄-open** in 0.5 M KOAc/HOAc buffer (pH 4.8) solution at a scan rate of 25 mV/s was similar to that of **β,β-Fe₄-open** (Figure S4). However, the reduction processes of **α,α-Fe₄-open** based on $W^{6+}$ were observed at a slightly more negative potential *i.e.*, (−0.704, −0.866, −0.936 V *vs.* Ag/AgCl) (Figure S4).

On the other hand, the redox waves observed at −0.212 and 0.380 V can be attributed to the redox processes of $Fe^{3+}$ centers, since the Zn-containing open-Wells–Dawson POM displayed no redox waves in this region [23]. The area ratio between the second reduction wave of $W^{6+}$ (two-electron process, $E$ = −0.794 V) and the reduction of $Fe^{3+}$ ($E$ = −0.212 V) is *ca.* 1:2, *i.e.*, 2:4 electrons. Based on the comparison of the area ratio, the wave is likely to be the simultaneous one-electron redox of the four-$Fe^{3+}$ center.

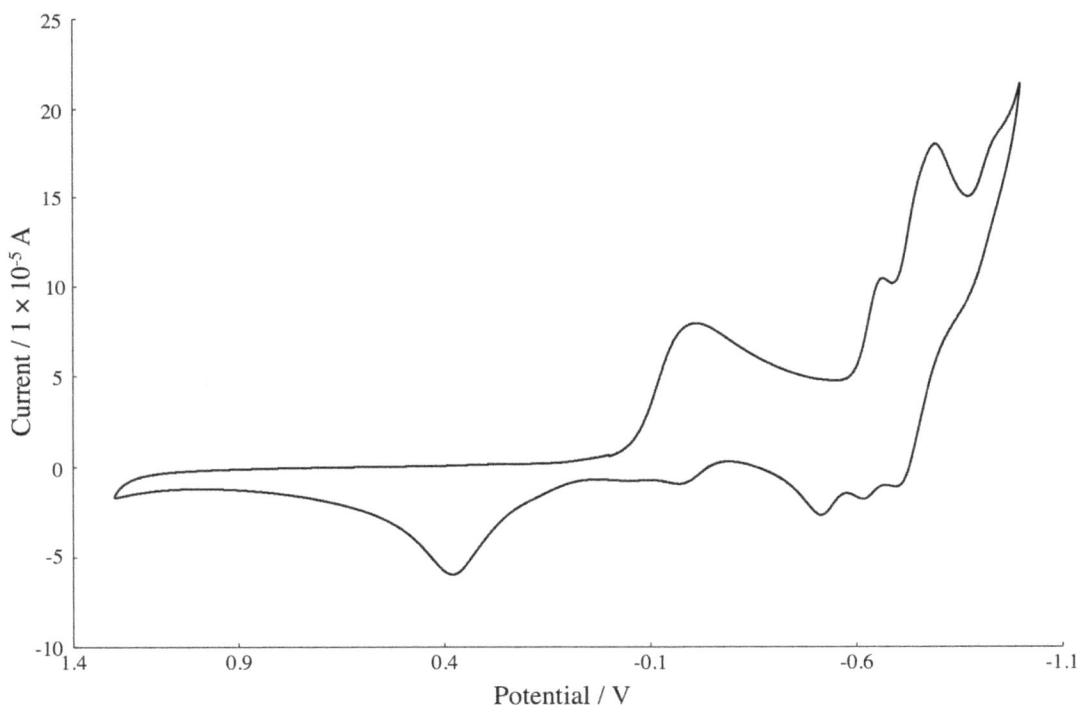

**Figure 5.** Cyclic voltammograms (CV) of 0.5 mM potassium salt of **β,β-Fe₄-open** in 0.5 M potassium acetate buffer, pH 4.8, scan rate 25 mV·s$^{-1}$, under $N_2$.

## 3. Experimental Section

### 3.1. Materials

The following reagents were used as received: $FeCl_3 \cdot 6H_2O$, HCl, KOH, and KCl (from Wako Pure Chemical Industries, Osaka, Japan). The sodium salt of trilacunary Keggin POM, *i.e.*, $Na_9H[A\text{-}\beta\text{-}SiW_9O_{34}] \cdot 23H_2O$, and **α,α-Fe₄-open** were prepared according to the literature method [19,40], and identified by X-ray crystallography, TG/DTA, FT-IR, UV/Vis absorption spectra, and cyclic voltammetry.

### 3.2. Instrumentation and Analytical Procedures

Elemental analyses were carried out by Mikroanalytisches Labor Pascher (Remagen, Germany). The sample was dried overnight at room temperature under $10^{-3}$–$10^{-4}$ Torr before analysis. Infrared spectra were recorded on a Jasco 4100 FTIR spectrometer (Jasco, Hachioji, Japan) using KBr disks at room temperature. Thermogravimetric (TG) and differential thermal analyses (DTA) were acquired using a Rigaku Thermo Plus 2 series TG8120 instrument (Rigaku, Akishima, Japan). TG/DTA measurements were run under air with a temperature ramp of 4.0 °C/min between 20 and 500 °C. Absorption spectra in $H_2O$ were obtained on a JASCO V-630 spectrophotometer (Jasco). Cyclic voltammetry was performed with ALS/CH Instruments (BAS, Sumida-ku, Japan), a Model 610E electrochemical analyzer with a three electrode cell in 0.5 M KOAc/HOAc buffer (pH 4.8) under $N_2$ atmosphere. A glassy carbon working electrode, a Pt auxiliary electrode and a Ag/AgCl reference electrode were employed. The scan rate was 25 mV·$s^{-1}$.

### 3.3. Synthesis of $K_9[\{Fe_4(H_2O)(OH)_5\}(\beta,\beta\text{-}Si_2W_{18}O_{66})]\cdot 17H_2O$ (Potassium Salt of $\beta,\beta\text{-}Fe_4\text{-}Open$)

$Na_9H[A\text{-}\beta\text{-}SiW_9O_{34}]\cdot 23H_2O$ (1.00 g, 0.351 mmol) was suspended in 50 mL of water. The pH was adjusted to 3.0 using 1.0 M $HCl_{aq}$, and 1.0 mL of saturated $KCl_{aq}$. was added to the solution. Upon addition of $FeCl_3\cdot 6H_2O$ (0.190 g, 0.703 mmol) to the solution, the pH dropped to 1.57. Subsequently, the pH was adjusted to 3.0 using 1.0 M $KOH_{aq}$. Thereafter, the solution was heated to 80 °C for 30 min and the pH was re-adjusted to 3.0 using 1.0 M $KOH_{aq}$. The resulting solution was left to stand undisturbed at room temperature for one week. The resulting yellow plate crystals were collected on a membrane filter (JG 0.2 μm), and dried *in vacuo* for 2 h to obtain a yield of 0.211 g (0.0390 mmol, 22.2% based on $Na_9H[A\text{-}\beta\text{-}SiW_9O_{34}]\cdot 23H_2O$).

The crystalline sample was soluble in water, but insoluble in organic solvents such as methanol, ethanol, and diethyl ether. Complete elemental analysis (%) calcd. for $H_{15}Fe_4K_9O_{76}Si_2W_{18}$ or $K_9[\{Fe_4(H_2O)(OH)_5\}(\beta,\beta\text{-}Si_2W_{18}O_{66})]\cdot 4H_2O$: H, 0.29; Fe, 4.32; K, 6.80; O, 23.51; Si, 1.09; W, 63.99. Found: H, 0.29; Fe, 4.25; K, 6.57; O, 23.6; Si, 1.23; W, 63.8; total ~99.74%. A weight loss of 4.23% (solvated water) was observed during overnight drying at room temperature, at $10^{-3}$–$10^{-4}$ Torr before analysis, suggesting the presence of 13 water molecules (calcd. 4.33%). TG/DTA under atmospheric conditions: a weight loss of 6.10% was observed below 500 °C; calc. 5.98% for $x = 18$ in $K_9[\{Fe_4(H_2O)(OH)_5\}(\beta,\beta\text{-}Si_2W_{18}O_{66})]\cdot xH_2O$. IR (KBr, $cm^{-1}$): 1615 (w), 1090 (w), 1008 (w), 962 (s), 906 (vs), 879 (vs), 810 (vs), 761 (vs), 716 (s), 643 (m), 520 (w), 483 (w), 407 (w).

### 3.4. X-ray Crystallography

For the potassium salt of $\beta,\beta\text{-}Fe_4\text{-}open$, a single crystal with dimensions of $0.11 \times 0.08 \times 0.07$ mm$^3$ was surrounded by liquid paraffin (Paratone-N) and analyzed at 100(2) K. Measurement was performed using a Bruker SMART APEX CCD diffractometer. The structure was solved by direct methods (SHELXS-97), followed by difference Fourier calculations and refinement by a full-matrix least-squares procedure on $F^2$ (program SHELXL-97) [41].

Crystal data: triclinic, space group $P{-}1$, $a = 12.560(2)$, $b = 19.013(3)$, $c = 20.359(4)$ Å, $\alpha = 92.925(3)$, $\beta = 98.002(3)$, $\gamma = 93.086(3)°$, $V = 4799.1(15)$ Å$^3$, $Z = 2$, $D_{calcd} = 3.694$ g·$cm^{-3}$, $\mu$(Mo-K$\alpha$) = 22.371 mm$^{-1}$. $R_1$ (I > 2.00σ(I)) = 0.0605, $R$ (all data) = 0.0751, $wR_2$ (all data) = 0.1642, GOF = 1.032. Most atoms in the main part of the structure were refined anisotropically, while the rest (as crystallization solvents) were refined isotropically, because of the presence of disorder. The composition and formula of the POM, containing countercations and crystalline water molecules, were determined by the complete elemental and TG analyses. Similar to other structural investigations of crystals of highly hydrated large polyoxometalate complexes, it was not possible to locate every countercation and hydrated water molecule, due to the extensive disorder of the cations and crystalline water molecules. Further details of the crystal structure investigations may be obtained from the Fachinformationszentrum Karlsruhe, 76344 Eggenstein-Leopoldshafen, Germany (Fax: +49-7247-808-666; E-Mail: crysdata@fiz-karlsruhe.de, http://www.fiz-karlsruhe.de/request_for_deposited_data.html?&L=1) on quoting the depository number CSD-431049 (Identification code; em1-6-3c).

## 4. Conclusions

In summary, we prepared a β,β-isomer of open-Wells–Dawson POM containing a tetra-iron(III) cluster, $K_9[\{Fe_4(H_2O)(OH)_5\}(\beta,\beta-Si_2W_{18}O_{66})] \cdot 17H_2O$ (potassium salt of **β,β-Fe₄-open**), by reacting trilacunary β-Keggin POM, $Na_9H[A-\beta-SiW_9O_{34}] \cdot 23H_2O$, with $FeCl_3 \cdot 6H_2O$. This compound is the first example containing the isomer of α,α-open-Wells–Dawson structural POM. Studies on the α,β-isomers of open-Wells–Dawson POMs, regarded as an open structure of the standard β-Wells–Dawson POM, are in progress.

**Supplementary Materials:**
Table S1: Bond valence sum (BVS) calculations of Fe and O atoms of the $\{Fe_4(H_2O)(OH)_5\}$ cluster moieties of **β,β-Fe₄-open**; Figure S1: FT-IR spectrum of potassium salt of **β,β-Fe₄-open** (KBr disk); Figure S2: TG/DTA data of potassium salt of **β,β-Fe₄-open**; Figure S3: UV/Vis absorption spectra of **α,α-** and **β,β-Fe₄-open** in $H_2O$; Figure S4: Cyclic voltammograms (CV) of 0.5 mM potassium salt of **α,α-** and **β,β-Fe₄-open** in 0.5 M potassium acetate buffer, pH 4.8, scan rate 25 $mV \cdot s^{-1}$, under $N_2$; checkCIF/PLATON report.

**Acknowledgments:** This study was supported by the Strategic Research Base Development Program for Private Universities of the Ministry of Education, Culture, Sports, Science and Technology of Japan, and also by a grant from Research Institute for Integrated Science, Kanagawa University (RIIS201505).

**Author Contributions:** Satoshi Matsunaga and Kenji Nomiya conceived and designed the experiments, and wrote the paper; Eriko Miyamae and Yusuke Inoue synthesized and characterized the compound.

## Abbreviations

The following abbreviations are used in this manuscript:

| | |
|---|---|
| POM | Polyoxometalate |
| TG/DTA | Thermogravimetric/differential thermal analyses |
| BVS | Bond valence sum |

## References

1. Pope, M.T.; Müller, A. Polyoxometalate chemistry: An old field with new dimensions in several disciplines. *Angew. Chem. Int. Ed.* **1991**, *30*, 34–48. [CrossRef]

2. Pope, M.T. *Heteropoly and Isopolyoxometalates*; Springer-Verlag: New York, NY, USA, 1983.

3. Hill, C.L.; Prosser-McCartha, C.M. Homogeneous catalysis by transition metal oxygen anion clusters. *Coord. Chem. Rev.* **1995**, *143*, 407–455. [CrossRef]

4. Neumann, R. Polyoxometalate complexes in organic oxidation chemistry. *Prog. Inorg. Chem.* **1998**, *47*, 317–370.

5. Proust, A.; Thouvenot, R.; Gouzerh, P. Functionalization of polyoxometalates: Towards advanced applications in catalysis and materials science. *Chem. Commun.* **2008**, 1837–1852. [CrossRef] [PubMed]

6. Hasenknopf, B.; Micoine, K.; Lacôte, E.; Thorimbert, S.; Malacria, M.; Thouvenot, R. Chirality in polyoxometalate chemistry. *Eur. J. Inorg. Chem.* **2008**, 5001–5013. [CrossRef]

7. Long, D.-L.; Tsunashima, R.; Cronin, L. Polyoxometalates: Building blocks for functional nanoscale systems. *Angew. Chem. Int. Ed.* **2010**, *49*, 1736–1758. [CrossRef] [PubMed]

8. Nomiya, K.; Sakai, Y.; Matsunaga, S. Chemistry of group IV metal ion-containing polyoxometalates. *Eur. J. Inorg. Chem.* **2011**, 179–196. [CrossRef]

9. Izarova, N.V.; Pope, M.T.; Kortz, U. Noble metals in polyoxometalates. *Angew. Chem. Int. Ed.* **2012**, *51*, 9492–9510. [CrossRef] [PubMed]

10. Song, Y.-F.; Tsunashima, R. Recent advances on polyoxometalate-based molecular and composite materials. *Chem. Soc. Rev.* **2012**, *41*, 7384–7402. [CrossRef] [PubMed]

11. Bijelic, A.; Rompel, A. The use of polyoxometalates in protein crystallography—An attempt to widen a well-known bottleneck. *Coord. Chem. Rev.* **2015**, *299*, 22–38. [CrossRef] [PubMed]

12. Wang, S.-S.; Yang, G.-Y. Recent advances in polyoxometalate-catalyzed reactions. *Chem. Rev.* **2015**, *115*, 4893–4962. [CrossRef] [PubMed]

13. Blazevic, A.; Rompel, A. The Anderson–Evans polyoxometalate: From inorganic building blocks via hybrid organic–inorganic structures to tomorrows "Bio-POM". *Coord. Chem. Rev.* **2016**, *307*, 42–64. [CrossRef]

14. Laronze, N.; Marrot, J.; Hervé, G. Synthesis, molecular structure and chemical properties of a new tungstosilicate with an open Wells–Dawson structure, α-[$Si_2W_{18}O_{66}$]$^{16-}$. *Chem. Commun.* **2003**, 2360–2361. [CrossRef]

15. Bi, L.-H.; Kortz, U. Synthesis and structure of the pentacopper(II) substituted tungstosilicate [$Cu_5(OH)_4(H_2O)_2(A$-α-$SiW_9O_{33})_2$]$^{10-}$. *Inorg. Chem.* **2004**, *43*, 7961–7962. [CrossRef] [PubMed]

16. Leclerc-Laronze, N.; Marrot, J.; Hervé, G. Cation-directed synthesis of tungstosilicates. 2. Synthesis, structure, and characterization of the open Wells–Dawson anion α-[{$K(H_2O)_2$}($Si_2W_{18}O_{66}$)]$^{15-}$ and its transiton-metal derivatives [{$M(H_2O)$}(μ-$H_2O)_2K(Si_2W_{18}O_{66})$]$^{13-}$ and [{$M(H_2O)$}(μ-$H_2O)_2K${$M(H_2O)_4$}($Si_2W_{18}O_{66}$)]$^{11-}$. *Inorg. Chem.* **2005**, *44*, 1275–1281. [PubMed]

17. Nellutla, S.; Tol, J.V.; Dalal, N.S.; Bi, L.-H.; Kortz, U.; Keita, B.; Nadjo, L.; Khitrov, G.A.; Marshall, A.G. Magnetism, electron paramagnetic resonance, electrochemistry, and mass spectrometry of the pentacopper(II)-substituted tungstosilicate [$Cu_5(OH)_4(H_2O)_2(A$-α-$SiW_9O_{33})_2$]$^{10-}$, A model five-spin frustrated cluster. *Inorg. Chem.* **2005**, *44*, 9795–9806. [CrossRef] [PubMed]

18. Leclerc-Laronze, N.; Haouas, M.; Marrot, J.; Taulelle, F.; Hervé, G. Step-by-step assembly of trivacant tungstosilicates: Synthesis and characterization of tetrameric anions. *Angew. Chem. Int. Ed.* **2006**, *45*, 139–142. [CrossRef] [PubMed]

19. Leclerc-Laronze, N.; Marrot, J.; Hervé, G. Dinuclear vanadium and tetranuclear iron complexes obtained with the open Wells–Dawson [$Si_2W_{18}O_{66}$]$^{16-}$ tungstosilicate. *C. R. Chim.* **2006**, *9*, 1467–1471. [CrossRef]

20. Sun, C.-Y.; Liu, S.-X.; Wang, C.-L.; Xie, L.-H.; Zhang, C.-D.; Gao, B.; Su, Z.-M.; Jia, H.-Q. Synthesis, structure and characterization of a new cobalt-containing germanotungstate with open Wells–Dawson structure: $K_{13}$[{$Co(H_2O)$}(μ-$H_2O)_2K(Ge_2W_{18}O_{66})$]. *J. Mol. Struct.* **2006**, *785*, 170–175. [CrossRef]

21. Wang, C.-L.; Liu, S.-X.; Sun, C.-Y.; Xie, L.-H.; Ren, Y.-H.; Liang, D.-D.; Cheng, H.-Y. Bimetals substituted germanotungstate complexes with open Wells–Dawson structure: Synthesis, structure, and electrochemical behavior of [{$M(H_2O)$}(μ-$H_2O)_2K${$M(H_2O)_4$}($Ge_2W_{18}O_{66}$)]$^{11-}$ (M = Co, Ni, Mn). *J. Mol. Struct.* **2007**, *841*, 88–95. [CrossRef]

22. Ni, L.; Hussain, F.; Spingler, B.; Weyeneth, S.; Patzke, G.R. Lanthanoid-containing open Wells–Dawson silicotungstates: Synthesis, crystal structures, and properties. *Inorg. Chem.* **2011**, *50*, 4944–4955. [CrossRef] [PubMed]

23. Zhu, G.; Geletii, Y.V.; Zhao, C.; Musaev, D.G.; Song, J.; Hill, C.L. A dodecanuclear Zn cluster sandwiched by polyoxometalate ligands. *Dalton Trans.* **2012**, *41*, 9908–9913. [CrossRef] [PubMed]

24. Zhu, G.; Glass, E.N.; Zhao, C.; Lv, H.; Vickers, J.W.; Geletii, Y.V.; Musaev, D.G.; Song, J.; Hill, C.L. A nickel containing polyoxometalate water oxidation catalyst. *Dalton Trans.* **2012**, *41*, 13043–13049. [CrossRef] [PubMed]

25. Zhu, G.; Geletii, Y.V.; Song, J.; Zhao, C.; Glass, E.N.; Bacsa, J.; Hill, C.L. Di- and tri-cobalt silicotungstates: Synthesis, characterization, and stability studies. *Inorg. Chem.* **2013**, *52*, 1018–1024. [CrossRef] [PubMed]

26. Ni, L.; Spingler, B.; Weyeneth, S.; Patzke, G.R. Trilacunary Keggin-type POMs as versatile building blocks for lanthanoid silicotungstates. *Eur. J. Inorg. Chem.* **2013**, 1681–1692. [CrossRef]

27. Guo, J.; Zhang, D.; Chen, L.; Song, Y.; Zhu, D.; Xu, Y. Syntheses, structures and magnetic properties of two unprecedented hybrid compounds constructed from open Wells–Dawson anions and high-nuclear transition metal clusters. *Dalton Trans.* **2013**, *42*, 8454–8459. [CrossRef] [PubMed]

28. Matsunaga, S.; Inoue, Y.; Otaki, T.; Osada, H.; Nomiya, K. Aluminum- and Gallium-Containing Open-Dawson Polyoxometalates. *Z. Anorg. Allg. Chem.* **2016**, *642*, 539–545. [CrossRef]

29. Minato, T.; Suzuki, K.; Kamata, K.; Mizuno, N. Synthesis of α-Dawson-type silicotungstate [α-$Si_2W_{18}O_{62}$]$^{8-}$ and protonation and deprotonation inside the aperture through intramolecular hydrogen bonds. *Chem. Eur. J.* **2014**, *20*, 5946–5952. [CrossRef] [PubMed]

30. Zhang, F.-Q.; Guan, W.; Yan, L.-K.; Zhang, Y.-T.; Xu, M.-T.; Hayfron-Benjamin, E.; Su, Z.-M. On the origin of the relative stability of Wells–Dawson isomers: A DFT study of α-, β-, γ-, α*-, β*-, and γ*-[($PO_4)_2W_{18}O_{54}$]$^{6-}$ anions. *Inorg. Chem.* **2011**, *50*, 4967–4977. [CrossRef] [PubMed]

31. Baker, L.C.W.; Figgis, J.S. New fundamental type of inorganic complex: Hybrid between heteropoly and conventional coordination complexes. Possibilities for geometrical isomerisms in 11-, 12-, 17-, and 18-heteropoly derivatives. *J. Am. Chem. Soc.* **1970**, *92*, 3794–3797. [CrossRef]

32. Dawson, B. The structure of the 9(18)-heteropoly anion in potassium 9(18)-tungstophosphate, $K_6(P_2W_{18}O_{62}) \cdot 14H_2O$. *Acta Crystalligr.* **1953**, *6*, 113–126. [CrossRef]

33. Neubert, H.; Fuchs, J. Crystal structures and vibrational spectra of two isomers of octadecatungsto-diarsenate $(NH_4)_6As_2W_{18}O_{62} \cdot nH_2O$. *Z. Naturforsch.* **1987**, *42b*, 951–958. [CrossRef]

34. Contant, R.; Thouvenot, R. A reinvestigation of isomerism in the Dawson structure: Syntheses and $^{183}$W NMR structural characterization of three new polyoxotungstates $[X_2W_{18}O_{62}]^{6-}$ (X = $P^V$, $As^V$). *Inorg. Chim. Acta* **1993**, *212*, 41–50. [CrossRef]

35. Richardt, P.J.S.; Gable, R.W.; Bond, A.M.; Wedd, A.G. Synthesis and redox characterization of the polyoxo Anion, $\gamma^*$-$[S_2W_{18}O_{62}]^{4-}$: A unique fast oxidation pathway determines the characteristic reversible electrochemical behavior of polyoxometalate anions in acidic media. *Inorg. Chem.* **2001**, *40*, 703–709. [CrossRef] [PubMed]

36. Zhang, J.; Bond, A.M. Voltammetric reduction of α- and $\gamma^*$-$[S_2W_{18}O_{62}]^{4-}$ and α-, β-, and $\gamma$-$[SiW_{12}O_{40}]^{4-}$: Isomeric dependence of reversible potentials of polyoxometalate anions using data obtained by novel dissolution and conventional solution-phase processes. *Inorg. Chem.* **2004**, *43*, 8263–8271. [CrossRef] [PubMed]

37. Sun, Y.-X.; Zhang, Z.-B.; Sun, Q.; Xu, Y. Syntheses, characterization and catalytic properties of two new Wells–Dawson molybdosulfates. *Chin. J. Inorg. Chem.* **2011**, *27*, 556–560.

38. Brown, I.D.; Altermatt, D. Bond-valence parameters obtained from a systematic analysis of the Inorganic Crystal Structure Database. *Acta Crystallogr.* **1985**, *B41*, 244–247. [CrossRef]

39. Zonnevijlle, F.; Tourné, C.M.; Tourné, G.F. Preparation and Characterization of Iron(III)- and Rhodium(III)-Containing Heteropolytungstates. Identification of Novel Oxo-Bridged Iron(III) Dimers. *Inorg. Chem.* **1982**, *21*, 2751–2757. [CrossRef]

40. Tézé, A.; Hervé, G. α-, β-, and $\gamma$-Dodecatungstosilicic acids: Isomers and related lacunary compounds. *Inorg. Synth.* **1990**, *27*, 85–96.

41. Sheldrick, G.M. A short history of SHELX. *Acta Crystallogr.* **2008**, *64*, 112–122. [CrossRef] [PubMed]

# 5

# Synthesis, Structure, and Characterization of In$_{10}$-Containing Open-Wells–Dawson Polyoxometalate

**Satoshi Matsunaga, Takuya Otaki, Yusuke Inoue, Kohei Mihara and Kenji Nomiya ***

Department of Chemistry, Faculty of Science, Kanagawa University, Hiratsuka, Kanagawa 259-1293, Japan; matsunaga@kanagawa-u.ac.jp (S.M.); toatkaukyia1@gmail.com (T.O.); r201470042gl@jindai.jp (Y.I.); r201203938gu@jindai.jp (K.M.)
* Correspondence: nomiya@kanagawa-u.ac.jp

Academic Editor: Greta Ricarda Patzke

**Abstract:** We have successfully synthesized $K_{17}\{[\{KIn_2(\mu\text{-}OH)_2\}(\alpha,\alpha\text{-}Si_2W_{18}O_{66})]_2[In_6(\mu\text{-}OH)_{13}(H_2O)_8]\}\cdot 35H_2O$ (potassium salt of **In$_{10}$-open**), an open-Wells–Dawson polyoxometalate (POM) containing ten indium metal atoms. This novel compound was characterized by X-ray crystallography, $^{29}$Si NMR, FTIR, complete elemental analysis, and TG/DTA. X-ray crystallography results for $\{[\{KIn_2(\mu\text{-}OH)_2\}(\alpha,\alpha\text{-}Si_2W_{18}O_{66})]_2[In_6(\mu\text{-}OH)_{13}(H_2O)_8]\}^{17-}$ (**In$_{10}$-open**) revealed two open-Wells–Dawson units containing two In$^{3+}$ ions and a K$^+$ ion, $[\{KIn_2(\mu\text{-}OH)_2\}(\alpha,\alpha\text{-}Si_2W_{18}O_{66})]^{11-}$, connected by an In$_6$-hydroxide cluster moiety, $[In_6(\mu\text{-}OH)_{13}(H_2O)_8]^{5+}$. **In$_{10}$-open** is the first example of an open-Wells–Dawson POM containing a fifth-period element. Moreover, to the best of our knowledge, it exhibits the highest nuclearity among the indium-containing POMs reported to date.

**Keywords:** polyoxometalates; open-Wells–Dawson structural polyoxometalate; Indium

---

## 1. Introduction

Polyoxometalates (POMs) are discrete metal oxide clusters that are of current interest as soluble metal oxides, as well as for their application in catalysis, medicine, and materials science [1–13]. Recently, open-Wells–Dawson POMs have been reported as an emerging class of POMs [14–28]. These compounds are a dimerized species of the trilacunary Keggin POMs, $[XW_9O_{34}]^{10-}$ (X = Si, Ge). Standard Wells–Dawson structural POMs are regarded as two trilacunary Keggin POM units assembled together via six W–O–W bonds. However, the electrostatic repulsion between the two units in $[XW_9O_{34}]^{10-}$ (X = Si and Ge), induced by the highly charged guest $XO_4^{4-}$ (X = Si, Ge) ion, is assumed to be so strong that it inhibits the assembly of the standard Wells–Dawson structure in aqueous media. Therefore, when the two trilacunary Keggin units comprise an $XO_4^{4-}$ (X = Si, Ge) ion, they are linked by only two W–O–W bonds. This results in the formation of an open-Wells–Dawson structural POM [29]. The open pocket of these POMs can accommodate multiple metal ions (one to six metal ions). Thus, this class of compounds may constitute a promising platform for the development of metal-substituted-POM-based materials and catalysts. To date, many compounds that contain various metal ions in their open pocket, e.g., V$^{5+}$ [19], Mn$^{2+}$ [16,21], Fe$^{3+}$ [19], Co$^{2+}$ [14,16,20,21,25,27], Ni$^{2+}$ [16,21,24,27], Cu$^{2+}$ [15–17], and Zn$^{2+}$ [23] have been reported. Some lanthanoid (Eu$^{3+}$, Gd$^{3+}$, Tb$^{3+}$, Dy$^{3+}$, and Ho$^{3+}$)-containing open-Wells–Dawson POMs have also been reported [22,26]. However, the large ionic radii of these lanthanoid atoms inhibit their complete insertion within the open pocket. This results in a weak coordination, similar to that of the K ions in K-containing open-Wells–Dawson POMs. Recently, we synthesized the Al$_4$- and Ga$_4$-containing open-Wells–Dawson POMs: $[\{Al_4(\mu\text{-}OH)_6\}(\alpha,\alpha\text{-}Si_2W_{18}O_{66})]^{10-}$ (**Al$_4$-open**) and $[\{Ga_4(\mu\text{-}OH)_6\}(\alpha,\alpha\text{-}Si_2W_{18}O_{66})]^{10-}$ (**Ga$_4$-open**), respectively, and successfully determined their molecular structures by single crystal X-ray crystallography [28]. X-ray structure analyses of **Al$_4$-** and

**Ga$_4$-open** revealed that the $\{M_4(\mu\text{-OH})_6\}^{6+}$ (M = Al$^{3+}$, Ga$^{3+}$) clusters are included in the open pocket of the open-Wells–Dawson unit.

In general, trivalent group 13 ions are found as various oligomeric hydroxide species in aqueous solution [30–32]. Synthetic and structural studies of group 13 ion-containing POMs provide informative and definitive molecular models of group 13 metal clusters in solution. However, among all the Al-, Ga-, and In-containing POMs, formed by the substitution of several tungsten ions in the parent POMs with trivalent group 13 ions [28,33–40], few well-characterized In-containing POMs have been reported to date [41–43]. Thus, In-containing POMs are intriguing target compounds from both a synthetic and a structural point of view.

In this study, we successfully synthesized an open-Wells–Dawson POM containing ten indium metal ions, $K_{17}\{[\{KIn_2(\mu\text{-OH})_2\}(\alpha,\alpha\text{-Si}_2W_{18}O_{66})]_2[In_6(\mu\text{-OH})_{13}(H_2O)_8]\}\cdot 35H_2O$ (potassium salt of **In$_{10}$-open**), and characterized it by X-ray crystallography, $^{29}$Si NMR, FTIR, complete elemental analysis, and thermogravimetric/differential thermal analyses (TG/DTA). In contrast to **Al$_4$-** and **Ga$_4$-open**, **In$_{10}$-open** showed a dimer structure bridged by a deca-indium-hydroxide cluster.

## 2. Results and Discussion

### 2.1. Synthesis

The crystalline sample of potassium salt of **In$_{10}$-open**, was afforded in 17.9% yield. This complex was prepared from a 1:5 molar ratio reaction of $K_{13}[\{K(H_2O)_3\}_2\{K(H_2O)_2\}(\alpha,\alpha\text{-Si}_2W_{18}O_{66})]\cdot 19H_2O$ with InCl$_3$. The sample was characterized using complete elemental analysis (H, In, K, O, Si, and W analyses), FTIR, TG/DTA, $^{29}$Si NMR in D$_2$O, and X-ray crystallography.

The FTIR spectrum of potassium salt of **In$_{10}$-open** (Figure S1) displays peaks at 1000 and 945 cm$^{-1}$ that correspond to $\nu_{as}$(Si–O) and $\nu_{as}$(W–O$_t$), respectively. The characteristic bands at 900–600 cm$^{-1}$ are associated with $\nu$(W–O$_c$), $\nu$(W–O$_b$), and $\nu$(W–O–W). The IR spectrum is very similar to those of the common open-Wells–Dawson POMs.

Before elemental analysis, the sample of **In$_{10}$-open** was dried overnight at room temperature under vacuum ($10^{-3}$–$10^{-4}$ Torr). All elements (H, In, K, O, Si, and W) were observed for a total analysis of 100.37%. The recorded data were in good accordance with the calculated values for the formula without water of crystallization, $K_{17}[\{KIn_2(\mu\text{-OH})_2\}(\alpha,\alpha\text{-Si}_2W_{18}O_{66})]_2[In_6(\mu\text{-OH})_{13}(H_2O)_8]$ (see Experimental section). The weight loss observed during drying, before analysis, was 5.28% corresponding to ca. 35 crystalline water molecules. On the other hand, during the TG/DTA measurements carried out under atmospheric conditions, a weight loss of 6.40%, observed at temperatures below 500 °C, corresponding to a total of ca. 42 water molecules, i.e., 8 coordinated water molecules and 34 molecules of water of crystallization (Figure S2). Thus, the elemental analysis and TG/DTA displayed a presence of a total of 34–35 water molecules for the sample under atmospheric conditions. The formula for potassium salt of **In$_{10}$-open** presented herein was determined as $K_{17}[\{KIn_2(\mu\text{-OH})_2\}(\alpha,\alpha\text{-Si}_2W_{18}O_{66})]_2[In_6(\mu\text{-OH})_{13}(H_2O)_8]\cdot 35H_2O$ based on the results of the complete elemental analysis.

### 2.2. Molecular Structure

The molecular structure of the polyoxoanion of potassium salt of **In$_{10}$-open** and its polyhedral representation are shown in Figure 1a,b, respectively. X-ray crystallographic data of **In$_{10}$-open** reveal that the two open-Wells–Dawson units that include two In$^{3+}$ ions and a K$^+$ ion, $[\{KIn_2(\mu\text{-OH})_2\}(\alpha,\alpha\text{-Si}_2W_{18}O_{66})]^{11-}$, are connected by a central In$_6$-hydroxide cluster moiety, $[In_6(\mu\text{-OH})_{13}(H_2O)_8]^{5+}$, to form a dimeric open-Wells–Dawson polyanion, $\{[\{KIn_2(\mu\text{-OH})_2\}(\alpha,\alpha\text{-Si}_2W_{18}O_{66})]_2[In_6(\mu\text{-OH})_{13}(H_2O)_8]\}^{17-}$ (Figure 1).

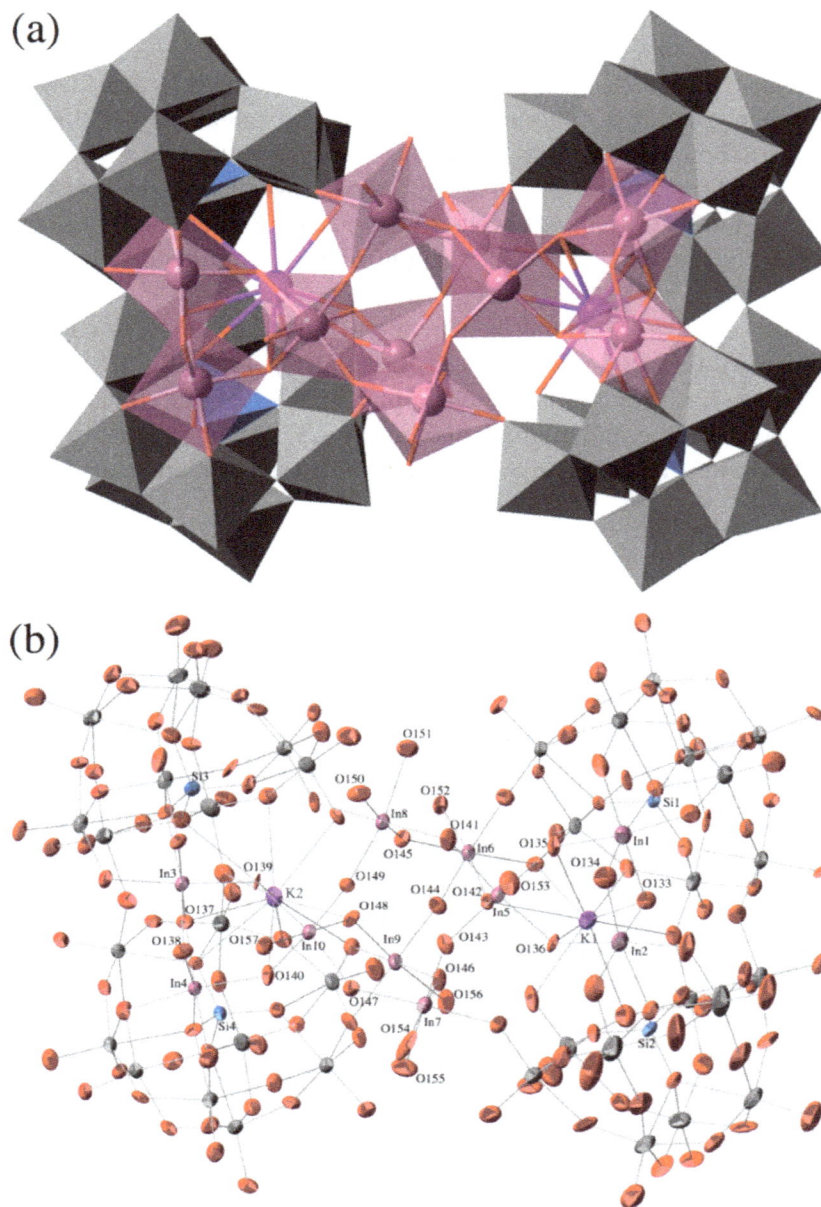

**Figure 1.** Molecular structure of the polyoxoanion, $\{[\{KIn_2(\mu\text{-}OH)_2\}(\alpha,\alpha\text{-}Si_2W_{18}O_{66})]_2[In_6(\mu\text{-}OH)_{13}(H_2O)_8]\}^{17-}$ of potassium salt of **In$_{10}$-open**. (a) Its polyhedral representation; and (b) thermal ellipsoidal plot. Color code: In, pink; K, purple; O, red; Si, blue; W, gray.

The two indium atoms in the open pocket of the open-Wells–Dawson POM units are connected through edge-sharing oxygen atoms (O133, O134 for In1, In2; O137, O138 for In3, In4) with In$\cdots$In distances of 3.266(2) (In1$\cdots$In2) and 3.299(2) (In3$\cdots$In4) (Figure 2a). Each Indium atom in the open pocket is bonded to three oxygen atoms of the lacunary site in the open-Wells–Dawson polyanion [In–O average = 2.2011 Å]. Open-Wells–Dawson POMs that include two metal atoms in the open pocket have been previously reported, e.g., $K_{11}[\{KV_2O_3(H_2O)_2\}(Si_2W_{18}O_{66})]\cdot 40H_2O$ (**V$_2$-open**) [19]. In contrast to our **In$_{10}$-open**, the two vanadium atoms in the open pocket of **V$_2$-open** are bound to only one half {SiW$_9$} of the open-Wells–Dawson unit, and are linked in a corner-sharing fashion (Figure 2b). Therefore, **In$_{10}$-open** is the first example of an open-Wells–Dawson POM that includes two metal atoms with edge-sharing fashion.

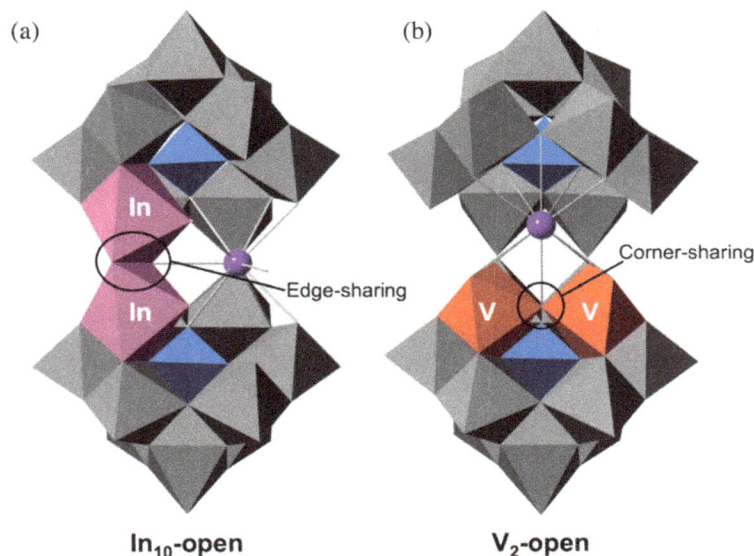

**Figure 2.** The metal arrangement of (**a**) **In$_{10}$-open** and (**b**) **V$_2$-open** in the open pocket.

In addition to the indium atoms in the open-pocket, the complex contains six indium atoms in the bridging hydroxide cluster. The indium atoms in the bridging cluster are connected to each other in a corner-sharing fashion. Moreover, the In atoms in the open-pocket and the W atoms of the open-Wells–Dawson POM units are also linked to the In atoms of the bridging cluster in a corner-sharing fashion. Thus, all the Indium atoms can be considered to be 6-coordinated. Bond valence sum (BVS) [44] calculations suggest that the corner- and edge-sharing oxygen atoms that are linked to the In atoms (corner-sharing: O135, O136, O139, O140, O141, O142, O143, O144, O145, O146, O147, O148, and O149; edge-sharing: O133, O134, O137, and O138) are protonated, *i.e.*, they are ascribed to the hydroxide groups. On the other hand, the terminal oxygen atoms on the indium atoms (O150, O151, O152, O153, O154, O155, O156, and O157) are ascribed to the water groups (Table S1).

**In$_{10}$-open** has a dimeric structure composed of two indium-containing open-Wells–Dawson POM moieties bridged by In$_6$ hydroxide clusters. Dimeric open-Wells–Dawson POMs, similar to **In$_{10}$-open**, have also been reported for $\{[Zn_6(\mu\text{-OH})_7(H_2O)(\alpha,\alpha\text{-Si}_2W_{18}O_{66})]_2\}^{22-}$ (**Zn$_{12}$-open**) by Hill *et al.* [23]. In this complex, the six zinc atoms are included in the open pocket of the open-Wells–Dawson unit, and the two Zn$_6$-containing open-Wells–Dawson units are connected through the two edge-sharing oxygen atoms. The arrangement and the number of metal ions in the open-pocket of the **In$_{10}$-open** are different from those of the **Zn$_{12}$-open** reported previously.

In open-Wells–Dawson POMs, the bite angle can be defined as the dihedral angle between the planes that pass through the six oxygen atoms of the lacunary site of each trilacunary Keggin unit. The bite angle varies, depending on the metal cluster included in the open pocket of the open-Wells–Dawson unit. The bite angles of **In$_{10}$-open** are 64.363° and 65.139° (Figure 3). These values are wider than those of other open-Wells–Dawson POMs, including other group 13 ions, such as **Al$_4$-** (54.274°) and **Ga$_4$-open** (56.110°) [28]. The difference between the bite angles of **Al$_4$-**, **Ga$_4$-**, and **In$_{10}$-open** is caused by the difference in the ionic radii of the Al (0.53 Å), Ga (0.76 Å), and In (0.94 Å) ions [45,46]. **In$_{10}$-open** displays the widest bite angles when compared to previously reported open-Wells–Dawson POMs, including the Co$_6$ (60.045°) [27], Zn$_6$ dimer (60.308°) [23], Ni$_5$ (58.925°) [24], and Cu$_5$ (61.663°) [15,17] clusters. The open-Wells–Dawson POM containing a Cu$_5$ cluster (**Cu$_5$-open**) exhibits a large bite angle (61.663°) due to the long bond lengths between the copper and the edge-sharing oxygen atom, caused by Jahn–Teller distortion [15,17]. The bite angles of **In$_{10}$-open** are ca. 3° wider than that of **Cu$_5$-open**. This increase appears to be caused by the large ionic radius of the indium ions incorporated in the open pocket.

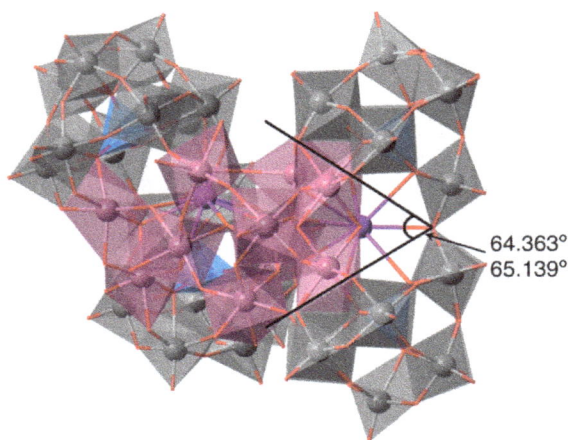

**Figure 3.** Bite angles of **In$_{10}$-open**.

The previously reported open–Wells–Dawson POMs mainly accommodated the fourth-period elements. Except for the lanthanoid-containing open–Wells–Dawson POMs, whose open-pockets weakly coordinate to the lanthanoid ions, open–Wells–Dawson POMs that accommodate the larger fifth- and sixth-period elements have not been reported to date. Thus, **In$_{10}$-open** indicates that elements (such as indium) having large ionic radii (0.94 Å) can be incorporated in the open-pocket of an open–Wells–Dawson unit.

## 2.3. Solution $^{29}$Si NMR

The solution $^{29}$Si NMR spectrum of **In$_{10}$-open** in D$_2$O displays a two-line spectrum at $-82.415$ and $-83.159$ ppm in a 1:1 ratio (Figure 4). The two Si atoms in one open-Wells–Dawson unit are nonequivalent due to the configuration of the other open-Wells–Dawson unit, even though the two units are equivalent. The adjacent two $^{29}$Si NMR peaks are consistent with the structure of **In$_{10}$-open** observed by X-ray crystallography. This suggests that **In$_{10}$-open** exists as a single species and maintains its structure in solution.

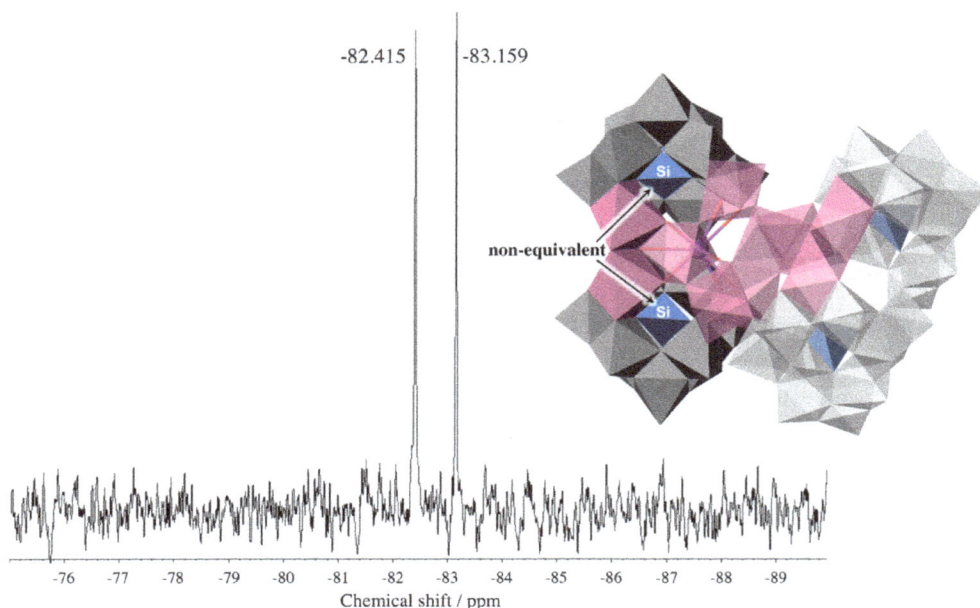

**Figure 4.** Solution $^{29}$Si NMR spectrum of **potassium salt of In$_{10}$-open** dissolved in D$_2$O.

## 3. Experimental Section

### 3.1. Materials

The following reagents were used as received: $InCl_3$ (from Sigma-Aldrich Japan, Shinagawa, Japan), KOH, KCl (from Wako Pure Chemical Industries, Osaka, Japan), and $D_2O$ (Kanto Chemical, Tokyo, Japan). The K ion-incorporating open-Wells–Dawson POM, $K_{13}[\{K(H_2O)_3\}_2\{K(H_2O)_2\}(\alpha,\alpha\text{-}Si_2W_{18}O_{66})]\cdot 19H_2O$, was prepared according to literature [14], and identified by TG/DTA and FT-IR analysis.

### 3.2. Instrumentation and Analytical Procedures

A complete elemental analysis was carried out by Mikroanalytisches Labor Pascher (Remagen, Germany). The sample was dried overnight at room temperature under a pressure of $10^{-3}$–$10^{-4}$ Torr before analysis. The $^{29}Si$ NMR (119.24 MHz) spectra in $D_2O$ solution were recorded in 5-mm outer diameter tubes, on a JEOL JNM ECP 500 FTNMR spectrometer with a JEOL ECP-500 NMR data-processing system (JEOL, Akishima, Japan). The $^{29}Si$ NMR spectrum was referenced to an internal standard of DSS. Infrared spectra were recorded on a Jasco 4100 FTIR spectrometer (Jasco, Hachioji, Japan) by using KBr disks at room temperature. TG/DTA measurements were performed using a Rigaku Thermo Plus 2 series TG8120 instrument (Rigaku, Akishima, Japan), under air flow with a temperature ramp of 4.0 °C per min at a temperature ranging between 26 and 500 °C.

### 3.3. Synthesis of $K_{17}\{[\{KIn_2(\mu\text{-}OH)_2\}(\alpha,\alpha\text{-}Si_2W_{18}O_{66})]_2[In_6(\mu\text{-}OH)_{13}(H_2O)_8]\}\cdot 35H_2O$ (Potassium Salt of $In_{10}$-open)

$InCl_3$ (398 mg, 1.80 mmol) was dissolved in water (40 mL). A separate solution of $K_{13}[\{K(H_2O)_3\}_2\{K(H_2O)_2\}(\alpha,\alpha\text{-}Si_2W_{18}O_{66})]\cdot 19H_2O$ (2.00 g, 0.361 mmol) dissolved in 100 mL of distilled water was added dropwise to the resulting solution. The pH of this solution was adjusted to 4.0 using 0.1 M $KOH_{aq}$. Next, 4 mL of saturated $KCl_{aq}$ were added to the solution. The resulting solution was left to stand undisturbed at room temperature for 3 days. The afforded colorless needle crystals were collected on a membrane filter (JG 0.2 μm), and dried in vacuo for 2 h. Yield = 0.381 g (0.0323 mmol, 17.9% based on $K_{13}[\{K(H_2O)_3\}_2\{K(H_2O)_2\}(\alpha,\alpha\text{-}Si_2W_{18}O_{66})]\cdot 19H_2O$).

The crystalline sample was soluble in water, but insoluble in organic solvents such as methanol, ethanol, and diethyl ether. Elemental analysis (%) calcd. for $H_{33}In_{10}K_{19}O_{157}Si_4W_{36}$ or $K_{17}\{[\{KIn_2(\mu\text{-}OH)_2\}(\alpha,\alpha\text{-}Si_2W_{18}O_{66})]_2[In_6(\mu\text{-}OH)_{13}(H_2O)_8]\}$: H 0.30, In 10.28, K 6.65, O 22.49, Si 1.01, W 59.27; Found: H 0.23, In 10.0, K 6.42, O 23.4, Si 1.02, W 59.3 (total 100.37%). A weight loss of 5.28% (solvated water) was observed during overnight drying at room temperature, at a pressure of $10^{-3}$–$10^{-4}$ Torr before analysis. This suggested the presence of 35 water molecules. From TG/DTA air flow, a weight loss of 6.40% was observed at a temperature below 500 °C; calc. 6.42% for a total of 42 water molecules, i.e., 34 solvated water molecules and 8 coordinated water molecules in $K_{17}\{[\{KIn_2(\mu\text{-}OH)_2\}(\alpha,\alpha\text{-}Si_2W_{18}O_{66})]_2[In_6(\mu\text{-}OH)_{13}(H_2O)_8]\}\cdot 34H_2O$; IR (KBr, $cm^{-1}$): 1622 (m), 1000 (w), 945 (m), 890 (vs), 786 (vs), 730 (vs), 649 (s), 548 (m), 523 (m); $^{29}Si$ NMR (50.0 °C, $D_2O$, DSS, ppm): $\delta = -82.415, -83.159$.

### 3.4. X-Ray Crystallography

For $In_{10}$-open, a single crystal with dimensions of $0.21 \times 0.06 \times 0.05$ mm$^3$ was surrounded by liquid paraffin (Paratone-N) and analyzed at 150(2) K. All measurements were performed on a Rigaku MicroMax-007HF with a Saturn CCD diffractometer (Rigaku). The structure was solved by direct methods (SHELXS-97), followed by difference Fourier calculations and refinement by a full-matrix least-squares procedure on $F^2$ (program SHELXL-97) [47].

Crystal data: monoclinic, space group P2(1)/a, $a = 23.525(5)$, $b = 32.926(7)$, $c = 25.424(6)$ Å, $\beta = 93.273(2)°$, $V = 19661(7)$ Å$^3$, $Z = 4$, $D_{calcd} = 3.857$ g cm$^{-3}$, $\mu$(Mo-Kα) = 22.491 mm$^{-1}$; $R_1[I > 2.00\sigma(I)] = 0.0829$, $R$ (all data) = 0.1012, $wR_2$ (all data) = 0.2296, GOF = 1.056. Most atoms in the main part of the structure were refined anisotropically, while the rest (as crystallization solvents) were refined isotropically, because of the presence of disorder. The composition and formula of the POM (containing many countercations and many crystalline water molecules) were determined from complete elemental and TG analyses. Similar to structural investigations of other crystals of highly hydrated large POM complexes, it was not possible to locate every countercation and hydrated water molecule in the complex. This frequently encountered situation is attributed to the extensive disorder of the cations and many of the crystalline water molecules. Further details on the crystal structure investigations may be obtained from: Fachinformationszentrum Karlsruhe, 76344 Eggenstein-Leopoldshafen, Germany (Fax: +49-7247-808-666; E-Mail: crysdata@fiz-karlsruhe.de, http://www.fiz-karlsruhe.de/request_for_deposited_data.html?&L=1) on quoting the depository number CSD-430962 (Identification code; 56_11_1).

## 4. Conclusions

In summary, we prepared and characterized an open-Wells–Dawson structural POM, potassium salt of $In_{10}$-open, containing ten indium ions, i.e., $K_{17}\{[\{KIn_2(\mu\text{-}OH)_2\}(\alpha,\alpha\text{-}Si_2W_{18}O_{66})]_2$ $[In_6(\mu\text{-}OH)_{13}(H_2O)_8]\}\cdot 35H_2O$ (potassium salt of $In_{10}$-open). Single-crystal X-ray analyses revealed that two open-Wells–Dawson units that include two $In^{3+}$ ions and a $K^+$ ion are connected by an $In_6$-hydroxide cluster moiety to form a dimeric open-Wells–Dawson polyanion. $In_{10}$-open displayed the widest bite angles among the previously reported open-Wells–Dawson POMs. This is mainly due to the large ionic radius of the indium ion. The solution $^{29}Si$ spectrum in $D_2O$ indicated that $In_{10}$-open was obtained as a single species and that its structure was maintained in solution. $In_{10}$-open is the first example of an open-Wells–Dawson POM containing a fifth-period element, and it exhibits the highest nuclearity of any indium-containing POM reported to date. This work can be extended to the future molecular design of novel open-Wells–Dawson POMs containing large fifth- and sixth-period elements, such as Ru, Rh, Pd, Pt. Studies of open-Wells–Dawson structural POMs containing larger metal atoms are in progress.

**Supplementary Materials:**
Figure S1: FT-IR spectrum of potassium salt of $In_{10}$-open (KBr disk), Figure S2: TG/DTA data of potassium salt of $In_{10}$-open (from 22 to 500 °C), Table S1: Bond valence sum (BVS) calculations of In and O atoms of the Indium-cluster moieties of $In_{10}$-open: checkCIF/PLATON report.

**Acknowledgments:** This study was supported by the Strategic Research Base Development Program for Private Universities of the Ministry of Education, Culture, Sports, Science and Technology of Japan, and also by a grant from the Research Institute for Integrated Science, Kanagawa University (RIIS201505).

**Author Contributions:** Satoshi Matsunaga and Kenji Nomiya conceived and designed the experiments, and wrote the paper; Takuya Otaki and Yusuke Inoue synthesized and characterized the compound; Kohei Mihara assisted characterization of the compound.

## Abbreviations

The following abbreviations are used in this manuscript:

| | |
|---|---|
| POM | Polyoxometalate |
| TG/DTA | Thermogravimetric/differential thermal analyses |
| BVS | Bond valence sum |

## References

1.    Pope, M.T.; Müller, A. Polyoxometalate chemistry: An old field with new dimensions in several disciplines. *Angew. Chem. Int. Ed.* **1991**, *30*, 34–48. [CrossRef]

2.    Pope, M.T. *Heteropoly- and Isopolyoxometalates*; Springer-Verlag: New York, NY, USA, 1983.

3.    Hill, C.L.; Prosser-McCartha, C.M. Homogeneous catalysis by transition metal oxygen anion clusters. *Coord. Chem. Rev.* **1995**, *143*, 407–455. [CrossRef]

4. Neumann, R. Polyoxometalate complexes in organic oxidation chemistry. *Prog. Inorg. Chem.* **1998**, *47*, 317–370.

5. Proust, A.; Thouvenot, R.; Gouzerh, P. Functionalization of polyoxometalates: Towards advanced applications in catalysis and materials science. *Chem. Commun.* **2008**, 1837–1852. [CrossRef] [PubMed]

6. Hasenknopf, B.; Micoine, K.; Lacôte, E.; Thorimbert, S.; Malacria, M.; Thouvenot, R. Chirality in polyoxometalate chemistry. *Eur. J. Inorg. Chem.* **2008**, *2008*, 5001–5013. [CrossRef]

7. Long, D.-L.; Tsunashima, R.; Cronin, L. Polyoxometalates: Building blocks for functional nanoscale systems. *Angew. Chem. Int. Ed.* **2010**, *49*, 1736–1758. [CrossRef] [PubMed]

8. Nomiya, K.; Sakai, Y.; Matsunaga, S. Chemistry of group IV metal ion-containing polyoxometalates. *Eur. J. Inorg. Chem.* **2011**, *2011*, 179–196. [CrossRef]

9. Izarova, N.V.; Pope, M.T.; Kortz, U. Noble metals in polyoxometalates. *Angew. Chem. Int. Ed.* **2012**, *51*, 9492–9510. [CrossRef] [PubMed]

10. Song, Y.-F.; Tsunashima, R. Recent advances on polyoxometalate-based molecular and composite materials. *Chem. Soc. Rev.* **2012**, *41*, 7384–7402. [CrossRef] [PubMed]

11. Bijelic, A.; Rompel, A. The use of polyoxometalates in protein crystallography—An attempt to widen a well-known bottleneck. *Coord. Chem. Rev.* **2015**, *299*, 22–38. [CrossRef] [PubMed]

12. Wang, S.-S.; Yang, G.-Y. Recent advances in polyoxometalate-catalyzed reactions. *Chem. Rev.* **2015**, *115*, 4893–4962. [CrossRef] [PubMed]

13. Blazevic, A.; Rompel, A. The Anderson–Evans polyoxometalate: From inorganic building blocks via hybrid organic–inorganic structures to tomorrows "Bio-POM". *Coord. Chem. Rev.* **2016**, *307*, 42–64. [CrossRef]

14. Laronze, N.; Marrot, J.; Hervé, G. Synthesis, molecular structure and chemical properties of a new tungstosilicate with an open Wells–Dawson structure, $\alpha$-$[Si_2W_{18}O_{66}]^{16-}$. *Chem. Commun.* **2003**, *21*, 2360–2361. [CrossRef]

15. Bi, L.-H.; Kortz, U. Synthesis and structure of the pentacopper(II) substituted tungstosilicate $[Cu_5(OH)_4(H_2O)_2(A-\alpha-SiW_9O_{33})_2]^{10-}$. *Inorg. Chem.* **2004**, *43*, 7961–7962. [CrossRef] [PubMed]

16. Leclerc-Laronze, N.; Marrot, J.; Hervé, G. Cation-directed synthesis of tungstosilicates. 2. Synthesis, structure, and characterization of the open Wells–Dawson anion $\alpha$-$[\{K(H_2O)_2\}(Si_2W_{18}O_{66})]^{15-}$ and its transiton-metal derivatives $[\{M(H_2O)\}(\mu-H_2O)_2K(Si_2W_{18}O_{66})]^{13-}$ and $[\{M(H_2O)\}(\mu-H_2O)_2K\{M(H_2O)_4\}(Si_2W_{18}O_{66})]^{11-}$. *Inorg. Chem.* **2005**, *44*, 1275–1281. [PubMed]

17. Nellutla, S.; Tol, J.V.; Dalal, N.S.; Bi, L.-H.; Kortz, U.; Keita, B.; Nadjo, L.; Khitrov, G.A.; Marshall, A.G. Magnetism, electron paramagnetic resonance, electrochemistry, and mass spectrometry of the pentacopper(II)-substituted tungstosilicate $[Cu_5(OH)_4(H_2O)_2(A-\alpha-SiW_9O_{33})_2]^{10-}$, A model five-spin frustrated cluster. *Inorg. Chem.* **2005**, *44*, 9795–9806. [CrossRef] [PubMed]

18. Leclerc-Laronze, N.; Haouas, M.; Marrot, J.; Taulelle, F.; Hervé, G. Step-by-step assembly of trivacant tungstosilicates: Synthesis and characterization of tetrameric anions. *Angew. Chem. Int. Ed.* **2006**, *45*, 139–142. [CrossRef] [PubMed]

19. Leclerc-Laronze, N.; Marrot, J.; Hervé, G. Dinuclear vanadium and tetranuclear iron complexes obtained with the open Wells–Dawson $[Si_2W_{18}O_{66}]^{16-}$ tungstosilicate. *C. R. Chim.* **2006**, *9*, 1467–1471. [CrossRef]

20. Sun, C.-Y.; Liu, S.-X.; Wang, C.-L.; Xie, L.-H.; Zhang, C.-D.; Gao, B.; Su, Z.-M.; Jia, H.-Q. Synthesis, structure and characterization of a new cobalt-containing germanotungstate with open Wells–Dawson structure: $K_{13}[\{Co(H_2O)\}(\mu-H_2O)_2K(Ge_2W_{18}O_{66})]$. *J. Mol. Struct.* **2006**, *785*, 170–175. [CrossRef]

21. Wang, C.-L.; Liu, S.-X.; Sun, C.-Y.; Xie, L.-H.; Ren, Y.-H.; Liang, D.-D.; Cheng, H.-Y. Bimetals substituted germanotungstate complexes with open Wells–Dawson structure: Synthesis, structure, and electrochemical behavior of $[\{M(H_2O)\}(\mu-H_2O)_2K\{M(H_2O)_4\}(Ge_2W_{18}O_{66})]^{11-}$ (M = Co, Ni, Mn). *J. Mol. Struct.* **2007**, *841*, 88–95. [CrossRef]

22. Ni, L.; Hussain, F.; Spingler, B.; Weyeneth, S.; Patzke, G.R. Lanthanoid-containing open Wells–Dawson silicotungstates: Synthesis, crystal structures, and properties. *Inorg. Chem.* **2011**, *50*, 4944–4955. [CrossRef] [PubMed]

23. Zhu, G.; Geletii, Y.V.; Zhao, C.; Musaev, D.G.; Song, J.; Hill, C.L. A dodecanuclear Zn cluster sandwiched by polyoxometalate ligands. *Dalton Trans.* **2012**, *41*, 9908–9913. [CrossRef] [PubMed]

24. Zhu, G.; Glass, E.N.; Zhao, C.; Lv, H.; Vickers, J.W.; Geletii, Y.V.; Musaev, D.G.; Song, J.; Hill, C.L. A nickel containing polyoxometalate water oxidation catalyst. *Dalton Trans.* **2012**, *41*, 13043–13049. [CrossRef] [PubMed]

25. Zhu, G.; Geletii, Y.V.; Song, J.; Zhao, C.; Glass, E.N.; Bacsa, J.; Hill, C.L. Di- and tri-cobalt silicotungstates: Synthesis, characterization, and stability studies. *Inorg. Chem.* **2013**, *52*, 1018–1024. [CrossRef] [PubMed]

26. Ni, L.; Spingler, B.; Weyeneth, S.; Patzke, G.R. Trilacunary Keggin-type POMs as versatile building blocks for lanthanoid silicotungstates. *Eur. J. Inorg. Chem.* **2013**, *2013*, 1681–1692. [CrossRef]

27. Guo, J.; Zhang, D.; Chen, L.; Song, Y.; Zhu, D.; Xu, Y. Syntheses, structures and magnetic properties of two unprecedented hybrid compounds constructed from open Wells–Dawson anions and high-nuclear transition metal clusters. *Dalton Trans.* **2013**, *42*, 8454–8459. [CrossRef] [PubMed]

28. Matsunaga, S.; Inoue, Y.; Otaki, T.; Osada, H.; Nomiya, K. Aluminum-and gallium-containing open-Dawson polyoxometalates. *Z. Anorg. Allg. Chem.* **2016**, *642*, 539–545. [CrossRef]

29. Zhang, F.-Q.; Guan, W.; Yan, L.-K.; Zhang, Y.-T.; Xu, M.-T.; Hayfron-Benjamin, E.; Su, Z.-M. On the origin of the relative stability of Wells–Dawson isomers: A DFT study of α-, β-, γ-, α*-, β*-, and γ*-$[(PO_4)_2W_{18}O_{54}]^{6-}$ anions. *Inorg. Chem.* **2011**, *50*, 4967–4977. [CrossRef] [PubMed]

30. Jordan, P.A.; Clayden, N.J.; Heath, S.L.; Moore, G.R.; Powell, A.K.; Tapparo, A. Defining speciation profiles of $Al^{3+}$ complexed with small organic ligands: The $Al^{3+}$-heidi system. *Coord. Chem. Rev.* **1996**, *149*, 281–309. [CrossRef]

31. Casey, W.H. Large aqueous aluminum hydroxide molecules. *Chem. Rev.* **2006**, *106*, 1–16. [CrossRef] [PubMed]

32. Mensinger, Z.L.; Wang, W.; Keszler, D.A.; Johnson, D.W. Oligomeric group 13 hydroxide compounds—A rare but varied class of molecules. *Chem. Soc. Rev.* **2012**, *41*, 1019–1030. [CrossRef] [PubMed]

33. Kikukawa, Y.; Yamaguchi, S.; Nakagawa, Y.; Uehara, K.; Uchida, S.; Yamaguchi, K.; Mizuno, N. Synthesis of a dialuminum-substituted silicotungstate and the diastereoselective cyclization of citronellal derivatives. *J. Am. Chem. Soc.* **2008**, *130*, 15872–15878. [CrossRef] [PubMed]

34. Kato, C.N.; Katayama, Y.; Nagami, M.; Kato, M.; Yamasaki, M. A sandwich-type aluminium complex composed of tri-lacunary Keggin-type polyoxotungstate: Synthesis and X-ray crystal structure of $[(A-PW_9O_{34})_2\{W(OH)(OH_2)\}\{Al(OH)(OH_2)\}\{Al(\mu-OH)(OH_2)_2\}_2]^{7-}$. *Dalton Trans.* **2010**, *39*, 11469–11474. [CrossRef] [PubMed]

35. Kikukawa, Y.; Yamaguchi, K.; Hibino, M.; Mizuno, N. Layered assemblies of a dialuminum-substituted silicotungstate trimer and the reversible interlayer cation-exchange properties. *Inorg. Chem.* **2011**, *50*, 12411–12413. [CrossRef] [PubMed]

36. Carraro, M.; Bassil, B.S.; Sorarù, A.; Berardi, S.; Suchopar, A.; Kortz, U.; Bonchio, M. A Lewis acid catalytic core sandwiched by inorganic polyoxoanion caps: Selective $H_2O_2$-based oxidations with $[Al^{III}_4(H_2O)_{10}(\beta-XW_9O_{33}H)_2]^{6-}$ (X = $As^{III}$, $Sb^{III}$). *Chem. Commun.* **2013**, *49*, 7914–7916. [CrossRef] [PubMed]

37. Kato, C.N.; Makino, Y.; Unno, W.; Uno, H. Synthesis, molecular structure, and stability of a zirconocene derivative with α-Keggin mono-aluminum substituted polyoxotungstate. *Dalton Trans.* **2013**, *42*, 1129–1135. [CrossRef] [PubMed]

38. Kato, C.N.; Kashiwagi, T.; Unno, W.; Nakagawa, M.; Uno, H. Syntheses and molecular structures of monomeric and hydrogen-bonded dimeric Dawson-type trialuminum-substituted polyoxotungstates derived under acidic and basic conditions. *Inorg. Chem.* **2014**, *53*, 4823–4832. [CrossRef] [PubMed]

39. Inoue, Y.; Matsunaga, S.; Nomiya, K. $Al_{16}$-hydroxide cluster-containing tetrameric polyoxometalate, $[\{\alpha-Al_3SiW_9O_{34}(\mu-OH)_6\}_4\{Al_4(\mu-OH)_6\}]^{22-}$. *Chem. Lett.* **2015**, *44*, 1649–1651. [CrossRef]

40. Allmen, K.; Car, P.-E.; Blacque, O.; Fox, T.; Müller, R.; Patzke, G.R. Structure and properties of new gallium-containing polyoxotungstates with hexanuclear and tetranuclear cores. *Z. Anorg. Allg. Chem.* **2014**, *640*, 781–789. [CrossRef]

41. Limanski, E.M.; Drewes, D.; Krebs, B. Sandwich-like polyoxotungstates with indium(III) as a heteroatom synthesis and characterization of the first examples of a new type of anions. *Z. Anorg. Allg. Chem.* **2004**, *630*, 523–528. [CrossRef]

42. Hussain, F.; Reicke, M.; Janowski, V.; Silva, S.D.; Futuwi, J.; Kortz, U. Some indium(III)-substituted polyoxotungstates of the Keggin and Dawson types. *C. R. Chim.* **2005**, *8*, 1045–1056. [CrossRef]

43. Zhao, D.; Ye, R.-H. Solvothermal synthesis and structure of a new indium-substituted polyoxotungstate: $[(CH_3)_2NH_2]_4[In_3(H_2O)_3(NO_3)(A-\alpha-H_3PW_9O_{34})_2]\cdot H_2O$. *J. Clust. Sci.* **2011**, *22*, 563–571. [CrossRef]

44. Brown, I.D.; Altermatt, D. Bond-valence parameters obtained from a systematic analysis of the Inorganic Crystal Structure Database. *Acta Crystallogr.* **1985**, *B41*, 244–247. [CrossRef]

45.  Shannon, R.D.; Prewitt, C.T. Effective ionic radii in oxides and fluorides. *Acta Cryst.* **1969**, *B25*, 925–946. [CrossRef]

46.  Shannon, R.D. Revised effective ionic radii and systematic studies of interatomic distances in halides and chalcogenides. *Acta Cryst.* **1976**, *A32*, 751–761. [CrossRef]

47.  Sheldrick, G.M. A short history of SHELX. *Acta Crystallogr.* **2008**, *64*, 112–122. [CrossRef] [PubMed]

# 6

# Silylation of Dinitrogen Catalyzed by Hydridodinitrogentris(Triphenylphosphine)Cobalt(I)

Wojciech I. Dzik

Van't Hoff Institute for Molecular Sciences (HIMS), Homogeneous, Supramolecular and Bio-Inspired Catalysis, Universiteit van Amsterdam, P.O. Box 94720, 1090 GS Amsterdam, The Netherlands; w.i.dzik@uva.nl

Academic Editors: Duncan H. Gregory and Gianfranco Pacchioni

**Abstract:** Recently, homogeneous cobalt systems were reported to catalyze the reductive silylation of dinitrogen. In this study the investigations on the silylation of dinitrogen catalyzed by $CoH(PPh_3)_3N_2$ are presented. We show that in the presence of the title compound, the reaction of $N_2$ with trimethylsilylchloride and sodium yields, on average, 6.7 equivalents of tris(trimethylsilyl)amine per Co atom in THF (tetrahydrofuran). The aim was to elucidate whether the active catalyst is: (a) the $[Co(PPh_3)_3N_2]^-$ anion formed after two-electron reduction of the title compound; or (b) a species formed via decomposition of $CoH(PPh_3)_3N_2$ in the presence of the highly reactive substrates. Time profile, and IR and EPR spectroscopic investigations show instability of the pre-catalyst under the applied conditions which suggests that the catalytically active species is formed through in situ modification of the pre-catalyst.

**Keywords:** nitrogen activation; silylamines; cobalt; homogeneous catalysis

---

## 1. Introduction

Effective catalytic activation of dinitrogen under ambient conditions remains one of the biggest challenges in synthetic chemistry. Although such a transformation is feasible, as evidenced by the biosynthesis of ammonia enabled by nitrogenase enzymes [1–4], the activation of dinitrogen on transition metal complexes is not trivial and the best man-made homogeneous catalysts for ammonia synthesis can perform only several dozens of catalytic turnovers [5–13].

In this context, most of the effort has been focused on molybdenum and iron systems [14–18]. Two major strategies for catalytic $N_2$ activation by homogeneous systems were developed: reductive protonation to yield ammonia and reductive silylation to yield silylamines (Figure 1).

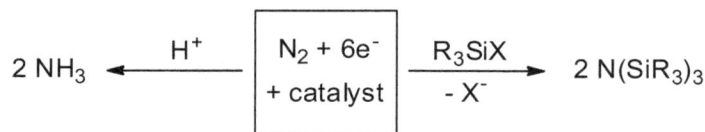

$$2\,NH_3 \xleftarrow{\quad H^+ \quad} \boxed{\begin{array}{c} N_2 + 6e^- \\ + \text{ catalyst} \end{array}} \xrightarrow[-\,X^-]{\quad R_3SiX \quad} 2\,N(SiR_3)_3$$

**Figure 1.** Strategies for catalytic reduction of $N_2$ at ambient conditions.

Catalytic formation of ammonia on well-defined systems was pioneered by the group of Schrock who developed a molybdenum-based system that catalyzed the formation of almost eight equivalents of $NH_3$ per metal atom [19]. The group of Nishibayashi developed binuclear catalysts which were able to catalyze the formation of up to 63 equivalents of $NH_3$ per Mo atom [20,21]. Iron also proved to be catalytically active as shown by the group of Peters, whose catalysts could produce up to 64 equivalents of $NH_3$ at very low temperatures [22,23].

However, these catalysts suffer from poisoning with ammonia and dihydrogen formed as side products. This is usually not the case in catalytic silylation reactions and therefore this reaction generally yields more turnovers of $N_2$ fixation (Figure 2). Catalytic reductive silylation of $N_2$ was pioneered by Shiina who used $CrCl_3$ as the pre-catalyst. This led to formation of 5.4 equivalents of $N(SiMe_3)_3$ when $N_2$ was reacted with lithium and $Me_3SiCl$ [24]. Other salts of various metals yielded much less of the desired product. For instance, the use of $CoCl_2$ which is relevant to the research presented here led to the formation of 1.2 equivalents of $N(SiMe_3)_3$ per metal centre. This chemistry was further developed for well-defined Mo and W systems by the groups of Hidai [25] and Nishibayashi [26]. Another well-defined molybdenum catalyst was recently disclosed by Mézailles and co-workers [27]. The group of Nishibayashi also reported that simple organometallic iron compounds can be used as pre-catalysts in this reaction and up to 34 equivalents of $N(SiMe_3)_3$ could be formed when a ferrocene derivative was used [28]. Interestingly, none of the studied iron compounds contain coordinated dinitrogen, and an approximately 1 h incubation period was observed, indicating that the catalytically active species is formed by reaction of the iron pre-catalyst with silylchlorides under reductive conditions. A two-coordinate iron(0) complex supported with a cyclic (alkyl)(amino)carbene ligand which revealed comparable activity was recently reported by Ung and Peters [29].

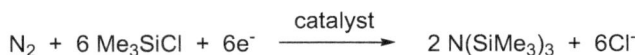

$$N_2 + 6\ Me_3SiCl + 6e^- \xrightarrow{\text{catalyst}} 2\ N(SiMe_3)_3 + 6Cl^-$$

catalysts:

TON = 24

TON = 15

TON = 195

TON = 226 (M = Mo)
60 (M = W)

TON = 34

TON = 24

$Co_2(CO)_8$/2bpy

TON = 49

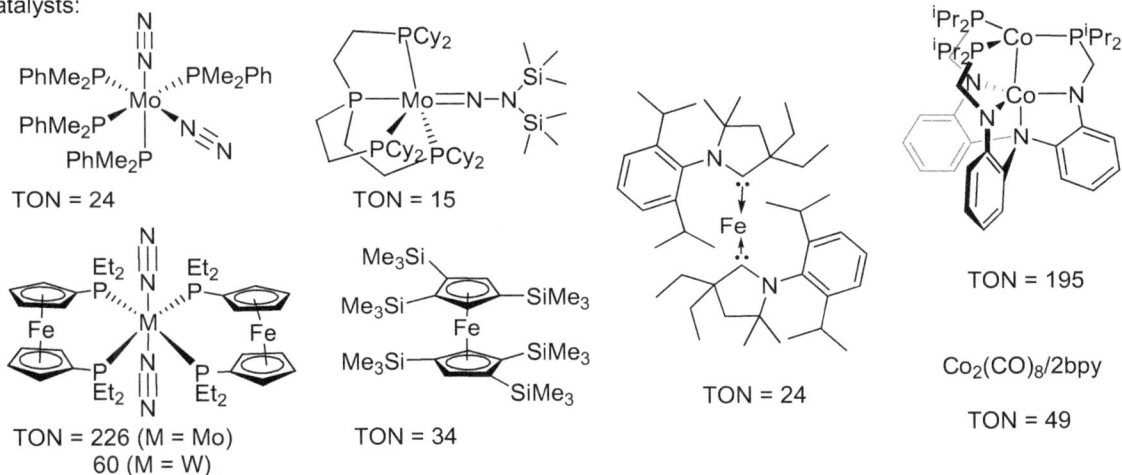

**Figure 2.** Selected (pre)catalysts for reductive trimethylsilylation of dinitrogen.

Despite the fact that cobalt-containing catalysts [30] are active in the Haber–Bosch reaction, only a handful of homogeneous systems based on this metal have been reported in the context of $N_2$ activation. Yamamoto and co-workers showed that treatment of $\{[Co(PPh_3)_3N_2]^-\}_xM^{x+}(THF)_{2+x}$ (M = Li, Na, Mg) complexes with acids leads to the formation of ammonia and hydrazine in sub-stoichiometric quantities [31]. Peters and co-workers reported methylation and trimethylsilylation of dinitrogen bound to $[(PhB(CH_2P^iPr_2)_3)CoN_2]_2Mg(THF)_4$ (Figure 3) [32]. Last year, Gagliardi, Lu and co-workers reported a dinuclear cobalt complex that catalyzes the formation of $N(SiMe_3)_3$ from dinitrogen with an impressive turnover number of 195 [33], while the group of Nishibayashi showed that in the presence of 2-2′-bipyridine, $Co_2(CO)_8$ can catalyze the formation of up to 49 equivalents of $N(SiMe_3)_3$ (Figure 2) [34]. Homogeneous cobalt complexes with tris(phosphine)borate ligands were also shown by Peters and co-workers to catalyze the formation of over two equivalents of ammonia [35].

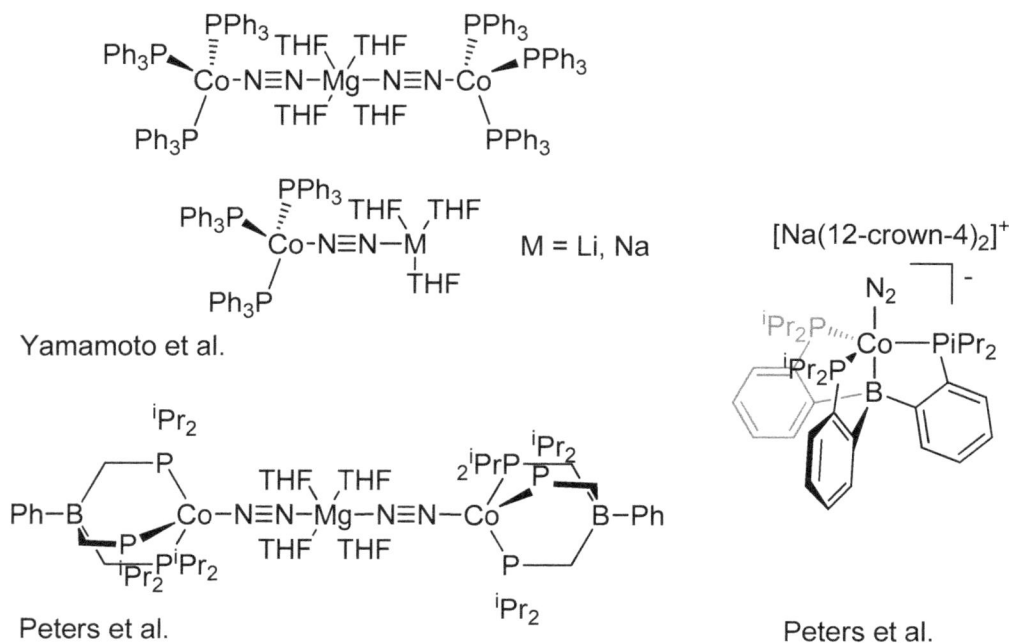

**Figure 3.** Cobalt dinitrogen complexes that allow reactivity of coordinated $N_2$ [31,32,35].

## 2. Results and Discussion

Since Yamamoto's $[Co(PPh_3)_3N_2]^-$ complex promoted the reduction of the coordinated $N_2$ in THF to ammonia and hydrazine, it seemed plausible that it could also promote the silylation of dinitrogen. Indeed, recently the group of Nishibayashi mentioned that $CoH(PPh_3)_3N_2$ catalyzes the reductive silylation of $N_2$ [34]. Given the fact that $[Co(PPh_3)_3N_2]^-$ can be formed, e.g., by the reduction of $CoH(PPh_3)_3N_2$ with alkali metals, we decided to investigate the activity of the cobalt triphenylphosphine system in silylation of $N_2$ in more detail.

One could expect that $[CoN_2(PPh_3)_3]^-$ should directly react with $Me_3SiCl$, resulting in the formation of Si–N bonds, similarly to the dinuclear system reported by the groups of Gagliardi and Lu [33]. However, as shown by Nishibayashi, Yoshizawa and co-workers, many simple cobalt complexes reveal long incubation periods before the onset of the catalytic reaction [34]. We were thus interested in the mechanism behind the catalytic dinitrogen silylation by the cobalt triphenylphosphine system. Is $[Co(PPh_3)_3N_2]^-$ the actual catalyst like the bimetallic Lu–Gagliardi complex or merely a pre-catalyst which undergoes a transformation into the catalytically active species through reaction with chlorosilane? To answer this question we started our investigations by studying the performance of the in-situ formed $[Co(PPh_3)_3N_2]^-$ anion in the catalytic silylation of dinitrogen.

Treating a 0.0025 M solution of $CoH(PPh_3)_3N_2$ in THF with 180 equivalents of Na and 190 equivalents of $Me_3SiCl$ led to the formation of $6.7 \pm 1.0$ equivalents (per Co atom) of $N(SiMe_3)_3$ as evidenced by GC. This corresponded to approximately 12% yield based on sodium (Table 1). The identity of $N(SiMe_3)_3$ was confirmed with mass spectrometry. Analysis of the ammonia content using indophenol method after saponification of the reaction mixture with 1 N $H_2SO_4$ was in accord with the GC analysis.

For comparison, we also investigated the behavior of cobalt compounds that do not have a coordinated dinitrogen molecule. The use of $Co(acac)_3$ led to the substoichiometric formation of $N(SiMe_3)_3$, while the use of $CoBr_2$ or Co powder did not result in the formation of $N(SiMe_3)_3$ (Table 1, entries 2–4). The use of cobaltocene led to the formation of $N(SiMe_3)_3$ in a slightly higher (but comparable) yield to the one obtained with $CoH(PPh_3)_3N_2$ (entry 5). $N(SiMe_3)_3$ was not formed when no cobalt was present (entry 6). $CoCl_2(PPh_3)_2$ which features the triphenylphosphine ligand had significantly lower activity (entry 7), while $CoCl(PPh_3)_3$ (which upon two-electron reduction under dinitrogen atmosphere could form $[Co(PPh_3)_3N_2]^-$) was less active than $CoH(PPh_3)_3N_2$ (entry 8). The use of diethyl ether as a solvent led to a slightly lower yield of $N(SiMe_3)_3$, and in benzene no formation of this compound was detected (entries 9,10). No significant changes in yield were measured when the amount of solvent varied (entries 11,12). The use of lithium as the reductant led to a comparable yield, while for potassium no desired product was formed (entries 13,14). Potassium gave good results with other systems [26] and the failure to obtain the desired product with this reductant is likely caused by a very fast reaction of potassium (disappearance of all solids within less than an hour) with all $Me_3SiCl$ before the catalytically active species could be formed (see below). The use of potassium graphite instead of metallic potassium led to the formation of 3.5 equivalents of $N(SiMe_3)_3$. The exact reason why $Me_3SiCl$ is consumed in unproductive reaction with metallic K at higher rates than with metallic Li or Na or $KC_8$ is currently not clear but remains outside of the scope of this work. For an example of a somewhat related low performance of metallic K compared to Na and $KC_8$ in reductive coupling of alkyl chlorides see e.g., [36].

**Table 1.** Screening of catalytic conditions of reductive silylation of $N_2$ [a].

| # | Catalyst | Solvent | Vol./mL | Reduct. | Yield/% [b] | TON [c] |
|---|---|---|---|---|---|---|
| 1 [d] | $CoH(PPh_3)_3N_2$ | THF | 10 | Na | $11.7 \pm 1.6$ | $6.7 \pm 1.0$ |
| 2 [e] | $Co(acac)_3$ | THF | 10 | Na | $0.9 \pm 0.6$ | $0.5 \pm 0.4$ |
| 3 [e] | $CoBr_2$ | THF | 10 | Na | 0 | 0 |
| 4 | Co | THF | 10 | Na | 0 | 0 |
| 5 [f] | $Co(Cp)_2$ | THF | 10 | Na | $15.9 \pm 3.4$ | $8.3 \pm 2.3$ |
| 6 [e] | - | THF | 10 | Na | 0 | 0 |
| 7 [e] | $CoCl_2(PPh_3)_2$ | THF | 10 | Na | $1.2 \pm 0.6$ | $0.7 \pm 0.3$ |
| 8 [e] | $CoCl(PPh_3)_3$ | THF | 10 | Na | $6.1 \pm 1.2$ | $3.7 \pm 0.7$ |
| 9 | $CoH(PPh_3)_3N_2$ | $Et_2O$ | 10 | Na | 9.1 | 5.8 |
| 10 | $CoH(PPh_3)_3N_2$ | $C_6H_6$ | 10 | Na | 0 | 0 |
| 11 | $CoH(PPh_3)_3N_2$ | THF | 5 | Na | 10.8 | 6.8 |
| 12 | $CoH(PPh_3)_3N_2$ | THF | 20 | Na | 11.6 | 7.3 |
| 13 | $CoH(PPh_3)_3N_2$ | THF | 10 | Li | 8.7 | 5.5 |
| 14 | $CoH(PPh_3)_3N_2$ | THF | 10 | K | 0 | 0 |
| 15 | $CoH(PPh_3)_3N_2$ | THF | 10 | $KC_8$ | 5.8 | 3.5 |

[a] Conditions: 0.025 mmol catalyst, 4.5 mmol reductant; 4.72 mmol $Me_3SiCl$, time = 48 h. $Me_3Si–SiMe_3$, $(Me_3Si)_2NH$ and $Me_3Si–C_4H_8O–SiMe_3$ were observed as the main side products; [b] Yield was determined with GC, using decane (10 μL) as an internal standard, and is based on the maximum theoretical yield of $(Me_3Si)_3N$ (1.5 mmol); [c] TON is the amount of $(Me_3Si)_3N$ formed per cobalt atom; [d] average of four runs; [e] average of two runs; [f] average of three runs.

Overall, $CoH(PPh_3)_3N_2$ showed a rather moderate activity, and although it performed better than its non-dinitrogen-containing analogs $CoCl(PPh_3)_3$ and $CoCl_2(PPh_3)_2$, it was still less effective than cobaltocene. Next, we undertook mechanistic studies of its catalytic activity, and attempted to detect any possible cobalt intermediates in the trimethylsilylation reaction. The time profile of the formation of $N(SiMe_3)_3$ reveals an induction period of ca. 4 h followed by a period of moderate activity (ca. 16 h) during which all the sodium is consumed (Figure 4). A somewhat shorter (1 h) induction period was reported by the group of Nishibayashi for the organometallic cobalt systems. This induction period was proposed to account for the formation of catalytically active species that features trimethylsilyl groups directly bound to the cobalt center [34].

We attempted to shed light on the possible initial transformations of $CoH(PPh_3)_3N_2$ using IR spectroscopy. Yamamoto et al. reported that the reduction of $CoH(PPh_3)_3N_2$ with sodium results in the formation of $[Co(PPh_3)_3N_2]Na(THF)_3$ [31]. Indeed, the addition of five equivalents of Na to a THF solution of $CoH(PPh_3)_3N_2$ resulted in the disappearance of its characteristic $N≡N$ stretch vibration at $2091 \text{ cm}^{-1}$ and the appearance of a new peak at $1912 \text{ cm}^{-1}$ attributable to $[Co(PPh_3)_3N_2]Na(THF)_3$. When three equivalents of $Me_3SiCl$ were added to a solution of $CoH(PPh_3)_3N_2$ in THF, the disappearance of the IR signal corresponding to the cobalt-dinitrogen complex was observed within 20 min and no peaks attributable to a reduced $N_2$ moiety could be observed. When the addition of $Me_3SiCl$ was performed in the presence of metallic sodium, the disappearance of the signal of $CoH(PPh_3)_3N_2$ was retarded, suggesting a possible concurrent reaction of $Me_3SiCl$ with sodium. Still, however, it seemed plausible that the attack of $Me_3SiCl$ on $[Co(PPh_3)_3N_2]^-$ would lead to the formation of the active catalyst. Therefore, we investigated whether the reaction of pre-formed $[Co(PPh_3)_3N_2]Na(THF)_3$ with six equivalents of trimethylchlorosilane would lead to the formation of $N(SiMe_3)_3$. GC-MS analysis of this reaction mixture did not reveal the formation of either $N(SiMe_3)_3$ or any organic compound containing an $N–SiMe_3$ moiety. The subsequent addition of 180 equivalents of sodium and $Me_3SiCl$ to the reaction mixture resulted in the formation of $N(SiMe_3)_3$, albeit an induction period still was observed.

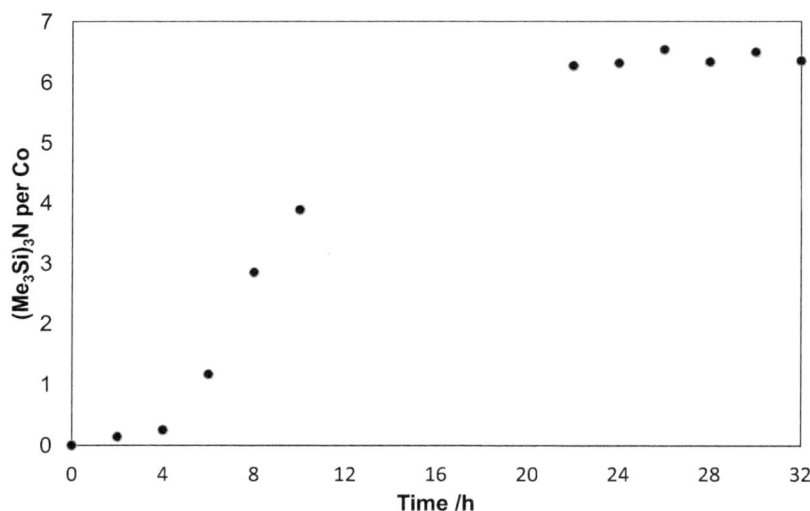

**Figure 4.** Formation of $(Me_3Si)_3N$ during $N_2$ reduction catalyzed by $CoH(PPh_3)_3N_2$.

Since in the first several hours of reaction no formation of $N(SiMe_3)_3$ is observed, and the stoichiometric reaction of $Me_3SiCl$ with $[Co(PPh_3)_3N_2]^-$ does not lead to the formation of N–Si bonds, it seemed rather unlikely that the catalytic reaction proceeds on $CoH(PPh_3)_3N_2$ (or $[Co(PPh_3)_3N_2]Na(THF)_3$). Indeed, IR spectroscopy analysis of an aliquot of a catalytically active mixture taken after 8 h of reaction revealed no peaks that could be assigned to any $Co–N_2$ species. This suggests that after the induction period, virtually all $CoH(PPh_3)_3N_2$ is decomposed and the bulk of the cobalt is not coordinated with a terminally bound $N_2$.

The majority of the coordination compounds of cobalt are paramagnetic, therefore we used EPR spectroscopy to probe the catalytic reaction mixture. Investigation of the reaction mixture after 3.5 h, i.e., before the onset of the formation of $N(SiMe_3)_3$, revealed the presence of a strong signal (Figure 5, red line) with clearly visible hyperfine couplings with the cobalt center and a very strong and broad signal spanning through the whole measurement window. Its spectral features ($g_x = 2.13$; $g_y = 2.12$; $g_z = 2.02$; $A_x = 215$; $A_y = 195$; $A_z = 180$ MHz) are indicative of a low-spin cobalt(II) species [35,37,38]. This species corresponded to approximately 8% of the total cobalt concentration. The amount of paramagnetic material was calculated from double integrals of the EPR signal of the catalytic reaction mixture and of a solution of Cobalt(II) meso-tetraphenylporphine used as the external standard (see Figure S1 in the supplementary materials). An aliquot from the reaction mixture taken after 48 h revealed the presence of the same paramagnetic species (with less resolved hyperfine structure, due to the lack of formation of good glass upon freezing) corresponding to approximately 5% of the total cobalt concentration (Figure 5, blue line). These data suggest that during the reaction of the $CoH(PPh_3)_3N_2$ with $Me_3SiCl$ in THF, a small amount of paramagnetic cobalt(II) species is formed which is not a catalytic intermediate in the $N_2$ fixation reaction. This species can be formed, e.g., by a transfer of a chloro radical from $Me_3SiCl$ to the cobalt(I) center, or by the attack of the trimethylsilyl radical formed after a one-electron reduction of Me₃SiCl. Formation of other cobalt species which could not be easily identified either using IR or EPR spectroscopy cannot be excluded; however, the reduction of the pre-catalyst to metallic cobalt has not been observed during the reaction.

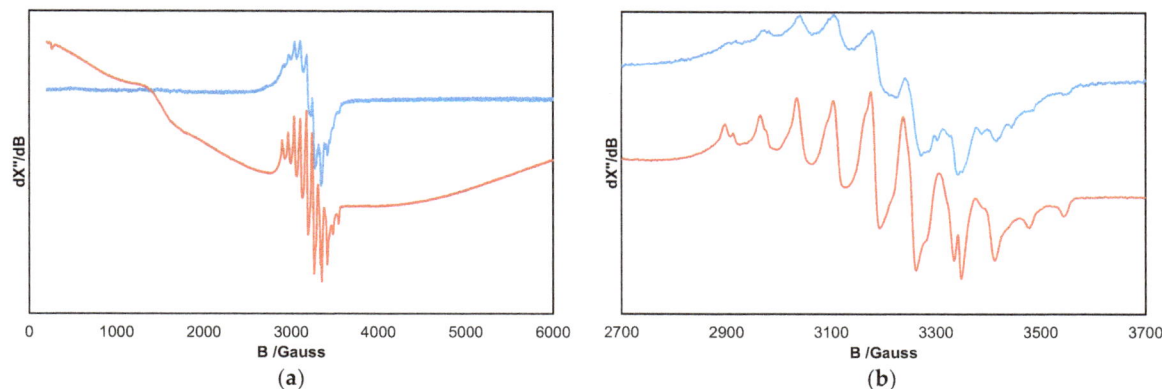

**Figure 5.** (**a**) X-band EPR spectra of the catalytic reaction mixture recorded at 20 K before the onset (red) and after reaching plateau (blue) of the formation of N(SiMe₃)₃. Frequency = 9.366 GHz, modulation amplitude: 4 Gauss, attenuation: 20 dB; (**b**) Selected part of the spectrum showing the hyperfine couplings of the cobalt(II) species.

Monitoring of the catalytic reaction with $^{31}$P NMR did not reveal any substantial signal of triphenylphosphine during the incubation period, and only after two days the signal at $-6.5$ ppm corresponding to free triphenylphosphine was observed, constituting less than 10% of the initial amount of CoH(PPh₃)₃N₂ (triphenylphosphine oxide added after the reaction stopped was used as an internal standard). This points to the possibility that triphenylphosphine remains bound to cobalt during catalysis; however, a fast exchange of triphenylphosphines on the cobalt center can also account for the lack of NMR signal.

To investigate whether cobalt nanoparticles are the active species, we analyzed the supernatant of the reaction mixture for the presence of nanoparticles using dynamic light scattering (DLS) after the formation of N(SiMe₃)₃ reached plateau. Measurements of the clear, brown solution obtained after sedimentation of the insoluble material formed during the catalytic reaction revealed a bimodal distribution of particles with an average size of 20 and 100 nm (see Figures S2 and S3 in the supplementary materials). This solution was, however, no longer active in the silylation of N₂ when additional sodium and Me₃SiCl were added. This suggests that even if cobalt nanoparticles are formed, they are an unlikely catalyst for the title reaction.

From the above results it is clear that neither CoH(PPh₃)₃N₂ nor its reduction product [Co(PPh₃)₃N₂]Na(THF)₃ activate N₂ towards the formation of N–Si bonds. Dinitrogen complexes of cobalt(II) are known [39]; however, it is unlikely that the observed paramagnetic complex is the active species given that it is already present in the reaction mixture before the onset of N(TMS)₃ formation is reached. It is possible that this species is being transformed into the actual catalyst during the incubation period. Since good molecular catalysts for the silylation of dinitrogen can reach over 200 turnovers [25,33], it is plausible that the actual very active catalyst is present only in minute amounts, which prevents its spectroscopic characterization. Based on the work of Nishibayashi, Yoshizawa and co-workers, an in-situ formed metallosilane species [28,34] can be responsible for the catalytic activity of CoH(PPh₃)₃N₂.

## 3. Materials and Methods

All manipulations were performed using Schlenk techniques under nitrogen or in a nitrogen-filled glovebox. CoH(PPh₃)₃N₂ was synthesized from Co(acac)₃ (99.99% trace metal basis, Sigma-Aldrich,

(St. Louis, MO, USA) (acac = acetylacetonato) as reported previously [40]. $CoCl(PPh_3)_3$ [41] and $CoCl_2(PPh_3)_2$ [42] were synthesized according to published procedures. THF and $Et_2O$ were distilled from sodium wire/benzoquinone, benzene and pentane was distilled from sodium wire. $Me_3SiCl$ (99%, purified by redistillation) was degassed using freeze-pump-thaw method (three cycles) and stored under nitrogen. Solvents used for catalytic runs were degassed using freeze-pump-thaw method (three cycles) immediately before use. GC measurements were performed on a Shimadzu GC-17A Gas Chromatograph (Shimadzu Corporation, Kyoto, Japan) with a Supelco SPB-1 fused silica capillary column. X-band EPR measurements were performed on a Bruker EMX spectrometer equipped with a He temperature control cryostat system (Oxford Instruments, Tubney Woods, UK). DLS measurements were performed on an ALV/LSE 5003 electronics and multiple Tau digital correlator (ALV-Laser Vertriebsgesellschaft m-b.H, Langen, Germany).

General procedure for catalytic silylation of dinitrogen: 320 mg of sodium (30%–35% in mineral oil, 4.5 mmol) was placed in a 100 mL Schlenk flask equipped with a glass stirring bar. Sodium was washed three times with 5 mL of pentane and a 22 mg (0.025 mmol) amount of $CoH(PPh_3)_3N_2$ was added followed by 10 mL of THF. The solution was vigorously stirred and 0.6 mL (4.7 mmol) of $Me_3SiCl$ was added with syringe. The initially orange-brown mixture turned gray/black after several hours of stirring. After 48 h, 10 μL of decane as an internal standard was added with a micropipette, and the mixture was centrifuged giving a purple precipitate and green-brown supernatant which was used for analysis of the amount of fixed nitrogen.

## 4. Conclusions

The cobalt-dinitrogen complex $CoH(PPh_3)_3N_2$ is a pre-catalyst in the reductive silylation of the dinitrogen reaction. This complex itself does not mediate the formation of N–Si bonds; however, under the applied catalytic conditions it transforms to the active species after approximately 4 h. Spectroscopic analysis of the reaction mixture did not reveal the presence of a resting state with a terminally bound $N_2$, and the nanoparticles formed during the course of the reaction are not catalytically active. It is therefore likely that in-situ formed homogeneous cobaltosilane species proposed by Nishibayashi, Yoshiawa and co-workers are responsible for the catalytic activity. Further studies should aim at an unambiguous determination of the structure of the catalytically active species generated from this mononuclear cobalt precursor.

**Acknowledgments:** We thank the Netherlands Organization for Scientific Research (NWO-CW, VENI grant 722.013.002) for funding, Mr. Christophe Rebreyend for measuring the EPR spectra, Dr. Saeed Raoufmoghaddam for measuring DLS, and Prof. Dr. Joost N. H. Reek for his constant encouragement and support.

## Abbreviations

The following abbreviations are used in this manuscript:

| | |
|---|---|
| acac | acetylacetonato |
| DLS | dynamic light scattering |
| EPR | electron paramagnetic resonance |
| GC | gas chromatography |
| IR | infrared |
| MS | mass spectrometry |
| THF | Tetrahydrofuran |
| TON | turnover number |

## References

1.  Bothe, H.; Schmitz, O.; Yates, M.G.; Newton, W.E. Nitrogen Fixation and Hydrogen Metabolism in Cyanobacteria. *Microbiol. Mol. Biol. Rev.* **2010**, *74*, 529–551. [CrossRef] [PubMed]
2.  Burgess, B.K.; Lowe, D.J. Mechanism of Molybdenum Nitrogenase. *Chem. Rev.* **1996**, *96*, 2983–3012. [CrossRef] [PubMed]
3.  Eady, R.R. Structure-Function Relationships of Alternative Nitrogenases. *Chem. Rev.* **1996**, *96*, 3013–3030. [CrossRef] [PubMed]
4.  Hoffman, B.M.; Lukoyanov, D.; Yang, Z.-Y.; Dean, D.R.; Seefeldt, L.C. Mechanism of Nitrogen Fixation by Nitrogenase: The Next Stage. *Chem. Rev.* **2014**, *114*, 4041–4062. [CrossRef] [PubMed]
5.  Hidai, M.; Mizobe, Y. Recent Advances in the Chemistry of Dinitrogen Complexes. *Chem. Rev.* **1995**, *95*, 1115–1133. [CrossRef]
6.  Bazhenova, T.A.; Shilov, A.E. Nitrogen fixation in solution. *Coord. Chem. Rev.* **1995**, *144*, 69–145.
7.  Fryzuk, M.D.; Johnson, S.A. The continuing story of dinitrogen activation. *Coord. Chem. Rev.* **2000**, *200–202*, 379–409. [CrossRef]
8.  MacKay, B.A.; Fryzuk, M.D. Dinitrogen coordination chemistry: On the biomimetic borderlands. *Chem. Rev.* **2004**, *104*, 385–401. [CrossRef] [PubMed]
9.  Hidai, M.; Mizobe, Y. Research inspired by the chemistry of nitrogenase. Novel metal complexes and their reactivity toward dinitrogen, nitriles, and alkynes. *Can. J. Chem.* **2005**, *83*, 358–374. [CrossRef]
10. Hinrichsen, S.; Broda, H.; Gradert, C.; Söncksen, L.; Tuczek, F. Recent developments in synthetic nitrogen fixation. *Ann. Rep. Prog. Chem. Sect. A Inorg. Chem.* **2012**, *108*, 17–47. [CrossRef]
11. Van der Ham, C.J.M.; Koper, M.T.M.; Hetterscheid, D.G.H. Challenges in reduction of dinitrogen by proton and electron transfer. *Chem. Soc. Rev.* **2014**, *43*, 5183–5191. [CrossRef] [PubMed]
12. Khoenkhoen, N.; de Bruin, B.; Reek, J.N.H.; Dzik, W.I. Reactivity of Dinitrogen Bound to Mid- and Late-Transition-Metal Centers. *Eur. J. Inorg. Chem.* **2015**, *567–598*. [CrossRef]
13. Bezdek, M.J.; Chirik, P.J. Expanding Boundaries: $N_2$ Cleavage and Functionalization beyond Early Transition Metals. *Angew. Chem. Int. Ed.* **2016**, *55*. [CrossRef] [PubMed]
14. Hazari, N. Homogeneous iron complexes for the conversion of dinitrogen into ammonia and hydrazine. *Chem. Soc. Rev.* **2010**, *39*, 4044–4056. [CrossRef] [PubMed]
15. Crossland, J.L.; Tyler, D.R. Iron-dinitrogen coordination chemistry: Dinitrogen activation and reactivity. *Coord. Chem. Rev.* **2010**, *254*, 1883–1894. [CrossRef]
16. MacLeod, K.C.; Holland, P.L. Recent developments in the homogeneous reduction of dinitrogen by molybdenum and iron. *Nat. Chem.* **2013**, *5*, 559–565.
17. Tanabe, Y.; Nishibayashi, Y. Developing more sustainable processes for ammonia synthesis. *Coord. Chem. Rev.* **2013**, *257*, 2551–2564. [CrossRef]
18. Jia, H.-P.; Quadrelli, E.A. Mechanistic aspects of dinitrogen cleavage and hydrogenation to produce ammonia in catalysis and organometallic chemistry: Relevance of metal hydride bonds and dihydrogen. *Chem. Soc. Rev.* **2014**, *43*, 547–564. [CrossRef] [PubMed]
19. Yandulov, D.V.; Schrock, R.R. Catalytic reduction of dinitrogen to ammonia at a single molybdenum center. *Science* **2003**, *301*, 76–78. [CrossRef] [PubMed]
20. Arashiba, K.; Miyake, Y.; Nishibayashi, Y. A molybdenum complex bearing PNP-type pincer ligands leads to the catalytic reduction of dinitrogen into ammonia. *Nat. Chem.* **2011**, *3*, 120–125. [CrossRef] [PubMed]
21. Arashiba, K.; Kinoshita, E.; Kuriyama, S.; Eizawa, A.; Nakajima, K.; Tanaka, H.; Yoshizawa, K.; Nishibayashi, Y. Catalytic Reduction of Dinitrogen to Ammonia by Use of Molybdenum–Nitride Complexes Bearing a Tridentate Triphosphine as Catalysts. *J. Am. Chem. Soc.* **2015**, *137*, 5666–5669. [CrossRef] [PubMed]
22. Anderson, J.S.; Rittle, J.; Peters, J.C. Catalytic conversion of nitrogen to ammonia by an iron model complex. *Nature* **2013**, *501*, 84–87. [CrossRef] [PubMed]
23. Del Castillo, T.J.; Thompson, N.B.; Peters, J.C. A Synthetic Single-Site Fe Nitrogenase: High Turnover, Freeze-Quench $^{57}$Fe Mössbauer Data, and a Hydride Resting State. *J. Am. Chem. Soc.* **2016**, *138*, 5341–5350. [CrossRef] [PubMed]
24. Shiina, K. Reductive silylation of molecular nitrogen via fixation to tris(trialkylsilyl)amine. *J. Am. Chem. Soc.* **1972**, *94*, 9266–9267. [CrossRef]

25. Komori, K.; Oshita, H.; Mizobe, Y.; Hidai, M. Catalytic conversion of molecular nitrogen into silylamines using molybdenum and tungsten dinitrogen complexes. *J. Am. Chem. Soc.* **1989**, *111*, 1939–1940. [CrossRef]

26. Tanaka, H.; Sasada, A.; Kouno, T.; Yuki, M.; Miyake, Y.; Nakanishi, H.; Nishibayashi, Y.; Yoshizawa, K. Molybdenum-Catalyzed Transformation of Molecular Dinitrogen into Silylamine: Experimental and DFT Study on the Remarkable Role of Ferrocenyldiphosphine Ligands. *J. Am. Chem. Soc.* **2011**, *133*, 3498–3506. [CrossRef] [PubMed]

27. Liao, Q.; Saffon-Merceron, N.; Mézailles, N. Catalytic Dinitrogen Reduction at the Molybdenum Center Promoted by a Bulky Tetradentate Phosphine Ligand. *Angew. Chem. Int. Ed.* **2014**, *53*, 14206–14210. [CrossRef] [PubMed]

28. Yuki, M.; Tanaka, H.; Sasaki, K.; Miyake, Y.; Yoshizawa, K.; Nishibayashi, Y. Iron-catalysed transformation of molecular dinitrogen into silylamine under ambient conditions. *Nat. Commun.* **2012**, *3*. [CrossRef] [PubMed]

29. Ung, G.; Peters, J.C. Low Temperature $N_2$ Binding to 2-coordinate $L_2Fe^0$ Enables Reductive Trapping of $L_2FeN_2^-$ and $NH_3$ Generation. *Angew. Chem. Int. Ed.* **2015**, *54*, 532–535.

30. Lloyd, L. *Handbook of Industrial Catalysts*; Springer: New York, NY, USA, 2011.

31. Yamamoto, A.; Miura, Y.; Ito, T.; Chen, H.; Iri, K.; Ozawa, F. Preparation, X-ray molecular structure determination, and chemical properties of dinitrogen-coordinated cobalt complexes containing triphenylphosphine ligands and alkali metal or magnesium. Protonation of the coordinated dinitrogen to ammonia and hydrazine. *Organometallics* **1983**, *2*, 1429–1436. [CrossRef]

32. Betley, T.A.; Peters, J.C. Dinitrogen Chemistry from Trigonally Coordinated Iron and Cobalt Platforms. *J. Am. Chem. Soc.* **2003**, *125*, 10782–10783. [CrossRef] [PubMed]

33. Siedschlag, R.B.; Bernales, V.; Vogiatzis, K.D.; Planas, N.; Clouston, L.J.; Bill, E.; Gagliardi, L.; Lu, C.C. Catalytic Silylation of Dinitrogen with a Dicobalt Complex. *J. Am. Chem. Soc.* **2015**, *137*, 4638–4641. [CrossRef] [PubMed]

34. Imayoshi, R.; Tanaka, H.; Matsuo, Y.; Yuki, M.; Nakajima, K.; Yoshizawa, K.; Nishibayashi, Y. Cobalt-Catalyzed Transformation of Molecular Dinitrogen into Silylamine under Ambient Reaction Conditions. *Chem. Eur. J.* **2015**, *21*, 8905–8909. [CrossRef] [PubMed]

35. Del Castillo, T.J.; Thompson, N.B.; Suess, D.L.M.; Ung, G.; Peters, J. Evaluating Molecular Cobalt Complexes for the Conversion of $N_2$ to $NH_3$. *Inorg. Chem.* **2015**, *54*, 9256–9262. [CrossRef] [PubMed]

36. Evans, W.J.; Workman, P.S. Accessing Lanthanide Diiodide Reactivity for Coupling Alkyl Chlorides to Carbonyl Compounds via the $NdI_3$/Alkali Metal Reduction System. *Organometallics* **2005**, *24*, 1989–1991. [CrossRef]

37. Jenkins, D.M.; Di Bilio, A.J.; Allen, M.J.; Betley, T.A.; Peters, J.C. Elucidation of a Low Spin Cobalt(II) System in a Distorted Tetrahedral Geometry. *J. Am. Chem. Soc.* **2002**, *124*, 15336–15350. [CrossRef] [PubMed]

38. Korstanje, T.J.; van der Vlugt, J.I.; Elsevier, C.J.; de Bruin, B. Hydrogenation of carboxylic acids with a homogeneous cobalt catalyst. *Science* **2015**, *360*, 298–302. [CrossRef] [PubMed]

39. Hojilla Atienza, C.C.; Milsmann, C.; Semproni, S.P.; Turner, Z.R.; Chirik, P.J. Reversible Carbon–Carbon Bond Formation Induced by Oxidation and Reduction at a Redox-Active Cobalt Complex. *Inorg. Chem.* **2013**, *52*, 5403–5417. [CrossRef] [PubMed]

40. Yamamoto, A.; Kitazume, S.; Pu, L.S.; Ikeda, S. Synthesis and properties of hydridodinitrogentris (triphenylphosphine)cobalt(I) and the related phosphine-cobalt complexes. *J. Am. Chem. Soc.* **1971**, *93*, 371–380. [CrossRef]

41. Aresta, M.; Rossi, M.; Sacco, A. Tetrahedral complexes of cobalt(I). *Inorg. Chim. Acta* **1969**, *3*, 227–231. [CrossRef]

42. Grutters, M.M.P.; Müller, C.; Vogt, D. Highly Selective Cobalt-Catalyzed Hydrovinylation of Styrene. *J. Am. Chem. Soc.* **2006**, *128*, 7414–7415. [CrossRef] [PubMed]

# 7

# Anion Ordering in Bichalcogenides

**Martin Valldor**

Physics of Correlated Matter, Max Planck Institute for Chemical Physics of Solids, Nöthnitzer Str. 40, 01187 Dresden, Germany; martin.valldor@cpfs.mpg.de

Academic Editors: Richard Dronskowski and Duncan H. Gregory

**Abstract:** This review contains recent developments and new insights in the research on inorganic, crystalline compounds with two different chalcogenide ions (bichalcogenides). Anion ordering is used as a parameter to form structural dimensionalities as well as local- and global-electric polarities. The reason for the electric polarity is that, in the heterogeneous bichalcogenide lattice, the individual bond-lengths between cations and anions are different from those in a homogeneous anion lattice. It is also shown that heteroleptic tetrahedral and octahedral coordinations offer a multitude of new crystal fields and coordinations for involved cations. This coordination diversity in bichalcogenides seems to be one way to surpass electro-chemical redox potentials: three oxidation states of a single transition metal can be stabilized, e.g., $Ba_{15}V_{12}S_{34}O_3$. A new type of disproportionation, related to coordination, is presented and results from chemical pressure on the bichalcogenide lattices of $(La,Ce)CrS_2O$, transforming doubly $[CrS_{3/3}S_{2/2}O_{1/1}]^{3-}$ (5+1) into singly $[CrS_{4/2}S_{2/3}]^{7/3-}$ (6+0) and $[CrS_{4/3}O_{2/1}]^{11/3-}$ (4+2) coordinations. Also, magnetic anisotropy is imposed by the anion ordering in BaCoSO, where magnetic interactions via S or O occur along two different crystallographic directions. Further, the potential of the anion lattice is discussed as a parameter for future materials design.

**Keywords:** anion; chalcogenide; superstructure; heteroleptic coordination; crystal field; charge ordering; coordination disproportionation; magnetic anisotropy

## 1. Introduction

In solid-state inorganic chemistry, the anion lattice is usually mentioned as a matrix where the cations dwell. The latter, positively charged, ions are chiefly in the research focus because their valence electrons are responsible for physical properties, like magnetism, optic activity, and conductivity. As a result, the fundamental aspects of the potentially interesting anion lattice receives relatively little attention. In this review, the elements in main group 16, the chalcogenides, will be used to display the possibilities of the anion lattice and how its design is related to new chemical situations that eventually lead to desirable properties. Focus is placed on describing the ordering of two anions, so solid-solutions will be disregarded.

According to Hume-Rothery [1], if the size difference of two atoms/ions, composing a common lattice, is larger than ~10%, an ordered superstructure is formed rather than a solid solution. Hence, by considering the sizes of the two-fold negatively charged chalcogenide ions ($r(O^{2-}) = 1.26$ Å, $r(S^{2-}) = 1.70$ Å, $r(Se^{2-}) = 1.84$ Å, and $r(Te^{2-}) = 2.07$ Å) [2], anion ordering is most probable for sulfide-oxides, selenide-oxides, telluride-oxides, and perhaps telluride-sulfides. These chemical prerequisites are chosen in this review to keep the focus on the phenomenon (anion superstructures) instead of on its vast possibilities.

Below, causes and consequences of involving two chalcogenides into one crystal lattice are presented. Each case is briefly introduced and subsequently discussed on the basis of existing compounds. At the end of this review, further possibilities are given, but only with a few practical examples.

So far, there are several known examples of bichalcogenides, but many of them were obtained by chance, often admitted in honesty by the discoverers themselves. By comparing existing compounds, this review aims to initiate an understanding in this research field.

## 2. Causes and Consequences of Chalcogenide Ordering

### 2.1. Partial Anionic Substitution

Despite a significant size difference between two anions, a stoichiometric composition of two anions might build up a common lattice with great similarities to that of an anion homogeneous one. For example, the ternary $La_2SO_2$ [3] is isostructural to the binary $La_2O_3$ [4], but exhibits an anion superstructure, i.e., S−O ordering. Also, by replacing half of $O^{2-}$ in $SrZnO_2$ [5] by $S^{2-}$ and at the same time substitute $Sr^{2+}$ for $Ba^{2+}$, the quaternary compound BaZnSO results [6]. As both compounds are structurally very similar, the latter sulfide oxide can be described as an anionic superstructure of the former. These kinds of anion substitutions are also expected to work both ways: for example, partly replacing S for O converts the anion lattice of $La_3CuSiS_7$ [7] (or $Ce_6Al_{10/3}S_{14}$ [8]) into that of $Ba_3V_2S_4O_3$ [9,10] and the latter is isostructural with the former, having even the same crystal symmetry. This means that there might be rational solubility of one chalcogenide in the matrix of another: These partial substitutions are possible if the anions in a candidate compound occupy two crystallographically unique sites, with preferably different environments.

### 2.2. Structural Dimensionality

The anion lattice dimensionality (D) is often inherited by the crystal structure as a whole. Decisive is the relative abundance of the two anions: the closer this ratio is to unity, the higher is the dimensionality. Hence, a 1:1 ration could constitute a 3-D lattice, a 2:1 ratio most likely a 2-D stacking, and a 6:1 probably leads to a 0-D structure e.g. a single anion surrounded by the other type of anion. For example, the sulfide–oxide $Eu_5V_3S_6O_7$ [11] contains a weaved net of O and S rows (O/S = 7/6 ≈ 1) and is best described as a 3-D lattice. Two-dimensional layers of Se and O are found in $La_2Fe_2Se_2O_3$ [12] where the ratio O/Se = 1.5. In $Bi_2SO_2$, stacked square layers of either O or S are observed [13] (Figure 1), with the ratio O/S = 2. Quasi 1-D chains of Se ions penetrate the crystal structure of $Sr_2Mn_2SeO_4$ [14], in correspondence to the composition O/Se = 4. Single O ions are separated by S in $La_4MnS_6O$ [15], which is a representative of a 0-D oxygen anionic order and the ratio S/O = 6.

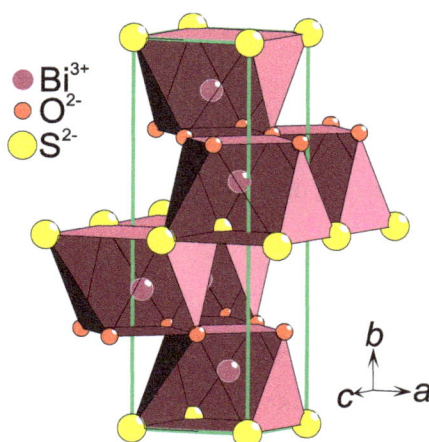

**Figure 1.** A part of the $Bi_2SO_2$ [13] crystal structure is shown as an example of a layered anion ordering. The heteroleptic coordination of Bi is emphasized with a polyhedron and the unit cell is drawn with green lines.

## 2.3. Charge Ordering

By combining the concepts of anion superstructures with that of hard–soft-acid–base (HSAB), which is often used for ions in solutions [16], it becomes apparent that a charge ordering might accompany an anion ordering. This means that higher (lower) oxidation states would be found close to anions with higher (lower) electronegativity. Assuming that the hypothetical compound $TM_3TeO_3$ ($TM$ = transition metal), with $TM^{2.67+}$ on average, exists and is charge ordered, it is most probable that $TM^{2+}$ coordinates mainly to $Te^{2-}$ and $TM^{3+}$ to $O^{2-}$. In $Eu_5V_3S_6O_7$, a charge order is proposed due to semiconducting behavior although the mean valence of V is +11/3 (+3.67) [11] and corresponds to $1 \times V^{3+}$ and $2 \times V^{4+}$. As the crystal structure contains two V sites, it is possible to order the charges although the authors chose differently: both sites are in first approximation octahedrally coordinated, $[VS_{4/2}O_{2/1}]^{13/3-}$, with similar interatomic distances, making it difficult to see any preference for the charges. However, the sites have different multiplicities, according to the Wyckoff definition (*4f* and *2b*), and probably should be divided as $V^{4+}$ (*4f*) and $V^{3+}$ (*2b*). In contrast, it can be expected from the stoichiometry that V is +4 in $Ba_3V_2S_4O_3$ [9,10], but the observations correspond to a charge disproportionation: $V^{3+}$ is found in an octahedral $[VS_{6/2}]^{3-}$ coordination and $V^{5+}$ in a $[VSO_3]^{3-}$ tetrahedron [10], agreeing with the HSAB principle. Similarly, tetrahedrally coordinated $V^{5+}$ and an intermediate valence of $V^{26/7+}$ in $[VS_{6/3}]^{2/7-}$-octahedra were reported in $Sr_6V_9S_{22}O_2$ but no magnetic moment was observed, agreeing with a lack of true charge ordering [17]. However, in $Ba_{15}V_{12}S_{34}O_3$ three oxidation states ($V^{3+}$, $V^{4+}$, $V^{5+}$) were proven experimentally and a suggested charge ordering was presented (Figure 2), as based on the existence of a short vandyl bond, $(V=O)^{2+}$, the HSAB principle, and Coulombic repulsion forces [18]. Obviously, the anion superstructure offers several different coordination environments for the transition metal, which seems to be an important prerequisite for the ordering of multiple oxidation states of one element on neighboring sites in a mutual crystal structure. In metal oxides, similar situations were suggested in $La_4Mn_5Si_4O_{22}$ [19] ($Mn^{2+}$, $Mn^{3+}$, and $Mn^{4+}$) and in $Ba_3Co_{10}O_{17}$ ($Co^{2+}$, $Co^{3+}$, and $Co^{4+}$) [20]. However, only the former was supported by spectroscopic data [19].

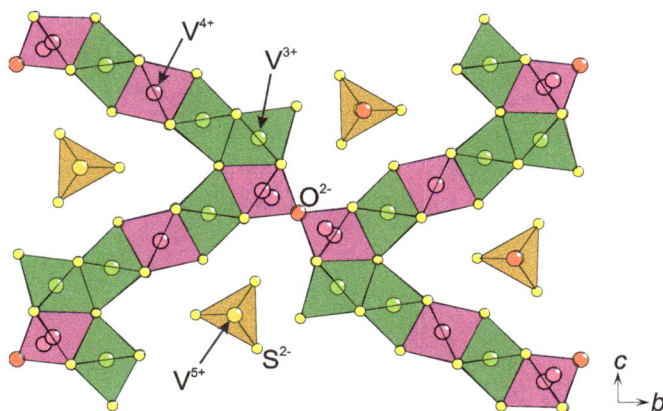

**Figure 2.** Part of the V—S—O lattice in the crystal structure of $Ba_{15}V_{12}S_{34}O_3$. The proposed charge ordering is emphasized with different colors: $V^{3+}$ (green), $V^{4+}$ (purple), and $V^{5+}$ (yellow) [18]. At the V—O—V junction, both half-occupied split-positions of V are indicated.

These, still rare, crystal structures prove that the lattice energy can surpass the electrochemical potential of redox pairs, which has to be studied further. Having several oxidation states of one element in one compound might induce novel optic and electronic properties.

## 2.4. Low Formal Oxidation States of Cations

A typical feature for the heavier chalcogens is the chemical non-stoichiometry towards reduced oxidation states of involved metals, exemplified by the non-stoichiometric $Fe_{1+x}Te$ having an $\alpha$-PbO type structure but with an additional partly occupied Fe site [21]. This is naturally also a consequence of the more covalent bonding character of Te—as compared to O, for example—and of the electronic itinerancy. Hence, to reach low formal oxidations states of transition metals, a combination of heavier chalcogenides with lighter ones can be the right strategy. Recently, the layered compound $CsV_2S_2O$ was reported with V having the formally intermediate charge +2.5 [22]. However, metallic properties were observed, meaning that new systems have to be found for the study of the low oxidation states. Still, it is fair to assume that the combination of hard and soft anions (i.e., according to HSAB) might stabilize rare charges of cations.

## 2.5. Polar Coordinations

A cation in a homogeneous anion lattice will bond similarly to its next neighbors, with a predictable coordination number (CN), as estimated from its ionic size relative to the anion size. However, within lattices consisting of two different anions, it is more challenging to make similar predictions. So far, no one has tried to settle any rules for the cations coordination in a heterogeneous anion lattice.

It is well established that the amount of covalency in the bonding nature between two elements is dependent on their differences in electronegativities. By introducing two different anions in the coordination of a common cation, the bonding will obtain irregularities (Figure 3). The bond distance between two specific atoms, e.g., Zn and O, varies somewhat depending on the other neighboring anions. Examples of this effect can be found in ternary bichalcogenides of actinides or lanthanides, like $Gd_2Se_2O$ [23] and $Ce_2TeO_2$ [24], but also in ZrSO [25], and $Ba_2TeO$ [26]. These four compounds—$Gd^{3+}$, $Ce^{3+}$, $Zr^{4+}$, and $Ba^{2+}$—have heteroleptic coordinations, and the bonding to either chalcogenide is clearly different from those in a homogeneous anion lattice (see Table 1). More specifically, heteroleptic coordination induces shorter $M-O$ ($M$ = metal) and longer $M-Ch$ ($Ch$ = chalcogen) distances in comparison to the corresponding bonds in homoleptic coordinations. This is also valid for ions of lighter elements like $Fe^{2+}$ in the more complex compounds $AEFe_2Ch_2O$ ($AE$ = $Ba^{2+}$, $Sr^{2+}$, $Ch = S^{2-}, Se^{2-}$) [27–30], where $Fe^{2+}$ has a quasi-tetrahedral $[FeS_{3/3}O_{1/2}]^-$ coordination and the Fe–S distance is surprisingly large, whereas Fe–O is correspondingly short in comparison to bonds in pure sulfides and oxides, respectively [30]. Naturally, this has strong influence on the involved magnetic coupling strengths, local electric polarizations, and crystal fields (see below). Similar situations can be observed in metal–organic complexes but those crystal structures are usually based on finite transition metal entities separated by organic molecules. On the contrary, in superstructures of anions, the lattices are extended and will cause global effects, so far, with unpredictable consequences.

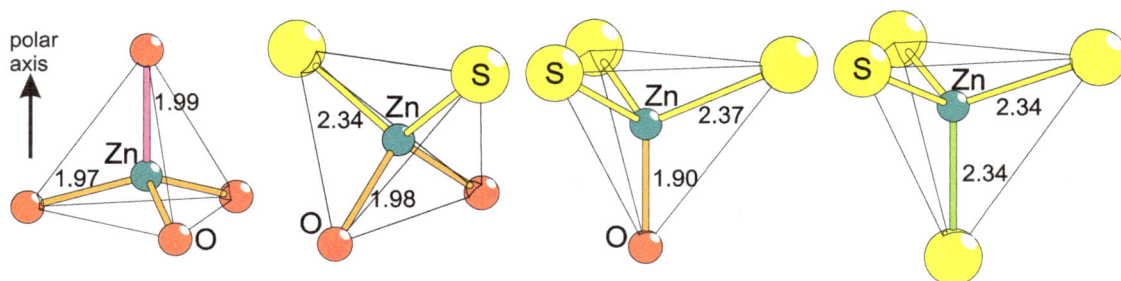

**Figure 3.** The principle of polar tetrahedral coordination is exhibited, as induced on Zn by S and O through heteroleptic coordination with comparisons to homoleptic ones. Displayed are, from left to right, the known examples ZnO [31], BaZnSO [6], CaZnSO [32], and 2H-ZnS [33]. The interatomic distances are differently colored to emphasize their different lengths (in Å).

**Table 1.** Comparisons of average bond lengths (in Å) for cations in homo- or heteroleptic coordination. Average distances are normalized by site multiplicities. The reference materials might contain two cations, but the cation of interest is chosen to have the same oxidation state and similar coordination number (CN) to validate the comparison. Bold fonts in formulas indicate the cations (**M**) in focus and *Ch* signifies chalcogens.

| **M**−*Ch*−O | CN $^1$ × **M**−O | CN $^1$ × **M**−S | CN $^1$ × **M**−Se | CN $^1$ × **M**−Te |
|---|---|---|---|---|
| **Gd**$_2$**Se**$_2$**O** [23] | 2 × 2.30 | | 5 × 2.98 | |
| **Gd**$_2$Ti$_2$O$_7$ [34], **Gd**$_2$**Se**$_3$ [35] | 8 × 2.46 | | 7.5 × 2.98 | |
| **Ce**$_2$**TeO**$_2$ [24] | 4 × 2.37 | | | 4 × 3.51 |
| **Ce**$_2$Zr$_2$O$_7$ [36], **Ce**$_{8/3}$**Te**$_4$ [37] | 8 × 2.47 | | | 8 × 3.31 |
| **ZrSO** [25] | 3 × 2.13 | 4 × 2.63 | | |
| **Zr**O$_2$ [38], Er$_2$**ZrS**$_5$ [39] | 8 × 2.21 | 7 × 2.59 | | |
| **Ba**$_2$**TeO** [26] | 1 × 2.45 | | | 5 × 3.59 |
| **Ba**O [40], **Ba**$_2$Mn**Te**$_3$ [41] | 6 × 2.76 | | | 7 × 3.54 |

$^1$ Average coordination number.

## 2.6. Structural Polarity

A local polarity, as induced by heterogeneous (heteroleptic) coordination can result in a global polarity, if the crystal structure is described with a non-centrosymmetric space group. For example, BaGeSe$_2$O contains GeSe$_2$O$_2$ tetrahedra, where the highly charged Ge$^{4+}$ is found far from the tetrahedral center due to the obvious size difference between Se and O [42]; the non-linear optic response from non-centrosymmetric BaGeSe$_2$O ($P2_12_12_1$) is surprisingly large. Another example is CaZnSO ($P6_3mc$) with strong electric polarity for tetrahedrally coordinated Zn$^{2+}$ ([ZnS$_{3/3}$O]$^{2-}$, Figure 3) [32]; Although the measured sample was polycrystalline, the second harmonic generation (SHG) effect is about 100 times stronger in CaZnSO than in $\alpha$-SiO$_2$. By replacing the non-magnetic Zn$^{2+}$ with the $S = 2$ entity high-spin Fe$^{2+}$, the isostructural CaFeSO was formed [43], where a magneto-electric coupling might result, although only a weak orbital moment is expected for a high-spin $d^6$ electronic configuration in tetrahedral coordination. However, according to the first report on this compound, a collinear anti-ferromagnetic long range order in each crystal structure layer was observed, which is further surprising because of the geometrical frustration from the hexagonal symmetry.

An indirect electric polarity is induced by non-magnetic [VSO$_3$]$^{3-}$ units on a quasi-1-D magnetic ($S = 1$) system in Ba$_3$V$_2$S$_4$O$_3$ (Figure 4): the trigonal packing of columns consisting of face-sharing [VS$_{6/3}$]$^{3-}$ octahedra are separated by Ba$^{2+}$ and polar [VSO$_3$]$^{3-}$ tetrahedra that all have a common polar direction [9,10]. The first investigations on its physical properties reveal two broad magnetic anomalies without long ranged spin-order, which might indicate thermal excitations of states that are induced by magneto-electric couplings [10].

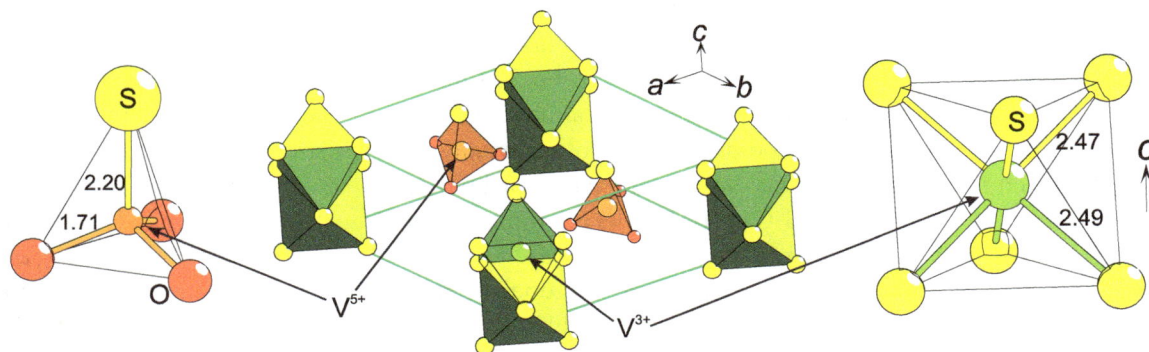

**Figure 4.** The polar [VSO$_3$]$^{3-}$ unit (**left**) imposes an indirect electric polarization on the quasi-1-D [VS$_{6/2}$]$^{3-}$ chain in Ba$_3$V$_2$S$_4$O$_3$ (**center**). This causes a shift of V$^{3+}$ within the chain (**right**). All distances are in Å and Ba$^{2+}$ ions were left out for simplification.

A similar 1-D magnetic sublattice is found in $La_3WTMS_3O_6$ ($TM$ = Cr, Mn, Fe, Co, Ni) [44]. By replacing $Ba^{2+}$ by $La^{3+}$ and the polar $[VSO_3]^{3-}$ unit by the non-polar $[WO_6]^{6-}$ trigonal prism, the quasi-1-D magnetic lattices can be obtained with ($P6_3$) or without ($P6_3/m$) structural polarity. To emphasize this, the compounds can be written as $Ba_3[VSO_3][TMS_{6/2}]$ and $La_3[WO_6][TMS_{6/2}]$, where the $[TMS_{6/2}]^{3-}$ columns are very similar even with the same +3 oxidation state of $TM$. This offers the unique situation to check the influence of the structural polarity on the material properties (magnetism, conductivity) on basically identical $TM$ sublattices. While this is not necessarily a result of anion ordering, the significant electric polarity in non-centrosymmetric $Ba_3[VSO_3][TMS_{6/2}]$ is assured by the heteroleptically coordinated $V^{5+}$ (Figure 4).

## 2.7. Coordination Configurations and Crystal Field Effects

The valence electrons of the first row $TMs$ (Sc−Zn) are strongly affected by the surrounding crystal field, which has consequences for energy differences between possible spin-states. With two different coordinating anions, there are more possibilities to design the crystal field strength and resulting relative energy of the orbitals on the $TM$. If the energy difference between two spin-states is relatively small, a transition between them can be achieved by means of temperature change and/or magnetic fields change (metamagnetism). The spin-state change is normally accompanied by a change in electrical conductivity, which is highly interesting for fundamental and applied science. For example, the spin-state transition in $Sr_{2-x}La_xCoO_4$ is related to the metal-insulator transition [45]. On the other hand, the spin-states of transition metals in compound with heterogeneous anionic lattice are far from understood.

As a consequence of (at least) two different coordinating anions, new coordination situations can be expected with novel crystal fields. Especially for the $TM$ ions, it is interesting to explore and to identify new electronic environments that influence the local polarity and the distribution of electrons among the $d$-orbitals. For tetrahedra, 3+1 and 2+2 coordinations are possible (Figure 3), of which a distorted type of the former is seen for $Fe^{2+}$ (HS, $d^6$) in $AEFe_2Ch_2O$ ($AE$ = $Ba^{2+}$, $Sr^{2+}$, $Ch$ = $S^{2-}$, $Se^{2-}$) [38–41] and in CaFeSO [43]. The latter 2+2 coordination is found for $TM^{2+}$ in $BaTMSO$ ($TM$ = $Zn^{2+}$ [8], $Co^{2+}$ HS, $d^7$ [46]) and $Fe^{2+}$ (HS, $d^6$) in CaFeSeO [47]. *Cis* and *trans* 2+2 distorted square planar coordinations of two electronically noble-gas-like anions have not been reported in bulk material to date, but should be possible. Numerous different five-fold coordinations are possible, but the two heteroleptic 4+1 and 3+2 configurations are most reasonable. A square pyramidal version of 4+1 is known for $Mn^{3+}$ (HS, $d^4$) in $Sr_2Mn_2SeO_4$ [14]. The 3+2 coordination has not been reported yet for bichalcogenides, but holds at least the possible *cis* and *trans* configurations that result in obviously different crystal fields. Moreover, a hypothetical trigonal bipyramidal *trans* coordination is likely to occur when the plane is constituted by the relatively smaller anions. The six-fold coordination has even more possibilities, namely 5+1, 4+2, and 3+3, of which all have several configurational names (Figure 5). So far, 5+1 has been reported for $Cr^{3+}$ ($d^3$) in $LaCrS_2O$ [48], for $Ti^{3+}$ ($d^1$) in $La_4Ti_2Se_5O_4$ [49], and for $Ti^{4+}$ ($d^0$) in $La_6Ti_3Se_9O_5$ [49]. A 4+2 coordination can be either *cis*, at least known for $Ti^{4+}$ ($d^0$) in $Pr_6Ti_2S_7O_6$ [50], or *trans* as for $Fe^{2+}$ (HS, $d^6$) in $La_2Fe_2Se_2O_3$ [12]. The facial (*fac*) and the meridional (*mer*) 3+3 coordinations are still very rare and were almost solely reported for metal-organic complexes, where polydental ligands make it possible to control the *fac* and *mer* isomerism. However, in extended crystals structure with monoatomic anions, the *mer* coordination in bichalcogenides only exists for non-magnetic $Ti^{4+}$ ($d^0$) in $La_6Ti_2S_7O_6$ [51]. Higher coordination numbers than six are mainly reported for lanthanides, actinides, and larger main group metals, see for example $Gd_2Se_2O$ [23], $Ce_2TeO_2$ [24], USeO [52], and BiCuSO [53]. Seven-fold capped trigonal antiprismatic coordinations are known for $Zr^{4+}$ ($d^0$) in ZrSO and $Nb^{5+}$ ($d^0$) in $Gd_3NbS_3O_4$. Further seven-fold (or higher) coordination might be possible at least for Mn or for $TMs$ of later periods ($4d$ or $5d$).

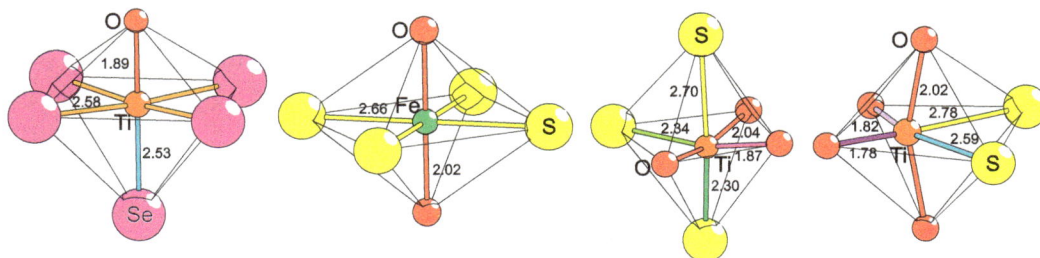

**Figure 5.** Different heteroleptic octahedral bichalcogenide coordinations of transition metals. From left to right: $[TiSe_{4/2}Se_{1/1}O_{1/1}]^{4-}$ in $La_4Ti_2Se_5O_4$ [49], *trans*-$[FeS_{4/4}O_{2/4}]^-$ in $La_2Fe_2Se_2O_3$ [12], *mer*-$[TiS_{3/1}O_{3/2}]^{5-}$ in $La_6Ti_2S_8O_5$ [51], and *cis*-$[TiS_{1/1}S_{1/2}O_{2/1}O_{2/2}]^{5-}$ in $Pr_6Ti_2S_7O_6$ [50] are shown. Interatomic bonds with the same color have the same bond lengths (in Å).

Clearly, the coordination flexibility is at hand, but a spin-state transition (HS−LS) has not been fully proven yet, although it was suggested for $Co^{3+}$ ($d^6$) in $La_3WCoS_3O_6$ [44], where Co is trigonal antiprismatically coordinated by $S^{2-}$ solely.

## 2.8. Coordination Disproportionation

$LaCrS_2O$ is a quasi-1-D system with zig-zag chains of $[CrS_{3/3}S_{2/2}O_{1/1}]^{3-}$ edge-sharing octahedra (Figure 6, left), where $La^{3+}$ ions separate the chains. By introducing the slightly smaller lanthanide ion $Ce^{3+}$ instead of $La^{3+}$, the lattice experiences a so-called "chemical pressure". As a result, the crystal structure of $CeCrS_2O$ consists of layers (Figure 6, right) built from two different octahedra that are edge- and vertex-sharing and are separated by $Ce^{3+}$ ions. The structural difference between the two chemically similar systems can be described as a coordination disproportionation: $LaCrS_2O$ contains only $[CrS_{3/3}S_{2/2}O_{1/1}]^{3-}$ entities [48] and $CeCrS_2O$ contains equal amount of $[CrS_{4/2}S_{2/3}]^{7/3-}$ and $[CrS_{4/3}O_{2/1}]^{11/3-}$ octahedra in an ordered fashion [54] (Figure 6). As it is clearly not a charge ordering phenomenon, this might be related to coordination order entropy that is rarely taken into consideration. An alternative explanation would be that the 5+1 coordination in $LaCrS_2O$ is relatively unstable, as indicated by the fact that the highly symmetrical electron distribution on $Cr^{3+}$ in an octahedron ($3t_{2g}0e_g$) does not agree with the relatively low symmetry of the Cr-site, i.e., four different bond lengths are found in the structure (Figure 6). Coordination disproportionation is also found in $CeCrSe_2O$ and even in $LaCrSe_2O$ [55], leaving $LaCrS_2O$ as only example, so far, with homogeneous 5+1 coordination of $Cr^{3+}$.

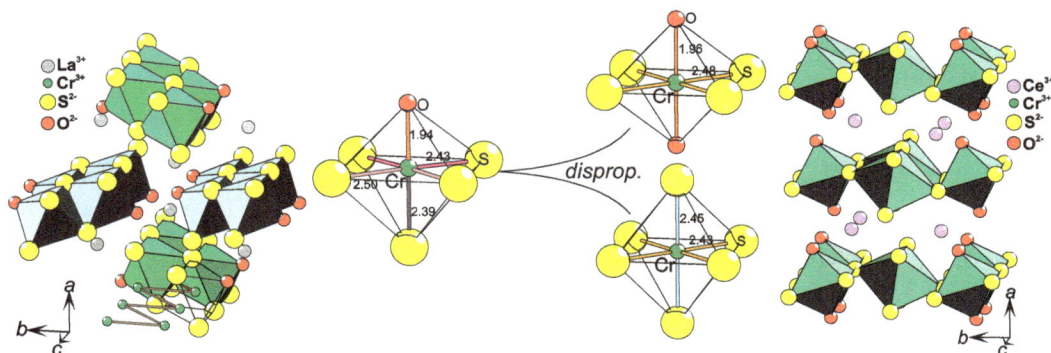

**Figure 6.** Parts of the $RECrS_2O$ ($RE$ = La, Ce) [48,54] crystal structures are displayed (**left** and **right**). In $LaCrS_2O$ (**left**), the 1-D rows of $[CrS_{3/3}S_{2/2}O_{1/1}]^{3-}$ edge-sharing octahedra are high-lighted, while the layers of edge- and vertex-sharing $[CrS_{4/3}O_{2/1}]^{11/3-}$−$[CrS_{4/2}S_{2/3}]^{7/3-}$ octahedra in $CeCrS_2O$ are exhibited (**right**). In the center, the coordination disproportionation is shown as changes in local coordination of Cr, where all distances (in Å) are added.

This type of polymorphism is probably described for the first time here and has to be studied further to understand its origin: The complex anionic lattice might be an ideal starting point for finding further examples of this phenomenon.

### 2.9. Magnetic Anisotropy

On an extended lattice, the magnetic interactions via super-exchange paths will be affected by heteroleptic coordinations of magnetic ions. The S—O ordering in BaCoSO [46] constitutes a new type of parameter causing magnetic anisotropy (Figure 7). The extraordinary bonding situation results in two different super-exchange paths, Co—S—Co and Co—O—Co, that run parallel to two orthogonal crystallographic directions, $a$ and $c$, respectively (Figure 7). The magnetic interactions between the $S = 3/2$ entities ($Co^{2+}$) are dependent on inter-atomic distances, bonding type (hopping integral), and bridges angle (Co—S—Co ($116°$) or Co—O—Co ($180°$)). All of these parameters, being different for the super-exchange via S or O, cause the magnetic coupling strengths to differ. Hence, the magnetic system BaCoSO is not obviously 2-D, as even the layers are anisotropic and carry features of a 1-D lattice.

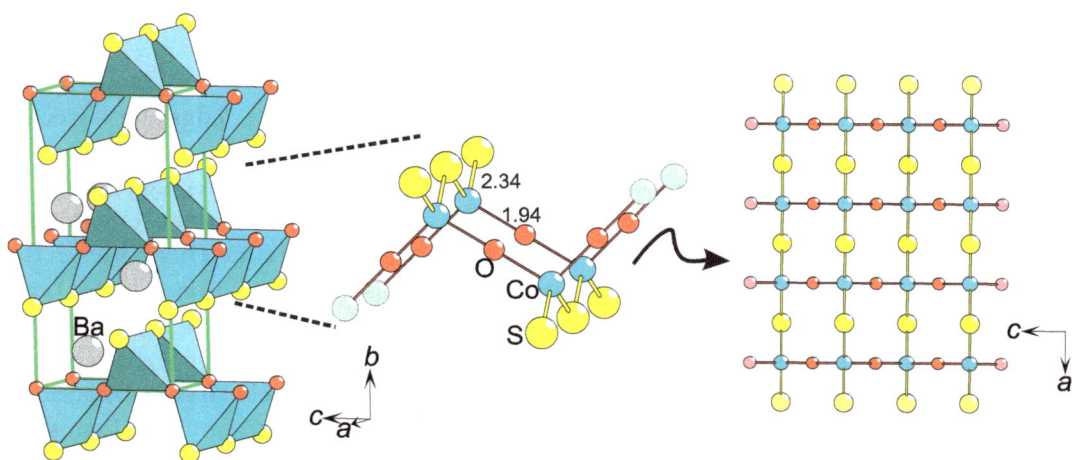

**Figure 7.** Magnetic anisotropy by chalcogen ordering in BaCoSO [46]. Displayed are: a part of the BaCoSO structure with unit cell in green (**left**), the local Co coordination (**center**), and the Co—S—O layer viewed along the $b$-axis. Interatomic distances are in Å.

Moreover, the low-dimensional surrounding, as formed by different sizes of next-neighbor ions, allows for orbital moments of magnetic ions or for a coupling between spin and structural polarity to evolve (see above).

It is clear that the ordering of anions can be counted as additional parameters for designing magnetic anisotropy.

### 2.10. Band-Gap Optimization

The covalent bond character of heavier anions to cations will decrease the electronic correlation strengths locally and decrease the band-gap globally: ZnO is a typical electronic insulator, ZnS and ZnSe are semiconductors, and ZnTe is metallic. The electronic properties of hypothetical compositions $Zn_2SO$ and $Zn_2TeO$ are, thus, difficult to predict and can be affected by the dimensionality of the anionic lattice. As the band-gap is important for several applicable materials, like in thermoelectrics and for photocatalytic splitting of water, compounds including heavier chalcogenides are receiving more attention lately. BiCuSeO is a prominent example with exceptional thermoelectric properties [56]. Its band-gap of about 0.8 eV increases proportionally to the S content in $BaCuSe_{1-x}S_xO$ up to about 1.1 eV [57], proving that this chalcogenide compound can be anion designed in a predictable way. The exceptional thermoelectric properties are naturally also a consequence of the low thermal conductivity, which is discussed as related to the 2-D layers of O and Se(S). Hence, the anion superstructure improves the applicable properties with its bonding anisotropy.

## 3. Further Notes

Anionic superstructures are only beginning to reveal their potential. In fact, there are several open problems in fundamental and applied sciences that might be solved by introducing the anionic superstructures as extra parameter in the design of related materials. For example, ionic conductors are often suffering from structural deterioration upon using the material, moving the ions; in cases where the moving ion is an anion, these compounds could be improved by introducing two anions, where one is moving and the other is holding the lattice together, perhaps in layers or in columns. On the other hand, if the moving ion is a cation, the anionic lattice can be designed with two ions to reach optimum diffusion paths. An already known example is $Ag_3SI$ with a combined anionic lattice and extremely high cationic mobility [58].

Even more intricate anionic lattices are known: $Ln_3SOF_5$ ($Ln$ = Nd, Sm, Gd–Ho) contain three different anions with only little mixing of $F^-$ and $O^{2-}$ [59]. The coordination is naturally more complex and the possibilities for the above mentioned parameters are extended. Heteroleptic coordinations by three anions in ordered layers of $U_2Te_2PO$ result in two different surroundings for U: either $UTe_5O_4$ or $UTe_5P_4$ [60]. Although the charge ordering was not stated in that report, the HSAB principle [16] suggests $U^{5+}$ in former and $U^{4+}$ in latter.

## 4. Final Words

Several of the above-mentioned phenomena are induced by the ordering of anions and the examples, presented here, only contained the bichalcogenides. Pnictogens in combination with chalcogens are already making the head-lines due to LaFePO [61] and related materials, but in that case Fe is homoleptically coordinated and does not have the same parameters of freedom as the materials in this review. However, the pnictogenide halogenide ZrNCl [62] does connect to the topic of this review and also exhibits superconductivity upon electronic doping. By combining a chalcogen with complex anions, further possibilities are obvious, and thereof $Bi_4(SO_4)S_2$ is a prominent example, which is reported to be superconducting [63]. Hence, further systematic investigations of other anion superstructures are needed, which probably will be as rewarding as this preliminary one for the bichalcogenides.

**Acknowledgments:** The author would like to thank Alexandra Zevalkink for improving the English language in this review.

## Abbreviations

The following abbreviations are used in this manuscript:

| | |
|---|---|
| Ch | Chalcogen |
| CN | Coordination Number |
| D | Dimensionality |
| HSAB | Hard–Soft-Acid–Base |
| HS | High Spin |
| LS | Low Spin |
| M | Metal |
| $r$ | radius |
| SHG | Second Harmonic Generation |
| TM | Transition Metal |

## References

1. Hume-Rothery, W.; Powell, H.M. On the Theory of Super-Lattice Structures in Alloys. *Z. Kristallogr.* **1935**, *91*, 23–47. [CrossRef]

2.     Shannon, R.D. Revised effective ionic radii and systematic studies of interatomic distances in halides and chalcogenides. *Acta Crystallogr.* **1976**, *A32*, 751–767. [CrossRef]

3.     Zachariasen, W.H. Crystal chemical studies of the 5*f*-series of elements. I. New structure types. *Acta Crystallogr.* **1948**, *1*, 265–268. [CrossRef]

4.     Zachariasen, W.H. Die Kristallstruktur der $\alpha$-Modifikation von den Sesquioxyden der seltenen Erdmetalle ($La_2O_3$, $Ce_2O_3$, $Pr_2O_3$, $Nd_2O_3$). *Z. Phys. Chem.* **1926**, *123*, 134–150.

5.     Von Schnering, H.G.; Hoppe, R. Die Kristallstruktur des $SrZnO_2$. *Z. Anorg. Allg. Chem.* **1961**, *312*, 87–98. [CrossRef]

6.     Broadley, S.; Gal, Z.A.; Cora, F.; Smura, C.F.; Clarke, S.J. Vertex-Linked $ZnO_2S_2$ Tetrahedra in the Oxysulfide BaZnOS: A New Coordination Environment for Zinc in a Condensed Solid. *Inorg. Chem.* **2005**, *44*, 9092–9096. [CrossRef] [PubMed]

7.     Guittard, M.; Julien-Pouzol, M. Les composes hexagonaux de type $La_3CuSiS_7$. *Bull. Soc. Chim. Fr.* **1970**, *1970*, 2467–2469.

8.     De Saint-Giniez, D.; Laruelle, P.; Flahaut, J. Structure cristalline du sulfure double de cerium et d'aluminium $Ce_6Al_{3.33}S_{14}$. *Comptes Rendus* **1968**, *267*, 1029–1032.

9.     Calvagna, F.; Zhang, J.-H.; Li, S.-J.; Zheng, C. Synthesis and structural analysis of $Ba_3V_2O_3S_4$. *Chem. Mater.* **2001**, *13*, 304–307. [CrossRef]

10.    Hopkins, E.J.; Prots, Y.; Burkhardt, U.; Watier, Y.; Hu, Z.; Kuo, C.-Y.; Chiang, J.-C.; Pi, T.-W.; Tanaka, A.; Tjeng, L.H.; et al. $Ba_3V_2S_4O_3$: A Mott insulating frustrated quasi-one-dimensional $S = 1$ magnet. *Chem. Eur. J.* **2015**, *21*, 7938–7943. [CrossRef] [PubMed]

11.    Meerschaut, A.; Lafond, A.; Palvadeau, P.; Deudon, C.; Cario, L. Synthesis and crystal structure of two new oxychalcogenides: $Eu_5V_3S_6O_7$ and $La_{10}Se_{14}O$. *Mater. Res. Bull.* **2002**, *37*, 1895–1905. [CrossRef]

12.    Mayer, J.M.; Schneemeyer, L.F.; Siegrist, T.; Waszczak, J.V.; van Dover, R.B. Neue Eisenlanthan-Oxidsulfid-und-Oxidselenid-Phasen mit Schichtstruktur: $Fe_2La_2O_3E_2$ ($E = S$, Se). *Angew. Chem.* **1992**, *104*, 1677–1678. [CrossRef]

13.    Koyama, E.; Nakai, I.; Nagashima, K. Crystal chemistry of oxide-chalcogenides. II. Synthesis and crystal structure of the first bismuth oxide-sulfide, $Bi_2O_2S$. *Acta Crystallogr.* **1984**, *B40*, 105–109. [CrossRef]

14.    Free, D.G.; Herkelrath, S.J.C.; Clarke, S.J. $Sr_2Mn_2O_4Se$: A New Oxychalcogenide with Antiferromagnetic Chains. *Z. Anorg. Allg. Chem.* **2012**, *638*, 2532–2537. [CrossRef]

15.    Ijjaali, I.; Bin, D.; Ibers, J.A. Seven new rare-earth transition-metal oxychalcogenides: Syntheses and characterization of $Ln_4MnOSe_6$ ($Ln$ = La, Ce, Nd), $Ln_4FeOSe_6$ ($Ln$ = La, Ce, Sm), and $La_4MnOS_6$. *J. Solid State Chem.* **2005**, *178*, 1503–1507. [CrossRef]

16.    Pearson, R.G. *Hard and Soft Acids and Bases*; Dowden, Hutchinson & Ross: Stroudsburg, PA, USA, 1973.

17.    Litteer, J.B.; Chen, B.-H.; Fettinger, J.C.; Eichhorn, B.W.; Ju, H.L.; Greene, R.L. Synthesis and magnetic and transport properties of $Sr_6V_9S_{22}O_2$: "$AM_2S_5$" phases revisited. *Inorg. Chem.* **2000**, *39*, 458–462. [CrossRef] [PubMed]

18.    Wong, C.J.; Hopkins, E.J.; Prots, Yu.; Hu, Z.; Kuo, C.J.; Pi, T.W.; Valldor, M. Anionic ordering in $Ba_{15}V_{12}S_{34}O_3$, affording three oxidation states of vanadium and a quasi-one-dimensional magnetic lattice. *Chem. Mater.* **2016**, *28*, 1621–1624. [CrossRef]

19.    Gueho, C.; Giaquinta, D.; Mansot, J.L.; Ebel, T.; Palvadeau, P. Structure and magnetism of $La_4Mn_5Si_4O_{22}$ and $La_4V_5Si_4O_{22}$: Two new rare-earth transition metal sorosilicates. *Chem. Mater.* **1995**, *7*, 486–492. [CrossRef]

20.    David, R.; Kabbour, H.; Bordet, P.; Pelloquin, D.; Leynaud, O.; Trentesaux, M.; Mentré, O. Triple $Co^{II, III, IV}$ charge ordering and spin states in modular cobaltites: A systemization through experimental and virtual compounds. *J. Mater. Chem. C* **2014**, *2*, 9457–9466. [CrossRef]

21.    Haraldsen, H.; Groenvold, F.; Vihovde, J. Faseforholdene i systemet jern-tellur. *Tidsskr. Kjem. Bergves.* **1944**, *4*, 96–98.

22.    Valldor, M.; Merz, P.; Prots, Y.; Schnelle, W. Bad-Metal-Layered Sulfide Oxide $CsV_2S_2O$. *Eur. J. Inorg. Chem.* **2016**, *2016*, 23–27. [CrossRef]

23.    Tougait, O.; Ibers, J.A. $Gd_2OSe_2$. *Acta Crystallogr.* **2000**, *C56*, 623–624.

24.    Benz, R. $Ce_2O_2Sb$ and $Ce_2O_2Bi$ crystal structure. *Acta Crystallogr.* **1971**, *B27*, 853–854. [CrossRef]

25.    McCullough, J.D.; Brewer, L.; Bromley, L.A. The crystal structure of zirconium oxysulfide, ZrOS. *Acta Crystallogr.* **1948**, *1*, 287–289. [CrossRef]

26. Besara, T.; Ramirez, D.; Sun, J.; Whalen, J.B.; Tokumoto, T.D.; McGill, S.A.; Singh, D.J.; Siegrist, T. $Ba_2TeO$: A new layered oxytelluride. *J. Solid State Chem.* **2015**, *222*, 60–65. [CrossRef]

27. Han, F.; Wan, X.; Shen, B.; Wen, H.-H. $BaFe_2Se_2O$ as an iron-based Mott insulator with antiferromagnetic order. *Phys. Rev.* **2012**, *B86*, 014411. [CrossRef]

28. Lei, H.; Ryu, H.; Ivanovski, V.; Warren, J.B.; Frenkel, A.I.; Cekic, B.; Yin, W.-G.; Petrovic, C. Structure and physical properties of the layered iron oxychalcogenide $BaFe_2Se_2O$. *Phys Rev.* **2012**, *B86*, 195133. [CrossRef]

29. Valldor, M.; Adler, P.; Prots, Y.; Burkhardt, U.; Tjeng, L.H. $S = 2$ Spin Ladders in the Sulfide Oxide $BaFe_2S_2O$. *Eur J. Inorg. Chem.* **2014**, *36*, 6150–6155. [CrossRef]

30. Huh, S.; Prots, Y.; Adler, P.; Tjeng, L.H.; Valldor, M. Synthesis and Characterization of Frustrated Spin Ladders $SrFe_2S_2O$ and $SrFe_2Se_2O$. *Eur. J. Inorg. Chem.* **2015**, *18*, 2982–2988. [CrossRef]

31. Aminoff, G. Über Lauephotogramme und Struktur von Zinkit. *Z. Kristallogr. Cryst. Mater.* **1921**, *56*, 495–505. [CrossRef]

32. Sambrook, T.; Smura, C.F.; Clarke, S.J. Structure and physical properties of the polar oxysulfide CaZnOS. *Inorg. Chem.* **2007**, *46*, 2571–2574. [CrossRef] [PubMed]

33. Aminoff, G. Untersuchungen über die Kristallstrukturen von Wurtzit und Rotnickelkies. *Z. Kristallogr. Cryst. Mater.* **1923**, *58*, 203–219. [CrossRef]

34. Knop, O.; Brisse, F.; Castelliz, L. Pyrochlores. V. Thermoanalytic, X-ray, neutron, infrared, and dielectric studies of $A_2Ti_2O_7$ titanates. *Can. J. Chem.* **1969**, *47*, 971–990. [CrossRef]

35. Guittard, M.; Benacerrat, A.; Flahaut, J. Les seleniures $L_2Se_3$ et $L_3Se_4$ des elements des terres rares. *Ann. Chim.* **1964**, *9*, 25–34.

36. Sasaki, T.; Ukyo, Y.; Kuroda, K.; Arai, S.; Muto, S.; Saka, H. Crystal structure of $Ce_2Zr_2O_7$ and β-($Ce_2Zr_2O_{7.5}$). *J. Ceram. Soc. Jpn.* **2004**, *112*, 440–444. [CrossRef]

37. Miller, J.F.; Matson, L.K.; Himes, R.C. Studies on the selenides and tellurides of selected rare-earth metals. *Proc. Conf. Rare Earth Res.* **1962**, *1962*, 233–248.

38. Böhm, J. Über das Verglimmen einiger Metalloxyde. *Z. Anorg. Allg. Chem.* **1925**, *149*, 217–222. [CrossRef]

39. Donohue, P.C.; Jeitschko, W. The preparation of $Ln_2MX_5$ where $Ln$ = rare earths, $M$ = Zr and Hf, and $X$ = S, Se. *Mater. Res. Bull.* **1974**, *9*, 1333–1336. [CrossRef]

40. Gerlach, W. Die Gitterstruktur der Erdalkalioxyde. *Z. Phys.* **1922**, *9*, 184–192. [CrossRef]

41. Matje, P.; Müller, W.; Schäfer, H. Zur Darstellung und Kristallstruktur von $Ba_2MnTe_3$. *Z. Naturforsch.* **1977**, *B32*, 835–836.

42. Liu, B.W.; Jiang, X.M.; Wang, G.E.; Zeng, H.Y.; Zhang, M.J.; Li, S.F.; Guo, W.H.; Guo, G.C. Oxochalcogenide $BaGeOSe_2$: Highly distorted mixed-anion building units leading to a large second-harmonic generation response. *Chem. Mater.* **2015**, *27*, 8189–8192. [CrossRef]

43. Jin, S.F.; Huang, Q.; Lin, Z.P.; Li, Z.L.; Wu, X.Z.; Ying, T.P.; Wang, G.; Chen, X.L. Two-dimensional magnetic correlations and partial long-range order in geometrically frustrated CaOFeS with triangle lattice of Fe ions. *Phys. Rev.* **2015**, *B91*, 094420. [CrossRef]

44. Bryhan, D.N.; Rakers, R.; Klimaszewski, K.; Patel, N.; Bohac, J.J.; Kremer, R.K.; Mattausch, H.; Zheng, C. $La_3TWS_3O_6$ ($T$ = Cr, Mn, Fe, Co, Ni): Quinary Rare Earth Transition-Metal Compounds Showing a Nonmagnetic/Magnetic Transition ($T$ = Co)—Synthesis, Structure and Physical Properties. *Z. Anorg. Allg. Chem.* **2010**, *636*, 74–78. [CrossRef]

45. Wu, H. Metal-insulator transition in $Sr_{2-x}La_xCoO_4$ driven by spin-state transition. *Phys. Rev.* **2012**, *B86*, 075120. [CrossRef]

46. Valldor, M.; Rößler, U.K.; Prots, Yu.; Kuo, C.-Y.; Chiang, J.-C.; Hu, Z.; Kniep, R.; Tjeng, L.H. Synthesis and Characterization of Ba[CoSO]: Magnetic Complexity in the Presence of Chalcogen Ordering. *Chem. Eur. J.* **2015**, *21*, 10821–10828. [CrossRef] [PubMed]

47. Han, F.; Wang, D.; Malliakas, C.D.; Sturza, M.; Chung, D.Y.; Wan, X.; Kanatzidis, M.G. (CaO)(FeSe): A Layered Wide-Gap Oxychalcogenide Semiconductor. *Chem. Mater.* **2015**, *27*, 5695–5701. [CrossRef]

48. Dugué, P.J.; Vovan, T.; Villers, J. Etude Structurale des Oxysulfures de Chrome(III) et de Terres Rares. I. Structure de l'Oxysulfure $LaCrOS_2$. *Acta Crystallogr.* **1980**, *B36*, 1291–1294. [CrossRef]

49. Tougait, O.; Ibers, J.A. Synthesis and crystal structures of the lanthanum titanium oxyselenides $La_4Ti_2O_4Se_5$ and $La_6Ti_3O_5Se_9$. *J. Solid State Chem.* **2001**, *157*, 289–295. [CrossRef]

50. Gardberg, A.S.; Ibers, J.A. Crystal structure of hexapraseodymium dititanium septasulfide hexaoxide, $Pr_6Ti_2S_7O_6$. *Z. Kristallogr.* **2001**, *216*, 491–492. [CrossRef]

51. Cody, J.A.; Ibers, J.A. Synthesis and characterization of the new rare-earth/transition-metal oxysulfides $La_6Ti_2S_8O_5$ and $La_4Ti_3S_4O_8$. *J. Solid State Chem.* **1995**, *114*, 406–412. [CrossRef]

52. Ferro, R. Über einige Selen-und Tellurverbindungen des Urans. *Z. Anorg. Allg. Chem.* **1954**, *275*, 320–326. [CrossRef]

53. Hiramatsu, H.; Yanagi, H.; Kamiya, T.; Ueda, K.; Hirano, M.; Hosono, H. Crystal structures, optoelectronic properties and electronic structures of layered oxychalcogenides *MCuOCh* (*M* = Bi, La; *Ch* = S, Se, Te): Effects of electronic configurations of $M^{3+}$ ions. *Chem. Mater.* **2008**, *20*, 326–334. [CrossRef]

54. Dugué, P.J.; Vovan, T.; Villers, J. Etude Structurale des Oxysulfures de Chrome(III) et de Terres Rares. II. Structure de I'Oxysulfure CeCrOS$_2$. *Acta Crystallogr.* **1980**, *B36*, 1294–1297. [CrossRef]

55. Van, T.V.; Huy Ng, D. Synthese et structure cristalld'une nouvelle famille d' oxyselenures de chrome III et de lanthanides legers, de formule generale *RCrSe$_2$O* (*R* = La, Ce). *Comptes Rendus* **1981**, *293*, 933–936.

56. Zhao, L.-D.; He, J.; Berardan, D.; Lin, Y.; Li, J.-F.; Nanc, C.-W.; Dragoe, N. BiCuSeO oxyselenides: New promising thermoelectric materials. *Energy Environ. Sci.* **2014**, *7*, 2900–2924. [CrossRef]

57. Berardan, D.; Li, J.; Amzallag, E.; Mitra, S.; Sui, J.; Cai, W.; Dragoe, N. Structure and Transport Properties of the BiCuSeO–BiCuSO Solid Solution. *Materials* **2015**, *8*, 1043–1058. [CrossRef]

58. Reuter, B.; Hardel, K. Über die Hochtemperaturmodifikation von Silbersulfidjodid. *Naturwiss* **1961**, *48*, 161–162. [CrossRef]

59. Grossholz, H.; Janka, O.; Schleid, T. Oxide fluoride sulfides of the lanthanoids with the formula $M_3OF_5S$ (*M* = Nd, Sm, Gd–Ho). *Z. Naturforsch.* **2011**, *B66*, 213–220. [CrossRef]

60. Stolze, K.; Isaeva, A.; Schwarz, U.; Doert, T. UPTe, ThPTe and U$_2$PTe$_2$O: Actinide pnictide chalcogenides with diphosphide anions. *Eur. J. Inorg. Chem.* **2015**, *2015*, 778–785. [CrossRef]

61. Kamihara, Y.; Hiramatsu, H.; Hirano, M.; Kawamura, R.; Yanagi, H.; Kamiya, T.; Hosono, H. Iron-Based Layered Superconductor: LaOFeP. *J. Am. Chem. Soc.* **2006**, *128*, 10012–10013. [CrossRef] [PubMed]

62. Shamoto, S.; Kato, T.; Ono, Y.; Miyazaki, Y.; Ohoyama, K.; Ohashi, M.; Yamaguchi, Y.; Kajitani, T. Structures of β-(ZrNCl) and superconducting Li$_{0.16}$ZrNCl: Double honeycomb lattice superconductor. *Physica C* **1998**, *306*, 7–14. [CrossRef]

63. Singh, S.K.; Kumar, A.; Gahtori, B.; Shruti, K.; Sharma, G.; Patnaik, S.; Awana, V.P.S. Bulk superconductivity in bismuth oxysulfide Bi$_4$O$_4$S$_3$. *J. Am. Chem. Soc.* **2012**, *134*, 16504–16507. [CrossRef] [PubMed]

# $\eta^1$:$\eta^2$-*P*-Pyrazolylphosphaalkene Complexes of Ruthenium(0)

**Victoria K. Greenacre and Ian R. Crossley** *

Department of Chemistry, University of Sussex, Brighton BN1 9QJ, UK
* Correspondence: i.crossley@sussex.ac.uk

Academic Editor: Lee J. Higham

**Abstract:** An extended range of novel ruthenium phosphaalkene complexes of the type $[Ru\{\eta^1\text{-}N\text{:}\eta^2\text{-}P,C\text{-}P(pz')=CH(SiMe_2R)\}(CO)(PPh_3)_2]$ (R = Tol, $C_6H_4CF_3$-*p*; pz' = $pz^{Me2}$, $pz^{CF3}$, $pz^{Me,CF3}$; R = Me, $C_6H_4CF_3$-*p*; pz' = $pz^{Ph}$) have been prepared from the respective ruthenaphosphaalkenyls $[Ru\{P=CH(SiMe_2R)\}Cl(CO)(PPh_3)_2]$ upon treatment with Lipz'. Where R = $C_6H_4CF_3$-*p* and pz' = $pz^{Me2}$ the complex is characterized by single crystal X-ray diffraction, only the second example of such species being structurally characterized. This indicates enhanced pyramidalisation of the alkenic carbon center when compared with precedent data (R = Me, pz' = pz) implying an enhanced $Ru\rightarrow\pi^*_{PC}$ contribution, which can be correlated with the greater donor power of $pz^{Me2}$. This is similarly reflected in spectroscopic data that reveal significant influence of the pyrazolyl substituents upon the phosphaalkene, stronger donors imparting significantly enhanced shielding to phosphorus; in contrast, a much lesser influence if noted for the silyl substituents.

**Keywords:** phosphorus; phosphaalkene; phosphaalkenyl; pi-complex; pyrazolate

## 1. Introduction

After almost half a century of study, the chemistry of low-coordinate phosphorus continues to fascinate both organic and inorganic chemists alike [1–8]. Dominated by the isolobal and isoelectronic relationship between phosphorus and the "CH" fragment, the chemistry of phosphacarbons is often familiar from their carbo-centric and nitrogenous counterparts, yet they simultaneously embody appreciable dichotomy in terms of their underlying electronic and chemical nature. Nowhere is this more apparent than in the chemistry of phosphaalkenes (RP=CR′R″) and phosphaalkynes (RC≡P), which rank among the most heavily studied classes of phosphacarbon. Reactivity is in each case dominated by the high-energy $\pi$-systems, though in the case of phosphaalkenes this is often competitive with the lone-pair, which lies close—albeit marginally lower—in energy. In contrast, the phosphaalkyne lone-pair is appreciably stabilized, but can be engaged chemically under appropriately directing conditions.

The study of these compounds is, however, often complicated by an intrinsic lack of stability, which restricts the range of available substrates and necessitates some synthetic ingenuity. Such difficulties are often addressed by the imposition of steric bulk, as a means of imparting kinetic stability, which has proven particularly effective in precluding homo-oligomerisation of phosphaalkenes. The same approach has typically been cited in the development of kinetically stabilized phosphaalkynes (e.g., $^tBuC≡P$, AdC≡P), however, in these cases the bulk is often sufficiently remote from the reactive $\pi$-system as to preclude it being the sole stabilizing influence. Moreover, even bulky phosphaalkynes, e.g., $R_3SiC≡P$ (R = Ph [9], Me [10,11]) often exhibit only limited stability, in some instances comparable to that of unencumbered systems (e.g., MeC≡P [12]). Significantly, the formally related phosphaethynyloate ion "O–C≡P⁻" is isolable as a sodium salt [13,14], which exhibits appreciable stability despite the lack of any steric "protection"; this fact is attributed to electronic influences, with the preferential adoption of an $O=C\equiv P^-$ type structure. Taken alongside computational studies of the isolated cyaphide ion ("$C\equiv P^-$"), which indicate an intrinsically unstable hypovalent "$[C=P]^-$" structure [15], this would imply that electronics serve an equally important role in imparting stability to low-coordinate phosphacarbons. Indeed, in the context of phosphaalkenes—which remain the longest and most heavily studied of the phosphacarbons—the incorporation of π-donor substituents (e.g., $NR_2$, OR) and/or conjugate π-systems has long been employed as an alternative means of stabilization [16].

In seeking to further stabilize, and chemically exploit phosphaalkenes, the incorporation of transition metal fragments has proven to be a particularly valuable tool. Prominently, the coordination of phosphaalkenes [17] in an $\eta^1$-fashion allows for selective reactivity of the π-system to be developed, by sequestering the lone-pair and precluding its competitive reaction [18]. Somewhat less extensively studied, $\eta^2$-coordination—common among carbo-centric analogues—has also been employed with phosphaalkenes [19,20], discrimination between $\eta^1$ and $\eta^2$ being achieved through judicious selection of the metal fragment, though bridging-$\eta^1$:$\eta^2$ complexes have also been described [21]. Moreover, both transition metal and, to a lesser extent, main-group fragments can be incorporated as discrete substituents on the phosphaalkenic core to afford a range of metallaphosphaalkenyl complexes (Figure 1) [22,23]. First described in 1985 [24], such systems remain relatively rare, though—with the exception of type **E**—all possible motifs have been realized, with P-metalla- (type **A**) and C-metalla- (type **B**) systems the most heavily studied.

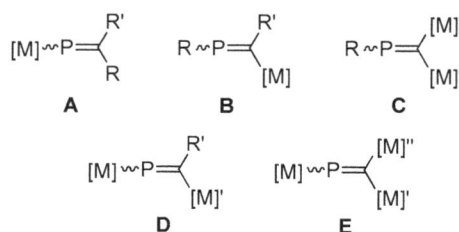

**Figure 1.** Metallaphosphaalkenyl motifs.

Recently, as part of an extended program investigating transition metal compounds featuring low-coordinate phosphacarbons with potential for conjugation [25–28], we have prepared and studied a range of ruthenaphosphaalkenyl complexes of the type $[Ru\{P=CH(SiMe_2R)\}Cl(CO)(PPh_3)_2]$ (R = Me **1a**, Ph **1b**, Tol **1c**) [29,30]. These are prepared by hydroruthenation of phosphaalkynes $R_3SiC\equiv P$, following from methodology developed initially by Hill and Jones for $^tBuC\equiv P$ and related systems [31–33], and superficially related to Nixon's independent reduction of $[(Ph_3P)_2Pt(\eta^2\text{-}P\equiv C^tBu)]$ with Schwartz's reagent [34]. Notably, we have described the first structural data for ruthenaphosphaalkenyls, which demonstrate the phosphaalkenyl moieties to behave as classical 1-electron donors, within a square-based pyramidal metal coordination sphere. Significantly, this precludes, in the ground state at least, any augmenting P→Ru donation from the lone-pair (i.e., phosphavinylidene character), which is consistent with precedent reactivity studies that demonstrate a nucleophilic phosphorus center [35–39]; indeed, in common with these precedent reports, we have found **1a–b** to engage in reaction with electrophilic fragments at phosphorus [30,40].

Notwithstanding, we have additionally observed an unusual reaction to proceed when **1a** or **1b** is exposed the lithium pyrazolates Li[pz'] (pz' = pz, $pz^{Me2}$, $pz^{CF3}$, $pz^{Me,CF3}$), giving rise to the unprecedented phosphaalkene complexes $[Ru\{\eta^1\text{-}N:\eta^2\text{-}P,C\text{-}P(pz')=CH(SiMe_2R)\}(CO)(PPh_3)_2]$ (Scheme 1) [29,30], which feature an apparent 3-member (Ru–P–C) metallacyclic core, bridged by a pyrazolyl moiety. Also replicated using Hill's $[Ru\{P=CH(^tBu)\}Cl(CO)(PPh_3)_2]$, this chemistry would perhaps imply ambiphilicty for the ruthenaphosphaalkenyl motif, if not directly of the alkenyl phosphorus center. On the basis of available data, we reasoned the resulting complexes to be best formulated as involving an $\eta^2$-phosphaalkene moiety, in line with the Dewar–Chatt–Duncanson model, with a dominant contribution from $d_\pi\rightarrow\pi^*_{(PC)}$ retro-donation to metal-ligand binding. The precise extent of the latter might reasonably be influenced by the nature of the phosphaalkene substituents, or indeed those of the pyrazolyl moiety, resulting variations in the atom-specific parameters (e.g., NMR shielding effects) offering a potential means of quantification. In seeking to assess the influence of such factors, and thus potentially to elaborate means of control over reactivity, we report herein the synthesis and characterization of an extended range of these novel pyrazolylphosphaalkene complex.

**Scheme 1.** Synthesis of $[Ru\{η^1-N:η^2-P,C-P(pz')=CH(SiMe_2R)\}(CO)(PPh_3)_2]$ (2–5) [29,30] (L = PPh_3). *Reagents and conditions:* (i) Lipz' (pz' = pz, pz$^{Me2}$, pz$^{CF3}$, pz$^{Me,CF3}$), THF.

## 2. Results and Discussion

### 2.1. Synthesis and Characterization of $η^2$-Pyrazolylphosphaalkene Complexes

As we have previously described [29,40], the ruthenaphosphaalkenyls **1c–d** were prepared via hydroruthenation of the respective phosphalkynes $RMe_2SiC≡P$ with $[RuHCl(CO)(PPh_3)_3]$ (Scheme 2). Each system is readily identified on the basis of characteristic spectroscopic signatures, *viz.* (i) the heavily deshielded $^{31}P\{^1H\}$-NMR resonance for the phosphaalkenyl moiety; (ii) consistently integrating resonances for two equivalent PPh_3 ligands; (iii) HMBC correlations ($^1H$–$^{29}Si$; $^1H$–$^{31}P$) confirming the presence of the silyl group and alkenic proton; (iv) infrared band for the retained carbonyl ($ν_{CO}$ 1930–1950 cm$^{-1}$), characteristic of a ruthenium(II) carbonyl complex.

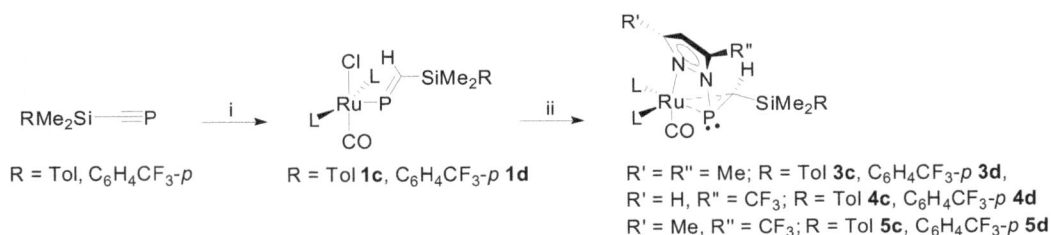

**Scheme 2.** Synthesis of $[Ru\{P=CH(SiMe_2R)\}Cl(CO)(PPh_3)_2]$ (R = Tol **1c**, $C_6H_4CF_3$-*p* **1d**) and their conversion to $[Ru\{η^1-N:η^2-P,C-P(pz')=CH(SiMe_2R)\}(CO)(PPh_3)_2]$ (2–5) (L = PPh_3). Reagents and conditions: (i) $[RuHCl(CO)(PPh_3)_3]$, $CH_2Cl_2$, 1 h.; (ii) Lipz' (pz' = pz$^{Me2}$, pz$^{CF3}$, pz$^{Me,CF3}$), THF, 1 h.

The reactions of **1c–d** with Li[pz$^{Me2}$] each afford a single complex (**3c–d** respectively), which is spectroscopically comparable to those obtained similarly from **1a** and **1b** [29,30]. Thus, the characteristic resonances for the phosphaalkenyl and PPh_3 ligands are lost, being replaced by three new, mutually coupling, resonances (1:1:1 ratio) in the region 50–30 ppm, which is commonly associated with saturated ($λ^3σ^3$) phosphorus centers. In each case, the lower frequency resonance is identified as that associated with the phosphacarbon, on the basis of a $^1H$–$^{31}P$-HMBC correlation to the, now appreciably shielded, "CH(SiMe_2R)" moiety (identified from $^1H$–$^{29}Si$-HMBC spectra and a significant $^1J_{PC}$ coupling (~70 Hz)). The latter is observed at $δ_H$ ~1.7 and $δ_C$ ~42, which, though appreciably lower frequency than for a classical alkenyl "CH", remains somewhat deshielded for a fully saturated "alkyl" moiety. Indeed, as was previously noted for **2a–b** and **3a–b**, the magnitude of the $^1J_{CH}$ coupling (~130–140 Hz) is intermediate between those of $C_2H_4$ (156 Hz) and $CH_4$ (125 Hz) [41], while $^1J_{PC}$ are essentially unperturbed from those of the parent phosphaalkenyls, consistent with the case of a π-bound phosphaalkene. This scenario is further supported by a significant reduction in $ν_{CO}$ for the retained carbonyl (1900–1910 cm$^{-1}$), indicative of increased density at metal and, on the basis of calculated force constants, consistent with reduction of the metal to Ru(0) [42–46].

Though augmented by computational data, the direct structural characterization of $\eta^2$-pyrazolylphosphaalkene complexes has been limited previously to the single crystal X-ray structure of pyrazole derivative **2a** and a low-quality structure of **3a** that served to confirm connectivity [30]. However upon prolonged standing at ambient temperature, a concentrated CDCl$_3$ solution of **3d** yielded X-ray quality single crystals of the chloroform solvate (Figure 2).

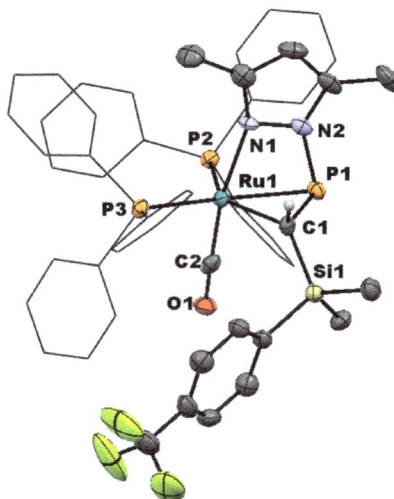

**Figure 2.** Molecular structure of [Ru{$\eta^1$-$N$:$\eta^2$-$P$,$C$-P(pz$^{Me2}$)=CH(SiMe$_2$C$_6$H$_4$CF$_3$-$p$)}(CO)(PPh$_3$)$_2$] (**3d**) in crystals of the chloroform solvate, with thermal ellipsoids at the 50% probability level. Ancillary phenyl rings are reduced, and hydrogen atoms omitted for clarity. Selected bond distance (Å) and angles (°): Ru–P1 2.381(1), Ru–P2 2.359(1), Ru–P3 2.369(1), Ru–C1 2.211(4), Ru–C2 1.831(5), Ru–N1 2.229(3), P1–C1 1.782(5), P1–N2 1.778(4), C1–Si 1.860(4), C2–O 1.155(6); Ru–C1–Si 128.2(2), Ru–C1–P1 72.3(2), Ru–P1–C1 62.2(1), Ru–P1–N2 82.5(1), Ru–C1–H1 112.1(4), P1–C1–Si 113.4(2), P1–C1–H1 112.1(4), C1–P1–N2 96.1(2), P1–Ru–P2 93.6(1), P2–Ru–P3 114.0(1), P3–Ru–C1 107.1(1), P1–Ru–C1 43.5(1), P2–Ru–C2 91.1(1), P3–Ru–C2 89.1(2), C2–Ru–N1 170.2(2).

In common with the precedent structure of **2a**, compound **3d** exhibits distorted trigonal-bipyramidal geometry about ruthenium, with PPh$_3$ ligands and the $\eta^2$-phosphaalkene lying in the equatorial plane. The carbonyl adopts an axial position, along with the metal-pyrazolyl bond, the latter being marginally distorted from linearity ($\angle C_{CO}$–Ru–N 170.2(2)°) due to the strain of bridging the Ru–phosphacarbon linkage. Internal angles for the phosphacarbon moiety indicate partial pyramidalization of the carbon center, while the unusually tight angles about phosphorus can be attributed to the constraint of the bridging pyrazolyl group. Taken together with the P–C linkage 1.782(5) Å, which is intermediate between a single and double bond [47,48], these data are wholly consistent with our previous conclusion of $\eta^2$-binding, rather than a discrete metallacyclic species. The C≡O linkage is marginally elongated relative to the structurally characterized ruthenium(II) phosphaalkenyls [30], consistent with a more electron rich metal center, and is comparable to those of other established Ru(0) carbonyls, including that previously reported for **2a**. Indeed, within the bounds of uncertainty, the molecular geometry of **3d** is fully comparable to that of **2a**, with the exception of a marginally contracted P–N linkage (1.788(4) Å, *cf.* 1.809(5) Å **2a**), which might reasonably be attributed to the increased donor strength of pz$^{Me2}$ relative to pz, as is reflected in the spectroscopic features (*vide infra*). Moreover, marginally greater pyramidalization about carbon is apparent in **3d** ($\angle$Si–C–P 113.4(2)° *cf.* 116.7(3)° **2a**), which might similarly reflect the enhanced donor strength of pz$^{Me2}$ leading to an increase in retro-donation from the metal.

In analogous fashion to the synthesis of **3**, treatment of phosphaalkenyls with Li[pz′] (pz′ = pz$^{CF3}$, pz$^{Me,CF3}$, pz$^{Ph}$) led to formation of the respective $\eta^2$-phosphaalkene complexes **4c–d**, **5c–d** (Scheme 2),

**6a** and **6d** (Scheme 3); the SiMe$_3$ (**a**) and SiMe$_2$Ph (**b**) derivatives of **4** and **5** have been previously described [29]. In contrast to **2** and **3**, the asymmetrically substituted pyrazolates offer potential for positional isomerism; however, only a single isomer is observed in each case. Though non-trivial in lieu of crystallographic data, assignment of the specific isomer obtained can be achieved with recourse to spectroscopic data. Thus, for each of **4** and **5** a notable coupling ($J_{FP}$ ~20 Hz) is apparent between the CF$_3$ substituent and the phosphacarbon only; this would imply proximity of the two moieties, strongly suggesting that the bulkier CF$_3$ substituent (*cf.* H or Me) is more favorably positioned away from the sterically encumbered metal center. This is supported by significant deshielding of the phosphacarbon center, consistent with proximity to an electron withdrawing substituent. Indeed, we find the resonant frequency of the phosphacarbon center to be very sensitive to the pyrazolyl substituents (*vide infra*), most significantly so for that at the proximal site.

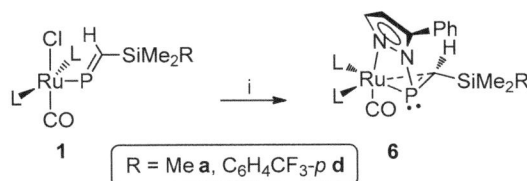

**Scheme 3.** Synthesis of [Ru{η$^1$-$N$:η$^2$-$P,C$-P(pz$^{Ph}$)=CH(SiMe$_2$R)}(CO)(PPh$_3$)$_2$] (R = Me **6a**, C$_6$H$_4$CF$_3$-$p$ **6d**) (L = PPh$_3$). *Reagents and conditions:* (**i**) Lipz$^{Ph}$, THF.

Assignment of the isomeric preference in **6a** and **6d** is a more complex undertaking, given the lack of any direct spectroscopic handle, though the same steric arguments can reasonably be applied. A more electron-withdrawing character is apparent, relative to pyrazole itself ($\delta_P$ 64.4 **6a**, *cf.* 58.7 **2a**), which conforms to expectation and is again consistent with the precedent systems, though the modest level of deshielding is inadequate to unequivocally confirm substituent proximity to the phosphacarbon. Nonetheless, we believe this to be the most likely scenario.

## 2.2. Spectroscopic Features and Trends

The key NMR spectroscopic data for all compounds **2**, **3**, **4** and **5**, along with **6a** and **6d** are summarized in Table 1. As one would anticipate, the pz$^{Me2}$ derivatives impart significant shielding to the phosphacarbon "P" center, which resonates around 25 ppm to lower frequency than for the pz systems; conversely, the presence of an electron-withdrawing CF$_3$ substituent leads to appreciable deshielding of this site ($\Delta\delta_P$ ~+20), though this is moderated in the pz$^{Me,CF3}$ derivatives by competitive induction from the methyl. All complexes featuring a methyl substituent proximal to the metal also exhibit additional shielding of one PPh$_3$ ligand ($\Delta\delta_P$ −3) compared to their non-alkylated analogues; significantly, the extent of this effect is quantitatively comparable whether the second substituent is methyl or trifluoromethyl, which would imply a localized effect, and also offers further support for the isomeric assignment. The presence of a methyl proximal to the metal also leads to an appreciable decrease in the magnitude of the alkenic C–H coupling constant. This implies reduced s-character in this linkage, consistent with an increased contribution from $d_\pi \rightarrow \pi^*_{(CP)}$ retrodonation (the result of increased electron density at the metal) and thus greater pyramidalization; indeed, this concurs with structural data for **3d** (*vide supra*).

Influence of the pyrazole upon the alkenic carbon center is also apparent, but exhibits a less defined trend; in contrast, the nature of the silyl substituent clearly holds significance, the SiMe$_3$ derivatives being notably more deshielded than their SiMe$_2$Ar analogues, distinctions between the latter being ill defined. This is consistent with $^{13}$C-NMR studies of vinyl- [49] and aryl-silanes [50], for which enhanced deshielding of the α-carbon follows from increasing methyl substitution at silicon, presumably reflecting the relative "inductive effect", given the greater electronegativity of carbon relative to silicon. While carbocentric systems show the reverse shielding trend for the β-center, in the case of **2–6** enhanced deshielding of phosphorus is again noted for the SiMe$_3$ derivatives. Though the relatively lower electronegativity of phosphorus (*cf.* carbon) might reasonably be invoked in accounting for this disparity, indirect effects via the synergic metal-ligand binding mode (such complexes have not been considered for the carbocentric silanes) cannot be discounted. Indeed, it is noted that analogues of **2** and **3** featuring a *tert*-butyl [29], rather than silyl, substituent exert a significant shielding effect upon the phosphorus center ($\delta_P$ 38.8 (pz); 14.2 (pz$^{Me2}$)), yet deshield the $\alpha$-carbon nucleus ($\delta_C$ 81.6 (pz), 79.8 (pz$^{Me2}$)) even more effectively than does SiMe$_3$, which would appear counter-intuitive on the basis of purely electronegativity arguments. The relative contributions of ligand→metal $\sigma$ donation, and $d_\pi$→$\pi^*_{(CP)}$ retrodonation for each center must thus be considered contributory, though inadequate data are currently available to quantify this.

**Table 1.** NMR Spectroscopic data for [Ru{$\eta^1$-$N$:$\eta^2$-$P,C$-P(pz$'$)=CH(SiMe$_2$R)}Cl(CO)(PPh$_3$)$_2$].

| Compound | | R | $\delta_P$[1] P=C | $\delta_P$[1] PPh$_3$ | $\delta_C$[2] P=C | $\delta_H$[2] P=CH ($^1J_{CH}$/Hz)[3] |
|---|---|---|---|---|---|---|
| **2**[4] | **2a** | Me | 58.7 | 46.6, 42.0 | 45.1 | 1.59 (137) |
| (pz$'$ = pz) | **2b** | Ph | 57.0 | 47.0, 41.7 | 42.6 | 1.72 (135) |
| | | | | | | |
| **3** | **3a** | Me[4] | 32.9 | 46.6, 39.2 | 44.9 | 1.62 (123) |
| (pz$'$ = pz$^{Me2}$) | **3b** | Ph[4] | 32.3 | 47.0, 38.9 | 41.8 | 1.77 (128) |
| | **3c** | C$_6$H$_4$Me-$p$[5] | 32.6 | 46.7, 39.1 | 41.8 | 1.75 (136) |
| | **3d** | C$_6$H$_4$CF$_3$-$p$[5] | 32.1 | 46.6, 38.6 | 39.8 | 1.66 (135) |
| | | | | | | |
| **4**[5] | **4a** | Me | 76.6 | 47.7, 41.5 | 47.1 | 1.78 (136) |
| (pz$'$ = pz$^{CF3}$) | **4b** | Ph | 74.9 | 48.0, 41.3 | 46.7 | 1.91 (136) |
| | **4c** | C$_6$H$_4$Me-$p$ | 75.0 | 47.9, 41.3 | 45.4 | 1.90 (134) |
| | **4d** | C$_6$H$_4$CF$_3$-$p$ | 73.8 | 47.8, 40.9 | 43.8 | 1.82 (134) |
| | | | | | | |
| **5**[5] | **5a** | Me | 64.6 | 46.9, 38.4 | 45.2 | 1.76 (129) |
| (pz$'$ = pz$^{Me,CF3}$) | **5b** | Ph | 62.7 | 47.2, 38.3 | 41.8 | 1.97 (131) |
| | **5c** | C$_6$H$_4$Me-$p$ | 61.6 | 47.2, 38.4 | 42.1 | 1.97 (135) |
| | **5d** | C$_6$H$_4$CF$_3$-$p$ | 62.0 | 47.1, 37.8 | 40.7 | 1.85 (133) |
| | | | | | | |
| **6**[5] | **6a** | Me | 64.4 | 47.4, 41.8 | 47.5 | 1.74 (137) |
| (pz$'$ = pz$^{Ph}$) | **6d** | C$_6$H$_4$CF$_3$-$p$ | 60.5 | 47.7, 41.3 | 43.7 | 1.78 (136) |

[1] Referenced to 85% H$_3$PO$_4$; [2] Referenced to SiMe$_4$; [3] Measured from coupled $^1$H–$^{13}$C HSQC spectra; [4] Recorded as solutions in CD$_2$Cl$_2$; [5] Recorded as solutions in CDCl$_3$.

## 3. Materials and Methods

### 3.1. General Methods

All manipulations were performed under anaerobic conditions using standard Schlenk line and glovebox (MBraun, Germany) techniques, working under an atmosphere of dry argon or dinitrogen respectively. Solvents were distilled from appropriate drying agents and stored over either molecular sieves (4 Å, for DCM and THF) or potassium mirrors. General reagents were obtained from Sigma-Aldrich (Gillingham, UK) or Fisher Scientific (Loughborough, UK) and purified by appropriate methods before use, precious metal salts were obtained from STREM (Cambridge, UK). [RuHCl(CO)(PPh$_3$)$_3$] [51], RMe$_2$SiCH$_2$PCl$_2$ (R = Me, Ph [52], Tol [29], C$_6$H$_4$CF$_3$ [39]), RMe$_2$SiC≡P [29,30,40] and [Ru{P=CH(SiMe$_2$R)}Cl(CO)(PPh$_3$)$_2$] (R = Me [29,30], Ph [29], Tol, C$_6$H$_4$CF$_3$-$p$ [40]) were prepared as previously described. Unless otherwise stated, NMR spectra were recorded at 303 K on a Varian VNMRS 400 ($^1$H 399.50 MHz, $^{13}$C 100.46 MHz, $^{19}$F 375.87, $^{31}$P 161.71 MHz, $^{29}$Si 79.37 MHz) spectrometer (Varian, Yarnton, UK). All spectra are referenced to external Me$_4$Si, 85% H$_3$PO$_4$ or CFCl$_3$ as appropriate. Carbon-13 spectra were assigned by recourse to the 2D (HSQC, HMBC) spectra; phosphaalkenic proton and silicon shifts were determined indirectly by $^1$H–$^{31}$P and $^1$H–$^{29}$Si correlation (HMBC). Elemental analyses were obtained by Mr. S. Boyer of the London Metropolitan University Elemental Analysis Service.

## 3.2. X-Ray Crystallography

Single crystal X-ray diffraction data were recorded on an Agilent Xcalibur Eos Gemini Ultra diffractometer (Agilent Technologies, Yarnton, UK) with CCD plate detector using Cu-K$\alpha$ ($\lambda$ = 1.54184) radiation. Structure solution and refinement were performed using SHELXS [53] and SHELXL [53] respectively, running under Olex2 [54].

## 3.3. Syntheses and Characterisation

[Ru{η$^1$-$N$:η$^2$-$P,C$-P(pz$^{Me2}$)=CH(SiMe$_2$Tol)}(CO)(PPh$_3$)$_2$] (**3c**): In a representative procedure, following literature precedent for compounds **3a** and **3b** [29,30], Hpz$^{Me2}$ (12.7 mg, 0.13 mmol) as solution in THF was treated, at ambient temperature, with a single equivalence of $^n$BuLi (0.06 cm$^3$, 2.5 M in hexanes) and the mixture stirred for 5 min. The resulting solution was added directly to a solution of [Ru{P=CH(SiMe$_2$Tol)}Cl(CO)(PPh$_3$)$_2$] (**1c**, 67 mg, 0.08 mmol) at ambient temperature and the mixture stirred for 1h. Volatiles were removed under reduced pressure, then the residue extracted with CH$_2$Cl$_2$ and the resulting solution filtered; the solvent was removed under reduced pressure and the resulting solid dried in vacuo, before being redissolved in CDCl$_3$ for spectroscopic analysis. $^1$H NMR (CDCl$_3$): $\delta_H$ 0.06 (s, 3H, SiMe$_2$), 0.15 (s, 3H, SiMe$_2$), 0.42 (s, 3H, Pz*–Me), 1.75 (m, 1H, CHSi ($^1J_{C-H}$ = 135.31 Hz), 1.95 (s, 3H, Pz*–Me), 2.34 (s, 3H, CH$_3$), 5.04 (s, 1H, Pz*–H4), 7.08–7.42 (m, 35H, Ar–H). $^{13}$C{$^1$H} NMR (CDCl$_3$): $\delta_C$ −0.9 (d, $^3J_{C-P}$ = 7.6 Hz, SiCH$_3$), 0.16 (d, $^3J_{C-P}$ = 8.5 Hz, SiCH$_3$), 9.6 (d, $^3J_{C-P}$ = 5.3 Hz, Pz–CH$_3$), 11.9 (s, Pz–CH$_3$), 21.6 (s, CH$_3$), 41.8 (ddd, $J_{C-P}$ = 4.5, 31.3, 78.6 Hz, CHSi ($^1J_{C-H}$ = 136.3 Hz)), 105.0 (d, $J_{C-P}$ = 2.7 Hz, Pz*–C4), 127.6 (d, $J_{C-P}$ = 8.87 Ha, Ar–C), 127.8 (d, $J_{C-P}$ = 8.87 Hz, ArC), 128.2 (s, ArC), 128.6 (m, ArC), 128.9 (d, $J_{C-P}$ = 6.7 Hz, ArC), 130.4 (d, $J_{C-P}$ = 13.0 Hz, ArC), 133.8 m, ArC), 133.6–134.4 (m, ArC), 137.4 (s, ArC), 138.2 (d, $J_{C-P}$ = 30.7 Hz, ArC), 138.4 (d, $J_{C-P}$ = 31.8 Hz, ArC), 145.2 (s, Pz*–C5), 152.4 (s, Pz*–C3), 209.5 (br, CO). $^{31}$P{$^1$H} NMR (CDCl$_3$): $\delta_P$ 46.7 (d, $J_{P-P}$ = 16.9 Hz), 39.1 (dd, $J_{P-P}$ = 50.1, 16.7 Hz), 32.6 (d, $J_{P-P}$ = 50.4 Hz, P=C). $^{29}$Si{$^1$H} NMR (CDCl$_3$): $\delta_{Si}$ −5.3.

[Ru{η$^1$-$N$:η$^2$-$P,C$-P(pz$^{Me2}$)=CH(SiMe$_2$C$_6$H$_4$CF$_3$-$p$)}(CO)(PPh$_3$)$_2$] (**3d**): In comparable fashion to **3c**, from the respective phosphaalkenyl (**1d**). $^1$H NMR (CDCl$_3$): $\delta_H$ −0.02 (s, 3H, SiCH$_3$), 0.18 (s, 3H, SiCH$_3$), 0.42 (s, 3H, Pz*–Me), 1.66 (m, 1H, CHSi ($^1J_{C-H}$ = 135.3 Hz), 1.95 (s, 3H, Pz*–Me), 5.07 (s, 1H, Pz*–H4), 7.09–7.34 (m, 35H, Ar–H), 7.45–7.56 (m, 4H, C$_6$H$_4$CF$_3$). $^{19}$F NMR (CDCl$_3$): $\delta_F$ −63.06 (s). $^{13}$C{$^1$H} NMR (CDCl$_3$): $\delta_C$ −1.3 (d, $^3J_{C-P}$ = 8 Hz, SiCH$_3$), 0.3 (d, $^3J_{C-P}$ = 8 Hz, SiCH$_3$), 9.6 (d, $^3J_{C-P}$ = 5.3 Hz, Pz–CH$_3$), 11.9 (s, Pz–CH$_3$), 39.8 (br. ($^1J_{C-H}$ = 135.3 Hz) CHSi), 105.2 (d, $J_{C-P}$ = 2.9 Hz, Pz*C4), 123.8 (q, $^2J_{C-F}$ = 3.9 Hz, CCF$_3$), 130.9 (q, $^1J_{C-F}$ = 240.0 Hz, CF$_3$), 127.6–129.1, 134.1–135.5, 137.2–138.1 (3 × m, PPh$_3$, C$_6$H$_4$), 145.3 (d, $^2J_{C-P}$ = 1.5 Hz, Pz*–C5), 152.7 (s, Pz*–C3), 209.7 (br, CO). $^{31}$P{$^1$H} NMR (CDCl$_3$): $\delta_P$ 46.6 (d, $^2J_{P-P}$ = 16.2 Hz, PPh$_3$), 38.6 (dd, $^2J_{P-P}$ = 54.4, 16.2 Hz, PPh$_3$), 32.1 (d, $^2J_{P-P}$ = 51.4 Hz, P=C). $^{29}$Si{$^1$H} NMR (CDCl$_3$): $\delta_{Si}$ −4.5. $\nu_{CO}$ = 1913 cm$^{-1}$. Anal. Found: C, 62.90; H, 5.01; N, 2.90; Calcd for C$_{52}$H$_{48}$F$_3$N$_2$OP$_3$RuSi: C, 62.71; H, 4.86; N, 2.81. Crystal data for **3d**: C$_{53}$H$_{48}$F$_3$N$_2$OP$_3$RuSi·CHCl$_3$, $M_w$ = 1115.42, triclinic, $P$-1 (No. 2), $a$ = 10.5494(5), $b$ = 11.5375(6), $c$ = 21.871(1) Å, $\alpha$ = 76.877(4), $\beta$ = 82.084(4), $\gamma$ = 85.941(4), $V$ = 2565.46(2) Å$^3$, $Z$ = 2, $D_c$ = 1.444 Mg m$^{-3}$, $\mu$(Cu-K$\alpha$) = 5.439 mm$^{-1}$, $T$ = 173(2) K, 9464 independent reflections, full-matrix $F^2$ refinement $R_1$ = 0.0567, $wR_2$ = 0.1894 on 8126 independent absorption corrected reflections [$I > 2\sigma(I)$; $2\theta_{max}$ = 143.4°], 607 parameters. The empirical absorption correction was conducted using spherical harmonics, as implemented in the SCALE3 ABSPACK scaling algorithm (CryAlisPro Version 1.171.38.41). CCDC 1502285.

[Ru{η$^1$-$N$:η$^2$-$P,C$-P(pz$^{CF3}$)=CH(SiMe$_2$Tol)}(CO)(PPh$_3$)$_2$] (**4c**): In comparable fashion to **3**, but commencing from Hpz$^{CF3}$. $^1$H NMR (CDCl$_3$) $\delta_H$ −0.1 (s, 3H, SiCH$_3$), 0.1 (s, 3H, SiCH$_3$), 1.90 (m, 1H, CHSi), 2.35 (br, 3H, CH$_3$), 5.30 (s, 1H, Pz–H3), 5.57 (1 H, s, Pz–H4), 7.04–7.48 (m, 35H, C$_6$H$_5$). $^{19}$F NMR (CDCl$_3$) $\delta_F$ −60.4 (d, $^4J_{F-P}$ = 17.6 Hz)).$^{13}$C{$^1$H} NMR (CDCl$_3$) $\delta_C$ −0.9 (d, $^3J_{C-P}$ = 10 Hz, SiCH$_3$), −0.25 (d, $^3J_{C-P}$ = 5 Hz, SiCH$_3$), 21.9 (s, CH$_3$), 45.4 (ddd, $J_{C-P}$ = 80.5, 31.0, 4.8 Hz, ($^1J_{C-H}$ = 134 Hz), SiCH), 105.2 (s, Pz–C$^4$), 119.1 (q, $^1J_{C-F}$ 269 Hz, CF$_3$), 128.2–128.6, 128.7–129.2, 133.7–134.4, 137.6–138.0 (4 × m, CH), 140.9 (s, Pz–C3), 210.5 (t, $^2J_{C-P}$ = 13 Hz, C≡O). $^{31}$P{$^1$H} NMR (CDCl$_3$) $\delta_P$ 75.0 (dq, $^2J_{P-P}$ = 44.5 Hz,

$^4J_{P-F}$ = 17.50 Hz, P=CH), 47.9 (d, $^2J_{P-P}$ = 17.5 Hz, PPh$_3$), 41.3 (dd, $^2J_{P-P}$ = 44.5, 17.5 Hz, PPh$_3$). $^{29}$Si{$^1$H} NMR (CDCl$_3$) δ −5.6. Anal. Found: C, 62.10; H, 4.85; N, 2.91; Calcd for C$_{51}$H$_{46}$F$_3$N$_2$OP$_3$RuSi: C, 62.39; H, 4.72; N, 2.85.

[Ru{η$^1$-$N$:η$^2$-$P,C$-P(pz$^{CF3}$)=CH(SiMe$_2$C$_6$H$_4$CF$_3$-$p$)}(CO)(PPh$_3$)$_2$] (**4d**): As for **4c**, commencing from **1d**. $^1$H NMR (CDCl$_3$) δ −0.06 (s, 3H, SiCH$_3$), 0.14 (s, 3H, SiCH$_3$), 1.85 (m, 1H, CHSi), 5.32 (s, 1H, Pz–H3), 5.60 (s, 1H, Pz–H4), 7.05–7.39 (m, 30H, C$_6$H$_5$), 7.5 (d, 2H, $J_{H-F}$ = 8.0 Hz, C$_6$H$_4$), 7.63 (d, 2H, $J_{H-F}$ = 7.6 Hz, C$_6$H$_4$). $^{19}$F NMR (CDCl$_3$) δ −60.1 (d ($^4J_{F-P}$ = 19.4 Hz)), −62.7 (s, C$_6$H$_4$CF$_3$). $^{13}$C{$^1$H} NMR (CDCl$_3$) δ 1.2 (s, SiCH$_3$), 43.8 (br. ($^1J_{C-H}$ = 134 Hz), SiCH), 104.9 (s, Pz–C4), 128.0 (t, $J_{C-P}$ = 4.3 Hz, CH), 128.7 (d, $J_{C-P}$ = 7.3 Hz, CH), 128.9 (s, CH), 129.5 (s, CH), 132.2–132.6, 133.5–134.1 (2 × m, CH), 142.8 (s, Pz–C5), 145.4 (br. Pz–C3). *CF$_3$ resonances obscured by aromatics*. $^{31}$P{$^1$H} NMR (CDCl$_3$) δ 37.8 (dd, $^2J_{P-P}$ = 47.4, 15.8 Hz), 47.1 (d, $^2J_{P-P}$ = 15.8 Hz, PPh$_3$), 62.0 (dq, $^2J_{P-P}$ = 47.4 Hz, $^4J_{P-F}$ = 19.40 Hz, P=CH). $^{29}$Si{$^1$H} NMR (CDCl$_3$) δ −4.2. $\nu_{CO}$ = 1913 cm$^{-1}$.

[Ru{η$^1$-$N$:η$^2$-$P,C$-P(pz$^{Me,CF3}$)=CH(SiMe$_2$Tol)}(CO)(PPh$_3$)$_2$] (**5c**): As for **4c**, but commencing from Hpz$^{Me,CF3}$. $^1$H NMR (CDCl$_3$): δ$_H$ −0.03 (s, 3H, SiCH$_3$), 0.17 (s, 3H, SiCH$_3$), 0.55 (s, 3H, Pz–CH$_3$), 1.97 (br, 1H, CHSi), 2.36 (s, 3H, CH$_3$), 5.52 (br. Pz–H4), 7.12–7.50 (m, 34 H, 2 × PPh$_3$, C$_6$H$_4$). $^{19}$F NMR (CDCl$_3$): δ$_F$ −59.9 (d, $J_{F-P}$ = 19.4 Hz). $^{13}$C{$^1$H} NMR (CDCl$_3$): δ$_C$ −1.1 (d, $^3J_{C-P}$ = 7.8 Hz, SiCH$_3$), 0.14 (d, $^3J_{C-P}$ = 8.8 Hz, SiCH$_3$), 11.85 (s, Pz–CH$_3$), 21.6 (m, CH$_3$), 42.1 (br. ddd, $J_{C-P}$ = 28.2, 78.8 Hz, $^1J_{C-H}$ = 137 Hz, SiCH), 105.7 (br m, Pz–C4), 119.5 (br, $^1J_{C-F}$ = 269 Hz, CF$_3$), 127.6–130.3 (m, PPh$_3$), 133.5–135.1 (m, PPh$_3$, C$_6$H$_4$) 137.6 (dd, $J$ = 1.49, 30.1 Hz, Pz–C5), 152.6 (br, Pz–C3), 209.0 (m, CO). $^{31}$P{$^1$H} NMR (CDCl$_3$): δ$_P$ 61.62 (dq, $^2J_{P-P}$ = 45.7, $^4J_{P-F}$ = 19.3 Hz, P=C), 47.2 (d, $^2J_{P-P}$ = 16.4 Hz, PPh$_3$), 38.4 (ddd, $^2J_{P-P}$ = 45.7, 16.4, $J_{P-F}$ = 1.4 Hz, PPh$_3$). $^{29}$Si{$^1$H} NMR (CDCl$_3$): δ$_{Si}$ −5.7. $\nu_{CO}$ = 1918 cm$^{-1}$. Anal. Found: C, 62.42; H, 5.05; N, 2.89; Calcd for C$_{52}$H$_{48}$F$_3$N$_2$OP$_3$RuSi: C, 62.71; H, 4.86; N, 2.81.

[Ru{η$^1$-$N$:η$^2$-$P,C$-P(pz$^{Me,CF3}$)=CH(SiMe$_2$C$_6$H$_4$CF$_3$-$p$)}(CO)(PPh$_3$)$_2$] (**5d**): As for **5c**, but commencing from **1d**. $^1$H NMR (CDCl$_3$): δ$_H$ 0.01 (s, 3H, SiCH$_3$), 0.21 (s, 3H, SiCH$_3$), 0.56 (s, 3H, Pz–CH$_3$) 1.85 (br, 1H, CHSi), 5.54 (s, 1H, Pz–H4), 7.13–7.28, 7.37–7.46, 7.47–7.53, 7.58–7.67 (4 × m, 34H, PPh$_3$, C$_6$H$_4$). $^{19}$F NMR (CDCl$_3$): δ$_F$ −59.9 (d, $J$ = 20.3 Hz, Pz–CF3), −62.7 (s, CF$_3$-$p$). $^{13}$C{$^1$H} NMR (CDCl$_3$): δ$_C$ −1.4 (d, $^3J_{C-P}$ = 7.4 Hz, SiCH$_3$), 0.16 (d, $^3J_{C-P}$ = 8.0 Hz, SiCH$_3$), 11.9 (s, Pz–CH$_3$), 40.7 (ddd, $J_{C-P}$ = 32.9, 79.8, 5.0 Hz, $^1J_{C-H}$ = 137 Hz, SiCH), 105.9 (br m, Pz–C4), 119.4 (q, $^1J_{C-F}$ = 270 Hz, CF$_3$), 124.6 (q, $^1J_{C-F}$ = 272 Hz, CF$_3$), 127.6–128.0 (m, CH), 128.6 (d, $J$ = 7 Hz, CH), 128.5–129.1 (m, CH), 133.6–134.3 (m, CH), 137.6 (dd, $J$ = 1.3, 3.2 Hz, Pz–C5), 152.8 (br, Pz–C3), 209.1 (m, CO). $^{31}$P{$^1$H} NMR (CDCl$_3$): δ$_P$ 61.99 (dq, $^2J_{P-P}$ = 57.0, $^4J_{P-F}$ = 18.8 Hz, PPh$_3$), 47.1 (d, $^2J_{P-P}$ = 15.8 Hz, PPh$_3$), 37.8 (dd, $^2J_{P-P}$ = 47.0, 15.9, P=C). $^{29}$Si{$^1$H} NMR (CDCl$_3$): δ$_{Si}$ −4.16. $\nu_{CO}$ 1917 cm$^{-1}$.

[Ru{η$^1$-$N$:η$^2$-$P,C$-P(pz$^{Ph}$)=CH(SiMe$_3$)}(CO)(PPh$_3$)$_2$] (**6a**): In a comparable fashion, commencing from **1a** and Hpz$^{Ph}$. $^1$H NMR (CDCl$_3$): δ$_H$ 0.15 (s, 9H, Si(CH$_3$)$_3$), 1.74 (m, 1H, CHSi), 5.39 (s, 1H, Pz–H3), 5.56 (br. Pz–H4), 7.05–7.55 (m, 35H, PPh$_3$, C$_6$H$_5$,). $^{13}$C{$^1$H} NMR (CDCl$_3$): δ$_C$ 1.4 (d, $^3J_{C-P}$ = 5.5 Hz, Si(CH$_3$)), 47.5 (ddd, $^2J_{P-P}$ = 4.4, 31.0, 80.2 Hz, $^1J_{C-H}$ = 135.9 Hz, SiCH), 103.3 (d, $J_{C-P}$ = 3.2 Hz, PzC4), 127.4 (d, $J_{C-P}$ = 6.6 Hz, CH), 127.7 (d, $J_{C-P}$ = 8.7 Hz, CH), 128.1 (d, $J_{C-P}$ = 8.6 Hz, CH), 128.6 (s, CH), 128.8 (d, $J_{C-P}$ = 19.0 Hz, CH), 129.1 (s, CH), 133.9 (m, CH), 137.3 (d, $J_{C-P}$ = 9.7 Hz, *ipso*-C), 137.7 (d, $J_{C-P}$ = 32.0 Hz, *ipso*-C), 138.0 (d, $J_{C-P}$ = 32.0 Hz, *ipso*-CH), 141.0 (s, Pz–C3), 147.7 (br, Pz–C5), 210.7 (m, CO). $^{31}$P{$^1$H} NMR (CDCl$_3$): δ$_P$ 64.4 (d, $J_{P-P}$ = 46.2 Hz, P=C), 47.4 (d, $J_{P-P}$ = 18.5 Hz, PPh$_3$), 41.8 (dd, $J_{P-P}$ = 46.2, 18.4 Hz, PPh$_3$). $^{29}$Si{$^1$H} NMR (CDCl$_3$): δ$_{Si}$ −0.5. $\nu_{CO}$ = 1908 cm$^{-1}$. Anal. Found: C, 65.71; H, 5.18; N, 3.03; Calcd for C$_{46}$H$_{47}$N$_2$OP$_3$RuSi: C, 65.62; H, 5.33; N, 2.99.

[Ru{η$^1$-$N$:η$^2$-$P,C$-P(pz$^{Ph}$)=CH(SiMe$_2$C$_6$H$_4$CF$_3$-$p$)}(CO)(PPh$_3$)$_2$] (**6d**): As for **6a**, but commencing from **1d**. $^1$H NMR (CDCl$_3$): δ$_H$ −0.03 (s, 3H, SiCH$_3$), 0.27 (s, 3H, SiCH$_3$), 1.86 (m, 1H, CHSi), 5.56 (1H, s, Pz–H4), 5.60 (s, 1H, Pz–H5), 7.11–7.68 (m, 39 H, 2 × PPh$_3$, C$_6$H$_5$, C$_6$H$_4$). $^{19}$F NMR (CDCl$_3$): δ$_F$ 62.62 (s, CF$_3$). $^{13}$C{$^1$H} NMR (CDCl$_3$): δ$_C$ −1.3 (d, $^3J_{C-P}$ = 8 Hz, SiCH$_3$), −0.9 (d, $^3J_{C-P}$ = 5 Hz, SiCH$_3$), 45.6 (ddd, $J_{C-P}$ = 5.1, 31.4, 79.5 Hz, $^1J_{C-H}$ = 133.9 Hz, SiCH), 103.5 (d, $J_{C-P}$ = 3.3 Hz, Pz–C4), 124.5 (q, $^1J_{C-F}$ = 272 Hz, CF$_3$), 127.4–130.2, 133.7–134.5 (2 × m, PPh$_3$, C$_6$H$_5$, C$_6$H$_4$), 141.2 (s, Pz–C3), 147.9 (br. Pz–C5), 210.5 (m, CO). $^{31}$P{$^1$H} NMR (CDCl$_3$): δ$_P$ 60.6 (d, $^2J_{P-P}$ = 46.9 Hz, P=C), 47.7 (d, $^2J_{P-P}$ = 17.4 Hz, PPh$_3$), 41.3 (dd, $^2J_{P-P}$ = 46.9, 17.4 Hz, PPh$_3$). $^{29}$Si{$^1$H} NMR (CDCl$_3$): δ$_{Si}$ −4.9. $\nu_{CO}$ = 1912. Anal. Found: C, 64.02; H, 4.69; N, 2.71; Calcd for C$_{53}$H$_{48}$F$_3$N$_2$OP$_3$RuSi: C, 64.11; H, 4.77; N, 2.80.

## 4. Conclusions

We have described the synthesis of an extended range of pyrazolyl-bridged $\eta^2$-phosphaalkene complexes of ruthenium(0), obtained by the reaction of pyrazolates with ruthenaphosphaalkenyl complexes. Spectroscopic and structural data suggest the nature of the pyrazolyl substituents in the 3/5 positions has a significant influence on the nature of the phosphaalkene fragment, which is a balance of direct shielding/deshielding of the phosphorus center, and an indirect influence resulting from donation to the metal, mediated through the extent of metal-ligand retro-donation. In contrast, the nature of the silyl substituents exerts a much smaller, albeit still noticeable, influence. Presently, the available data pool is inadequate to formulate an unequivocally quantitative description of these effects, which are the subject of on-going investigations.

**Acknowledgments:** We thank the Royal Society and University of Sussex (studentship to Victoria K. Greenacre) for financial support. Ian R. Crossley gratefully acknowledges the award of a Royal Society University Research Fellowship. We thank Dr. L. Higham (guest editor) for the invitation to contribute to this issue.

**Author Contributions:** Victoria K. Greenacre identified the specific targets and conducted all experimental work and crystallographic determinations. Ian R. Crossley conceived and led the over-arching research project, contributed to interpretation and wrote the paper.

## References

1.   Waterman, R. Phosphorus chemistry: Discoveries and advances. *Dalton Trans.* **2016**, *45*, 1801–1803. [CrossRef] [PubMed]

2.   Mathey, F. Phospha-Organic Chemistry: Panorama and Perspectives. *Angew. Chem. Int. Ed.* **2003**, *42*, 1578–1604. [CrossRef] [PubMed]

3.   Dillon, K.B.; Mathey, F.; Nixon, J.F. *Phosphorus: The Carbon Copy*; Wiley: Chichester, UK, 1998.

4.   Nixon, J.F. Recent developments in the organometallic chemistry of phospha-alkynes, RC-P. *Coord. Chem. Rev.* **1995**, *145*, 201–258.

5.   Appel, R. *Multiple Bonds and Low Coordination in Phosphorus Chemistry*; Regitz, M., Scherer, O.J., Eds.; Thieme: Stutgart, Germany, 1990.

6.   Markovski, L.N.; Romanenko, V.D. Phosphaalkynes and phosphaalkenes. *Tetrahedron* **1989**, *45*, 6019–6090. [CrossRef]

7.   Nixon, J.F. Coordination chemistry of compounds containing phosphorus–carbon multiple bonds. *Chem. Rev.* **1988**, *88*, 1327–1362. [CrossRef]

8.   Appel, R.; Knoll, F.; Ruppert, I. Phospha-alkenes and Phospha-alkynes, Genesis and Properties of the (p–p) π-Multiple Bond. *Angew. Chem. Int. Ed. Engl.* **1981**, *20*, 731–744. [CrossRef]

9.   Cordaro, J.G.; Stein, D.; Rüegger, H.; Grützmacher, H. Making the True "CP" Ligand. *Angew. Chem. Int. Ed.* **2006**, *45*, 6159–6162. [CrossRef] [PubMed]

10.  Mansell, S.M.; Green, M.; Kilby, R.J.; Murry, M.; Russell, C.A. Facile preparation of trimethylsilylphosphaalkyne and its conversion to polyphospholide anions. *C. R. Chim.* **2010**, *13*, 1073–1081. [CrossRef]

11.  Mansell, S.M.; Green, M.; Russell, C.A. Coordination chemistry of trimethylsilylphosphaalkyne: A phosphaalkyne bearing a reactive substituent. *Dalton Trans.* **2012**, *41*, 14360–14368. [CrossRef] [PubMed]

12.  Jones, C.; Schulten, C.; Stasch, A. The first complexes and cyclodimerisations of methylphosphaalkyne (P ≡CMe). *Dalton Trans.* **2006**, *31*, 3733–3735. [CrossRef] [PubMed]

13.  Chen, X.; Alidori, S.; Puschmann, F.F.; Santiso-Quinones, G.; Benkö, Z.; Li, Z.; Becker, G.; Grützmacher, H.-F.; Grützmacher, H. Sodium Phosphaethynolate as a Building Block for Heterocycles. *Angew. Chem. Int. Ed.* **2014**, *53*, 1641–1645. [CrossRef] [PubMed]

14.  Jupp, A.R.; Goichoechea, J.M. The 2-Phosphaethynolate Anion: A Convenient Synthesis and [2+2] Cycloaddition Chemistry. *Angew. Chem. Int. Ed.* **2013**, *52*, 10064–10067. [CrossRef] [PubMed]

15. Mo, O.; Yanez, M.; Guillemin, J.-C.; Riague, E.H.; Gal, J.-F.; Maria, P.-C.; Poliart, C.D. The Gas-Phase Acidity of HCP, CH$_3$CP, HCAs, and CH$_3$CAs: An Unexpected Enhanced Acidity of the Methyl Group. *Chem. Eur. J.* **2002**, *8*, 4919–4924. [CrossRef]

16. Markovski, L.N.; Romanenko, V.D. Phosphaalkynes and phosphaalkenes. *Tetrahedron* **1989**, *45*, 6019–6090. [CrossRef]

17. Le Floch, P. Phosphaalkene, phospholyl and phosphinine ligands: New tools in coordination chemistry and catalysis. *Coord. Chem. Rev.* **2006**, *250*, 627–681. [CrossRef]

18. De Vaumes, R.; Marinetti, A.; Mathey, F. Catalytic hydrogenation of the phosphorus–carbon double bond in phosphaalkene complexes. *J. Organomet. Chem.* **1991**, *413*, 411–417. [CrossRef]

19. Van der Knaap, T.A.; Bickelhaupt, F.; Krasykamp, J.G.; van Koten, G.; Bernards, J.P.C.; Edzes, H.T.; Veeman, W.S.; de Boer, E.; Baerends, E.J. The η$^1$- and η$^2$-coordination in a (phosphaalkene)platinum(0) complex. *Organometallics* **1984**, *3*, 1804–1811. [CrossRef]

20. Kraajkamp, J.G.; van Koten, G.; van der Knaap, T.A.; Bickelhaupt, F.; Stam, C.H. Influence of steric factors on the coordination mode (η$^1$ or η$^2$) of phosphaalkenes to zerovalent Pt(0)L$_2$ centers. X-ray structure of bis(triphenylphosphine)[(2,6-dimethylphenyl)-9-fluorenylidenephosphine]platinum(0)-toluene. *Organometallics* **1986**, *5*, 2014–2020. [CrossRef]

21. Appel, R.; Casser, C.; Knoch, F. Über niederkoordinierte phosphorverbindungen: XXXX. 2,4,6-tri-t-Butylphenylmethylenphosphan,ein vielseitiger ligand in übergangsmetall-komplexen. *J. Organomet. Chem.* **1985**, *293*, 213–217.

22. Weber, L. Recent developments in the chemistry of metallophosphaalkenes. *Coord. Chem. Rev.* **2005**, *249*, 741–763. [CrossRef]

23. Weber, L. Metallophosphaalkenes—from Exotics to Versatile Building Blocks in Preparative Chemistry. *Angew. Chem. Int. Ed. Engl.* **1996**, *35*, 271–288. [CrossRef]

24. Weber, L.; Reizig, K.; Boese, R.; Polk, M. Z-[(η$^5$-C$_5$H$_5$)(CO)$_2$Fe-P=C(OSiMe$_3$)($^t$Bu)], a Phosphaalkenyl-Complex with FeP Single Bond. *Angew. Chem. Int. Ed. Engl.* **1985**, *24*, 604–605. [CrossRef]

25. Saunders, A.J.; Crossley, I.R.; Roe, S.M. Aroylphosphanes: Base-Free Synthesis and Their Coordination Chemistry with Platinum-Group Metals. *Eur. J. Inorg. Chem.* **2016**, *25*, 4076–4082. [CrossRef]

26. Saundersa, A.J.; Crossley, I.R. Synthesis of 3-stannyl and 3-silyl propargyl phosphanes and the formation of a phosphinoallene. *Dalton Trans.* **2016**, *45*, 2148–2155. [CrossRef] [PubMed]

27. Trathen, N.; Leech, M.C.; Crossley, I.R.; Greenacre, V.K.; Roe, S.M. Synthesis and electronic structure of the first cyaphide-alkynyl complexes. *Dalton Trans.* **2014**, *43*, 9004–9007. [CrossRef] [PubMed]

28. Saunders, A.J.; Crossley, I.R.; Coles, M.P.; Roe, S.M. Facile self-assembly of the first diphosphametacyclophane. *Chem. Commun.* **2012**, *48*, 5766–5768. [CrossRef] [PubMed]

29. Greenacre, V.K.; Trathen, N.; Crossley, I.R. Ruthenaphosphaalkenyls: Synthesis, Structures, and Their Conversion to η$^2$-Phosphaalkene Complexes. *Organometallics* **2015**, *34*, 2533–2542. [CrossRef]

30. Trathen, N.; Greenacre, V.K.; Crossley, I.R.; Roe, S.M. Ambiphilic Reactivity of a Ruthenaphosphaalkenyl: Synthesis of P-Pyrazolylphosphaalkene Complexes of Ruthenium(0). *Organometallics* **2013**, *32*, 2501–2504. [CrossRef]

31. Bedford, R.B.; Hill, A.F.; Jones, C. Phosphaalkyne Hydrometalation: Synthesis of [RuCl(P=CH$^t$Bu)(CO)(PPh$_3$)$_2$]. *Angew. Chem. Int. Ed. Engl.* **1996**, *35*, 547–549. [CrossRef]

32. Bedford, R.B.; Hill, A.F.; Jones, C.; White, A.J.P.; Williams, D.J.; Wilton-Ely, J.D.E.T. Phosphaalkyne Hydrometalation: Synthesis and Reactivity of the Complexes [Ru(PCHCMe$_3$)Cl(CA)(PPh$_3$)$_2$] (A = O, S). *Organometallics* **1998**, *17*, 4744–4753. [CrossRef]

33. Brym, M.; Jones, C. Synthesis, characterisation and reactivity of the first diphosphaalkyne. *Dalton Trans.* **2003**, 3665–3667. [CrossRef]

34. Benuvenutti, M.H.A.; Cenac, N.; Nixon, J.F. Hydrozirconation of an η$^2$-ligated phosphaalkyne. A new synthetic route to η$^2$-ligated phosphaalkenes. *Chem. Commun.* **1997**, 1327–1328. [CrossRef]

35. Bedford, R.B.; Hibbs, D.E.; Hill, A.F.; Hursthouse, M.B.; Abdul Malik, K.M.; Jones, C. Complete metal-mediated reduction of the triple bond of a phosphaalkyne: X-ray structure of [Ru(PHFCH$_2$Bu$^t$)Cl(CO)(CNC$_6$H$_3$Me$_2$-2,6)(PPh$_3$)$_2$]BF$_4$·CH$_2$Cl$_2$. *Chem. Commun.* **1996**, 1895–1896. [CrossRef]

36. Bedford, R.B.; Hill, A.F.; Jones, C.; White, A.J.P.; Williams, D.J.; Wilton-Ely, J.D.E.T. Novel syntheses of heterodinuclear phosphaalkenyl complexes: X-ray structure of [Ru{P(AuPPh$_3$)=CHBu$^t$}Cl$_2$(CO)(PPh$_3$)$_2$]. *Chem. Commun.* **1997**, 179–180. [CrossRef]

37.  Hill, A.F.; Jones, C.; White, A.J.P.; Williams, D.J.; Wilton-Ely, J.D.E.T. A metallacyclic $\lambda^5$-phosphaalkenyl complex of ruthenium(II): X-ray structure of $[Ru\{\kappa^2\text{-}P(=O)CBu^tC(=O)\}(CNBu^t)_2(PPh_3)_2]$. *Chem. Commun.* **1998**, 367–368. [CrossRef]

38.  Bedford, R.B.; Hill, A.F.; Jones, C.; White, A.J.P.; Williams, D.J.; Wilton-Ely, J.D.E.T. Co-ordinative activation of phosphaalkynes: Methyl neopentylidene phosphorane complexes of ruthenium(II); crystal structure of $[Ru(MeP=CHBu^t)Cl(I)(CO)(PPh_3)_2]$. *J. Chem. Soc. Dalton Trans.* **1997**, 139–140. [CrossRef]

39.  Hill, A.F.; Jones, C.; White, A.J.P.; Williams, D.J.; Wilton-Ely, J.D.E.T. Mercuriophosphaalkene-P complexes: Crystal structure of $[Ru\{P(=CHBu^t)HgC_5H_4Fe(\eta\text{-}C_5H_5)\}Cl_2(CO)(PPh_3)_2]$. *J. Chem. Soc. Dalton Trans.* **1998**, 1419–1420. [CrossRef]

40.  Greenacre, V.K.; Crossley, I.R. Hydrochlorination of Ruthenaphosphaalkenyls: Unexpectedly Facile Access to Alkylchlorohydrophosphane Complexes. *Organometallics* **2016**. Submitted.

41.  Maciel, G.E.; McIver, J.W., Jr.; Ostlund, N.S.; Pople, J.A. Approximate self-consistent molecular orbital theory of nuclear spin coupling. I. Directly bonded carbon–hydrogen coupling constants. *J. Am. Chem. Soc.* **1970**, *92*, 1–11. [CrossRef]

42.  Bohanna, C.; Esteruelas, M.A.; Lahoz, F.J.; Onate, E.; Oro, L.A.; Sola, E. Synthesis of Butadiene-Osmium(0) and -Ruthenium(0) Complexes by Reductive Carbon–Carbon Coupling of Two Alkenyl Fragments. *Organometallics* **1995**, *14*, 4825–4831. [CrossRef]

43.  Bolton, P.D.; Grellier, M.; Vautravers, N.; Vendier, L.; Sabo-Etienne, S. Access to Ruthenium(0) Carbonyl Complexes via Dehydrogenation of a Tricyclopentylphosphine Ligand and Decarbonylation of Alcohols. *Organometallics* **2008**, *27*, 5088–5093. [CrossRef]

44.  Hill, A.F.; Owen, G.R.; White, A.J.P.; Williams, D.J. The Sting of the Scorpion: A Metallaboratrane. *Angew. Chem. Int. Ed.* **1999**, *38*, 2759–2761. [CrossRef]

45.  Christian, D.F.; Roper, W.R. Isocyanide complexes of zerovalent ruthenium. Proton addition to a transition-metal base offering alternative base sites. *J. Chem. Soc. D Chem. Commun.* **1971**, 1271–1272. [CrossRef]

46.  Herberhold, M.; Hill, A.F. The coordination chemistry of iminooxosulphuranes VI. Factors affecting coordination geometry in complexes of tosyliminooxosulphurane. *J. Organomet. Chem.* **1990**, *395*, 195–206. [CrossRef]

47.  Allen, F.H. The Cambridge Structural Database: A quarter of a million crystal structures and rising. *Acta Crystallogr. Sect. B Struct. Sci.* **2002**, *58*, 380–388. [CrossRef]

48.  Allen, F.H.; Kennard, O.; Watson, D.G.; Brammer, L.; Orpen, A.G.; Taylor, R. Tables of bond lengths determined by X-ray and neutron diffraction. Part 1. Bond lengths in organic compounds. *J. Chem. Soc. Perkin Trans. II* **1987**, *12*, S1–S19. [CrossRef]

49.  Rakita, P.E.; Worsham, L.S. $^{13}$C NMR studies of organosilanes: IV. Vinyl- and allyl-silanes. *J. Organomet. Chem.* **1977**, *139*, 135–142. [CrossRef]

50.  Rakita, P.E.; Srebro, J.P.; Worsham, L.S. $^{13}$C NMR studies of organosilanes: I. Substituent-chemical-shift parameters for phenylsilanes. *J. Organomet. Chem.* **1976**, *104*, 27–37. [CrossRef]

51.  Boniface, S.M.; Clark, G.R.; Collins, T.J.; Roper, W.R. Preparation of octahedral hydrido-aquo-ruthenium(II) complexes, and structural characterisation of hydridoaquodicarbonylbis(triphenylphosphine)-ruthenium(II) tetrafluoroborate. *J. Organomet. Chem.* **1981**, *206*, 109–117. [CrossRef]

52.  Averre, C.E.; Coles, M.P.; Crossley, I.R.; Day, I.J. The open-chain triphosphanes $RMe_2SiCH_2P(PR'_2)_2$ (R = Me, Ph; R' = SiMe_3, Cy, Ph). *Dalton Trans.* **2012**, *41*, 278–284. [CrossRef] [PubMed]

53.  Sheldrick, G.M. A short history of SHELX. *Acta Crystallogr. Sect. A* **2008**, *64*, 112–122. [CrossRef] [PubMed]

54.  Dolomanov, O.V.; Bourhis, L.J.; Gildea, R.J.; Howard, J.A.K. Puschmann, H. OLEX2: A complete structure solution, refinement and analysis. *J. Appl. Crystallogr.* **2009**, *42*, 339–341. [CrossRef]

# 9

# Ammonothermal Synthesis and Crystal Structures of Diamminetriamidodizinc Chloride [Zn$_2$(NH$_3$)$_2$(NH$_2$)$_3$]Cl and Diamminemonoamidozinc Bromide [Zn(NH$_3$)$_2$(NH$_2$)]Br

**Theresia M. M. Richter [1], Sabine Strobel [1], Nicolas S. A. Alt [2], Eberhard Schlücker [2] and Rainer Niewa [1,*]**

[1]  Institute of Inorganic Chemistry, Universität Stuttgart, Pfaffenwaldring 55, 70569 Stuttgart, Germany; richter.theresia@yahoo.de (T.M.M.R.); sabine.strobel@iac.uni-stuttgart.de (S.S.)

[2]  Institute of Process Machinery and Systems Engineering, University of Erlangen-Nuremberg, Cauerstraße 4, 91058 Erlangen, Germany; nicolas.alt@hswt.de (N.S.A.A.); sl@ipat.uni-erlangen.de (E.S.)

*  Correspondence: rainer.niewa@iac.uni-stuttgart.de

Academic Editors: Helmut Cölfen and Duncan H. Gregory

**Abstract:** The treatment of excess zinc in the presence of ammonium chloride under ammonothermal conditions of 873 K and 97 MPa leads to diamminetriamidodizinc chloride [Zn$_2$(NH$_3$)$_2$(NH$_2$)$_3$]Cl with a two-dimensionally μ-amido-interconnected substructure. Similar reaction conditions using ammonium bromide instead of the chloride (773 K, 230 MPa) result in diamminemonoamidozinc bromide [Zn(NH$_3$)$_2$(NH$_2$)]Br with one-dimensional infinite μ-amido-bridged chains. Both compounds were obtained as colorless, very moisture sensitive crystals. Crystal structures and hydrogen bond schemes are analyzed. Raman spectroscopic data of the chloride are reported.

**Keywords:** ammonothermal synthesis; amides; ammoniates; zinc

---

## 1. Introduction

Synthesis of high-quality nitride materials presents a challenge for various applications. In particular, semiconductor nitride materials are currently the focus of crystal growth, one example being GaN wafers as superior substrates for high-performance blue and white LEDs [1]. A promising synthesis and crystal growth technique for such materials is the ammonothermal method, utilizing supercritical ammonia under either ammonoacidic or ammonobasic conditions. However, a number of unresolved issues remain, like the fundamental understanding of the chemistry of dissolution, materials transport and recrystallization processes. Additionally, the technique may not only provide superior GaN crystals, but further interesting nitride materials.

Recently, we have focused on the ammonothermal zinc nitride synthesis. In this respect, we have presented the ammonothermal synthesis and characterization of Zn(NH$_3$)$_3$F$_2$ and Zn(NH$_3$)$_2$F$_2$, which show five-fold coordination at Zn [2]. In contrast, the few further examples of ammoniates of zinc halides exclusively exhibit the tetrahedral environment of Zn, namely in [Zn(NH$_3$)$_2$Cl$_2$], [Zn(NH$_3$)$_2$Br$_2$] [3], [Zn(NH$_3$)$_4$]Br$_2$ and [Zn(NH$_3$)$_4$]I$_2$ [4], all obtained from reaction of either aqueous solutions or of the solid zinc halides with ammonia at ambient pressure. With [Zn$_2$(NH$_3$)$_2$(NH$_2$)$_3$]Cl and [Zn(NH$_3$)$_2$(NH$_2$)]Br, we present two ammineamidozinc halides, synthesized from nominally ammonoacidic conditions, which additionally show increased condensation within their cationic substructures. The formation of these compounds from supercritical ammonia may indicate their role as intermediates in a conceivable ammonothermal synthesis and crystal growth of the interesting semiconductor material $Zn_3N_2$ analogously to the already commercially available ammonothermally grown GaN crystals [1].

## 2. Results and Discussion

We present the synthesis of $[Zn_2(NH_3)_2(NH_2)_3]Cl$ and $[Zn(NH_3)_2(NH_2)]Br$, two ammoniates of zinc halide amides, both with tetrahedral coordination by ammonia and amide ligands at the Zn central atom. Similar simultaneous coordination by amide ions and ammonia molecules was previously reported in, for example, $KNH_2 \cdot 2NH_3$ [5], or more relevant for the discussion, $[Cr_2(NH_2)_3(NH_3)_6]I_2$ [6] and $InF_2(NH_2)(NH_3)$ [7], both with octahedral coordination at the metal atom and obtained from the respective halides under ammonothermal conditions.

Both title compounds were obtained in the colder zone of the autoclave, while additional $Zn(NH_2)_2$ was observed in the hot zone. According to literature, $Zn(NH_2)_2$ is insoluble in liquid ammonia at ambient conditions [8]; however, we have frequently observed that Zn reacts under various ammonoacidic as well as ammonobasic conditions to form the diamide and crystallizes in the colder zone of the autoclave in large crystals, indicating the ammonoamphoteric character of Zn and an enhanced solubility of the binary amide at elevated temperatures and pressures. However, it is apparently possible to adjust the solubility, respectively its temperature dependence, by the addition of halide ions and thus favor the formation of the supposedly less soluble title compounds in the zone with lower temperature. The deposition at the hotter or colder zone in the reaction vessel is usually dictated by the temperature dependence of the solubility. This temperature dependence can fundamentally change with the type of mineralizer and therefore with the nature of the dissolved species, as is well known, for example, for the ammonothermal synthesis of GaN [1].

The underlying thermodynamics governing these processes follow very similar principles to those that are well established for the so-called Chemical Vapor Transport [9].

It is interesting to note that we were only able to synthesize the title compounds in the presence of platinum used as liner material to minimize contact of the solution with the autoclave wall and thus minimize corrosion, known to be severe in ammonoacidic solutions at elevated temperatures [10]. A catalytic action of both $NH_4Cl$ and Pt, for example, for the formation of alkali- and alkaline-earth metal amides from liquid ammonia is well established [11].

### 2.1. Crystal Structure Description

$[Zn_2(NH_3)_2(NH_2)_3]Cl$ crystallizes in the chiral space group $P2_12_12$ with two formula units in the unit cell. A Flack parameter close to $\frac{1}{2}$ indicates the presence of a racemic inversion twin. Table 1 gives further selected information on the crystal structure and its determination. Table 2 presents positional parameters and Table A1 anisotropic displacement parameters. In the crystal structure, the ammonia molecules, the amide and the chloride ions together form the motif of a hexagonal closed packing with stacking of hexagonal layers along [100]. In addition, 1/6 of the tetrahedral voids exclusively built by ammonia molecules and amide groups are occupied by Zn. These tetrahedra are linked via amide vertices to layers $^2_\infty \left[ Zn(NH_3)(NH_2)_{3/2}^{1/2+} \right]$ orientated within the $a/b$ plane (Figure 1). As may be expected, the distances Zn–N to the amide groups with 199.1(5) pm and 201.2(6) pm are significantly shorter as compared to the distance to the $NH_3$ ligand (213.5(7) pm). Similar distance relations were earlier observed, e.g., in $KNH_2 \cdot 2NH_3$ [5] or $[Cr_2(NH_2)_3(NH_3)_6]I_3$ [6]. Angles around Zn are in the range of 101.8° to 116.4°, thus close to the ideal tetrahedral angle. Further selected distances and angles are summarized in Table 3.

The arrangement of tetrahedra within the layer $^2_\infty \left[ Zn(NH_3)(NH_2)_{3/2}^{1/2+} \right]$ is reminiscent of the very similar interconnection within the structure of binary $Zn(NH_2)_2$, where six tetrahedra are condensed to rings, however, interconnected within a three-dimensional framework. According to the nomenclature after Liebau, developed to classify oxosilicate structures, $^2_\infty \left[ Zn(NH_3)(NH_2)_{3/2}^{1/2+} \right]$ represents an unbranched *vierer* single-layer with a molar ratio $n(Zn):n(N)$ of 2:5 [12].

**Table 1.** Selected crystallographic data and information for the structure determination of $[Zn_2(NH_3)_2(NH_2)_3]Cl$ and $[Zn(NH_3)_2(NH_2)]Br$.

| Formula | $[Zn_2(NH_3)_2(NH_2)_3]Cl$ | $[Zn(NH_3)_2(NH_2)]Br$ [a] |
|---|---|---|
| Crystal system | orthorhombic | monoclinic |
| Space group | $P2_12_12$ | $P2_1/n$ |
| $a/pm$ | 577.15(4) | 760.55(4) |
| $b/pm$ | 1023.59(6) | 597.72(4) |
| $c/pm$ | 654.56(4) | 1257.22(8) |
| $\beta/deg.$ | - | 93.475(4) |
| $V/10^6\ pm^3$ | 386.69(5) | 570.48(6) |
| X-ray density $\rho_{XRD}/g/cm^3$ | 2.13 | 2.28 |
| $Z$ | 2 | 4 |
| $T$ | 298 K | 100 K |
| Radiation | Mo-$K\alpha$ | Mo-$K\alpha$ |
| $F(000)$ | 496 | 376 |
| Range $h/k/l$ | $\pm4/-13-14/-21-22$ | $\pm8/\pm11/-17-16$ |
| $\theta_{max}/deg.$ | 53.9 | 55.0 |
| Absorption coefficient/$mm^{-1}$ | 12.93 | 11.17 |
| Reflect. meas./indep. | 844/533 | 1298/751 |
| $R_{int}/R_\sigma$ | 0.030/0.038 | 0.030/0.036 |
| $R_1(|F_o| \geq 4\sigma(F_o))$ | 0.038 | 0.034 |
| $R_1/wR_2/GooF$(all refl.) | 0.044/0.098/1.021 | 0.045/0.086/1.047 |
| Parameters | 58 | 71 |
| Restraints | 6 | 8 |
| Extinction coefficient | 0.017(6) | 0.003(1) |
| Flack $x$ | 0.49(4) | - |
| Residual e$^-$-density/$10^{-6}\cdot pm^{-3}$ | 1.19/$-0.54$ | 1.23/$-1.13$ |

[a] Refined as twin by partial merohedry.

**Table 2.** Positional and isotropic displacement parameters $U_{iso}$ ($10^4\ pm^2$) for $[Zn_2(NH_3)_2(NH_2)_3]Cl$.

| Atom | Site | $x$ | $y$ | $z$ | $U_{iso}$ |
|---|---|---|---|---|---|
| Zn | 4c | 0.1009(1) | 0.33764(8) | 0.5374(1) | 0.0231(3) |
| Cl | 2a | 1/2 | 1/2 | 0.0434(5) | 0.0247(5) |
| N(1) | 4c | 0.449(1) | 0.3291(7) | 0.553(1) | 0.028(1) |
| H(1A) | 4c | 0.35(2) | 0.248(8) | 0.67(1) | $1.2U_{iso}$(N(1)) |
| H(1B) | 4c | 0.50(2) | 0.354(1) | 0.67(1) | $1.2U_{iso}$(N(1)) |
| N(2) | 4c | 0.007(1) | 0.3377(8) | 0.853(1) | 0.032(1) |
| H(2A) | 4c | 0.00(2) | 0.259(5) | 0.91(1) | $1.2U_{iso}$(N(2)) |
| H(2B) | 4c | 0.09(2) | 0.36(1) | 0.96(1) | $1.2U_{iso}$(N(2)) |
| H(2C) | 4c | 0.41(2) | 0.118(9) | 0.08(1) | $1.2U_{iso}$(N(2)) |
| N(3) | 2b | 0 | 1/2 | 0.395(2) | 0.027(2) |
| H(3A) | 4c | 0.354(7) | 0.01(1) | 0.65(1) | $1.2U_{iso}$(N(3)) |

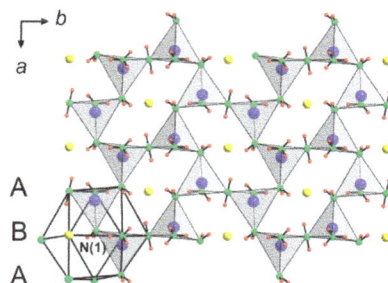

**Figure 1.** Section of the crystal structure of $[Zn_2(NH_3)_2(NH_2)_3]Cl$: Layers $^2_\infty\left[Zn(NH_3)(NH_2)_{3/2}^{1/2+}\right]$ built by occupation of tetrahedral voids exclusively formed by amide and ammonia molecules by Zn within the motif of an *hcp* of chloride (closed packed layers in stacking direction [100] are indicated by letters A and B), amide ions and ammonia molecules (**yellow** spheres: Cl, **green** spheres: N, **red** spheres: H, and **blue** spheres surrounded by **grey** tetrahedra: Zn). The *hcp* stacking motif is indicated by an anticuboctahedron and capital letters.

**Table 3.** Selected interatomic distances (pm) and angles (deg.) in the crystal structure of $[Zn_2(NH_3)_2(NH_2)_3]Cl$.

| Bond | | Distance | Occurence | Arrangement | Angle |
|---|---|---|---|---|---|
| Zn | –N(1) | 200.9(7) | | N(1)–Zn–N(1) | 114.5(2) |
| Zn | –N(1) | 201.2(6) | | N(1)–Zn–N(2) | 101.8(3) |
| Zn | –N(2) | 213.5(7) | | N(1)–Zn–N(2) | 110.1(3) |
| Zn | –N(3) | 199.1(5) | | N(1)–Zn–N(3) | 110.6(2) |
| | | | | N(1)–Zn–N(3) | 116.4(3) |
| | | | | N(2)–Zn–N(3) | 112.1(4) |
| N(1) | –H(1A) | 88(3) | | H(1A)–N(1)–H(1B) | 107(9) |
| N(1) | –H(1B) | 89(3) | | Zn–N(1)–H(1A) | 98(7) |
| | | | | Zn–N(1)–H(1B) | 102(7) |
| | | | | Zn–N(1)–Zn | 117.3(2) |
| N(2) | –H(2A) | 88(3) | | H(2A)–N(2)–H(2B) | 87(9) |
| N(2) | –H(2B) | 88(3) | | H(2A)–N(2)–H(2C) | 102(10) |
| N(2) | –H(2C) | 89(3) | | H(2B)–N(2)–H(2C) | 70(8) |
| | | | | Zn–N(2)–H(2A) | 113(7) |
| | | | | Zn–N(2)–H(2B) | 130(7) |
| | | | | Zn–N(2)–H(2C) | 139(7) |
| N(3) | –H(3A) | 90(3) | 2x | H(3A)–N(3)–H(3A) | 144(9) |
| | | | | Zn–N(3)–H(3A) | 105(7) |
| | | | | Zn–N(3)–Zn | 124.4(5) |
| ØN | –H | 89 | | | |
| ØZn | –N | 204 | | | |

Hydrogen atoms of the amide groups point towards the chloride ions located between the layers and form hydrogen bonds (Figure 2). Every chloride ion connects to twelve hydrogen atoms, where half of the hydrogen atoms belong to each neighboring layer. Table 4 summarizes donor–acceptor distances and angles.

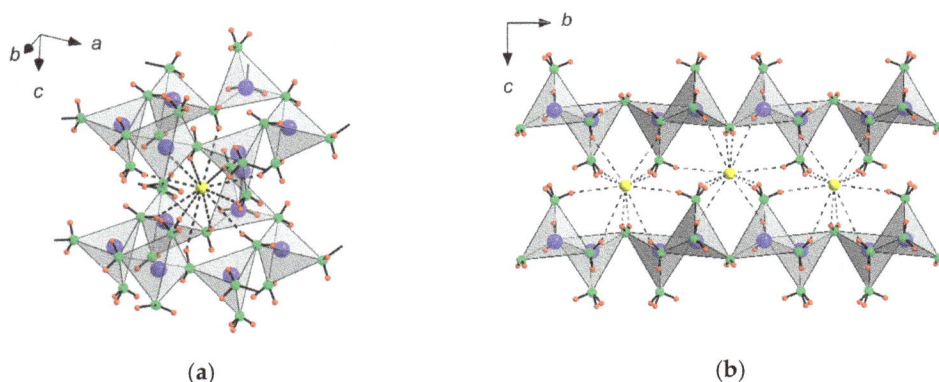

(a)                                    (b)

**Figure 2.** Sections of the crystal structure of $[Zn_2(NH_3)_2(NH_2)_3]Cl$: (a) every chloride ion employs twelve H-bonds and (b) connects two adjacent layers $^2_\infty \left[ Zn(NH_3)(NH_2)_{3/2}^{1/2+} \right]$ within the third dimension (**yellow** spheres: Cl, **green** spheres: N, **red** spheres: H, and **blue** spheres surrounded by **grey** tetrahedra: Zn).

**Table 4.** Donor–acceptor distances (pm) and angles (deg.) in $[Zn_2(NH_3)_2(NH_2)_3]Cl$.

| Arrangement | $d$(N–H) | $d$(H ... Cl) | <(NHCl) | $d$(N ... Cl) |
|---|---|---|---|---|
| N(1)–H(1A) ... Cl | 88 | 304 | 143.0 | 378 |
| N(1)–H(1B) ... Cl | 89 | 287 | 149.9 | 367 |
| N(2)–H(2A) ... Cl | 89 | 267 | 163.2 | 352 |
| N(2)–H(2B) ... Cl | 88 | 284 | 135.1 | 352 |
| N(2)–H(2C) ... Cl | 89 | 274 | 161.7 | 359 |
| N(3)–H(3A) ... Cl | 90 | 287 | 152.7 | 369 |

[Zn(NH$_3$)$_2$(NH$_2$)]Br suffers from twinning by partial merohedry via two-fold rotation around the face diagonal [101] in space group $P12_1/n1$ setting. Due to application of the respective twin law, the reliability factors significantly drop (see the experimental part). Table 1 summarizes selected information for the final structure refinement, Table 5 gives positional parameters and Table A2 anisotropic displacement parameters.

**Table 5.** Positional and isotropic displacement parameters $U_{iso}$ ($10^4$ pm$^2$) for [Zn(NH$_3$)$_2$(NH$_2$)]Br.

| Atom | Site | $x$ | $y$ | $z$ | $U_{iso}$ |
|------|------|-----|-----|-----|-----------|
| Br | 4$e$ | 0.1858(4) | 0.22474(7) | 0.4382(4) | 0.0214(3) |
| Zn | 4$e$ | 0.1606(4) | 0.1440(2) | 0.7794(4) | 0.0166(3) |
| N(1) | 4$e$ | 0.163(3) | 0.243(5) | 0.9347(2) | 0.021(2) |
| H(1A) | 4$e$ | 0.054(9) | 0.28(5) | 0.95(2) | 1.5$U_{iso}$(N(1)) |
| H(1B) | 4$e$ | 0.20(3) | 0.14(2) | 0.98(1) | 1.5$U_{iso}$(N(1)) |
| H(1C) | 4$e$ | 0.23(2) | 0.36(2) | 0.94(1) | 1.5$U_{iso}$(N(1)) |
| N(2) | 4$e$ | 0.419(4) | 0.247(5) | 0.206(2) | 0.023(8) |
| H(2A) | 4$e$ | 0.83(3) | 0.22(2) | 0.75(2) | 1.5$U_{iso}$(N(2)) |
| H(2B) | 4$e$ | 0.89(3) | 0.38(2) | 0.67(1) | 1.5$U_{iso}$(N(2)) |
| H(2C) | 4$e$ | 0.86(3) | 0.13(2) | 0.67(2) | 1.5$U_{iso}$(N(2)) |
| N(3) | 4$e$ | 0.859(3) | 0.190(1) | 0.221(3) | 0.015(3) |
| H(3A) | 4$e$ | 0.95(1) | 0.26(2) | 0.257(8) | 1.2$U_{iso}$(N(3)) |
| H(3B) | 4$e$ | 0.90(1) | 0.20(2) | 0.157(4) | 1.2$U_{iso}$(N(3)) |

Very similar to [Zn$_2$(NH$_3$)$_2$(NH$_2$)$_3$]Cl, the ammonia molecules, amide and bromide ions in the crystal structure of [Zn(NH$_3$)$_2$(NH$_2$)]Br, form the motif of a hexagonal closed packing with stacking along [010] (Figure 3). According to the composition, only 1/8 of the tetrahedral voids exclusively formed by two ammonia molecules and two amide ions are occupied by Zn. As discussed above for the chloride, the distances to the terminal ammonia ligands with 204(3) pm and 211(3) pm are significantly longer than those to the bridging amide ligands with 198.4(8) pm and 200.2(8) pm. Angles around Zn with 104°–123.8° deviate little from to the ideal tetrahedral angle. Table 6 collects selected interatomic distances and angles. Vertex-sharing via amide ligands leads to $\frac{1}{\infty}\left[Zn(NH_3)_2(NH_2)_{2/2}^+\right]$ zigzag chains running along [010]. According to the Liebau nomenclature, these chains are classified as unbranched *zweier* single-chains with molar ratio of $n(Zn):n(N) = 1:3$ [12]. Once again, these chains may be viewed as sections from the three-dimensional crystal structure of binary zinc amide, formally broken up by the addition of HBr.

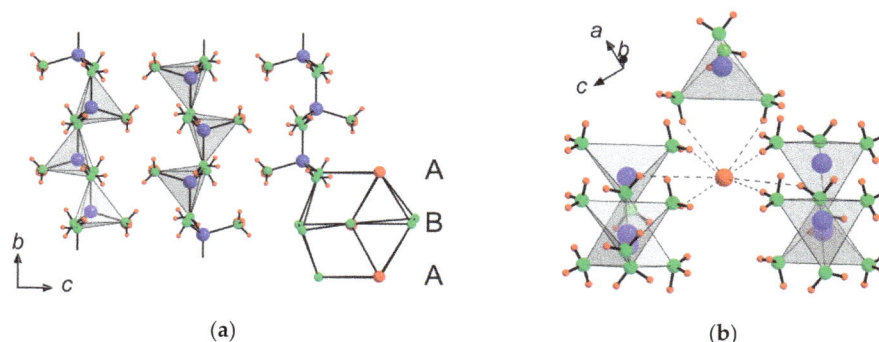

(a)             (b)

**Figure 3.** Sections of the crystal structure of [Zn(NH$_3$)$_2$(NH$_2$)]Br: (**a**) infinite $\frac{1}{\infty}\left[Zn(NH_3)_2(NH_2)_{2/2}^+\right]$ zigzag chains result from occupation of tetrahedral voids exclusively built by amide ions and ammonia molecules by Zn within the motif of an *hcp* formed by bromide and amide ions together with ammonia molecules (the *hcp* stacking motif is indicated by an anticuboctahedron and capital letters); (**b**) every bromide ion employs eight H-bonds and interconnects three chains (large **red** spheres: Cl, **green** spheres: N, small **red** spheres: H, and **blue** spheres surrounded by **grey** tetrahedra: Zn).

**Table 6.** Selected interatomic distances (pm) and angles (deg.) in the crystal structure of $[Zn(NH_3)_2(NH_2)]Br$.

| Bond | | Distance | Arrangement | Angle |
|---|---|---|---|---|
| Zn | –N(1) | 204(3) | N(1)–Zn–N(2) | 106.7(6) |
| Zn | –N(2) | 211(3) | N(1)–Zn–N(3) | 104(1) |
| Zn | –N(3) | 198.4(8) | N(1)–Zn–N(3) | 107(1) |
| Zn | –N(3) | 200.2(8) | N(2)–Zn–N(3) | 110(1) |
| | | | N(2)–Zn–N(3) | 105(1) |
| N(1) | –H(1A) | 91(7) | N(3)–Zn–N(3) | 123.8(2) |
| N(1) | –H(1B) | 89(3) | H(1A)–N(1)–H(1B) | 103(10) |
| N(1) | –H(1C) | 89(3) | H(1A)–N(1)–H(1C) | 109(10) |
| | | | H(1B)–N(1)–H(1C) | 110(10) |
| N(2) | –H(2A) | 92(4) | Zn–N(1)–H(1A) | 112(10) |
| N(2) | –H(2B) | 90(4) | Zn–N(1)–H(1B) | 116(10) |
| N(2) | –H(2C) | 90(4) | Zn–N(1)–H(1C) | 108(10) |
| | | | H(2A)–N(2)–H(2B) | 112(10) |
| N(3) | –H(3A) | 89(3) | H(2A)–N(2)–H(2C) | 83(10) |
| N(3) | –H(3B) | 89(3) | H(2B)–N(2)–H(2C) | 113(10) |
| | | | Zn–N(2)–H(2A) | 107(10) |
| $\varnothing$Zn | –NH_3 | 207.5 | Zn–N(2)–H(2B) | 131(10) |
| $\varnothing$Zn | –NH_2 | 199.3 | Zn–N(2)–H(2C) | 100(10) |
| $\varnothing$Zn | –N | 203.4 | H(3A)–N(3)–H(3B) | 97(9) |
| | | | Zn–N(3)–H(3A) | 97(9) |
| | | | Zn–N(3)–H(3A) | 123(10) |
| | | | Zn–N(3)–H(3B) | 128(8) |
| | | | Zn–N(3)–H(3B) | 97(8) |
| | | | Zn–N(3)–Zn | 116.2(5) |

Bromide ions interconnect three $\frac{1}{\infty}\left[Zn(NH_3)_2(NH_2)_{2/2}{}^+\right]$ zigzag chains each via eight H-bonds (see Figure 3). Table 7 gives relevant donor–acceptor distances and angles. For both title compounds, there are no indications for a rotational disorder of the ammonia ligands, prohibited by an involvement in hydrogen bonding networks.

**Table 7.** Donor–acceptor distances (pm) and angles (deg.) in $[Zn_2(NH_3)_2(NH_2)]Br$.

| Arrangement | $d$(N–H) | $d$(H ... Br) | <(NHBr) | $d$(N ... Br) |
|---|---|---|---|---|
| N(1)–H(1A) ... Br | 91 | 280 | 153.9 | 364 |
| N(1)–H(1B) ... Br | 89 | 268 | 159.2 | 352 |
| N(1)–H(1C) ... Br | 89 | 280 | 146.1 | 357 |
| N(2)–H(2A) ... Br | 91 | 268 | 150.4 | 351 |
| N(2)–H(2B) ... Br | 89 | 265 | 178.6 | 355 |
| N(2)–H(2C) ... Br | 90 | 268 | 173.6 | 357 |
| N(3)–H(3A) ... Br | 89 | 283 | 144.3 | 360 |
| N(3)–H(3B) ... Br | 89 | 313 | 127.4 | 373 |

## 2.2. Raman Spectroscopy

The Raman spectrum (Figure 4) of a $[Zn_2(NH_3)_2(NH_2)_3]Cl$ single crystal can be interpreted according to those of $Zn(NH_3)_2Br_2$ [13], $Zn(NH_3)_6Cl_2$ [14], $[Zn(NH_3)_4]Br_2$ [4], $[Zn(NH_3)_4]I_2$ [4,15] as well as calculations for $[Zn(NH_3)_4]^{2+}$ [16] and is in agreement with general trends for ammoniates of transition metal halides [14,17,18]. Four groups of signals appear in the spectrum, of which three groups are due to modes of $NH_3$ and $NH_2{}^-$ molecules (3167–3493 cm$^{-1}$, 1301–1621 cm$^{-1}$, 737–1031 cm$^{-1}$) and one group evokes from Zn–N skeletal vibrations (134–476 cm$^{-1}$).

In the range of 3167–3493 cm$^{-1}$, three asymmetric and three symmetric valence vibrations appear in agreement with the three crystallographic different nitrogen sites of ammonia and amide ligands. The broadening of this group of signals indicates the presence of hydrogen bonds of relevant strength, which is in perfect agreement with short NH ... Cl distances (see above) [18,19]. These hydrogen bonds are also the reason for a shift to smaller wave numbers as compared to the respective modes of the free ammonia molecule (3337 and 3450 cm$^{-1}$ [20]). For ammoniates of zinc fluoride $Zn(NH_3)_3F_2$ and $Zn(NH_3)_2F_2$ (3093–3337 cm$^{-1}$ [2]), a similar, but even larger shift was observed, due to the higher electronegativity of fluorine compared to chlorine, while the bromide $[Zn(NH_3)_4]Br_2$ (3194 cm$^{-1}$ [4]) and the iodide $[Zn(NH_3)_4]I_2$ (3177 cm$^{-1}$ [21]) exhibit smaller shifts in the symmetric valence mode.

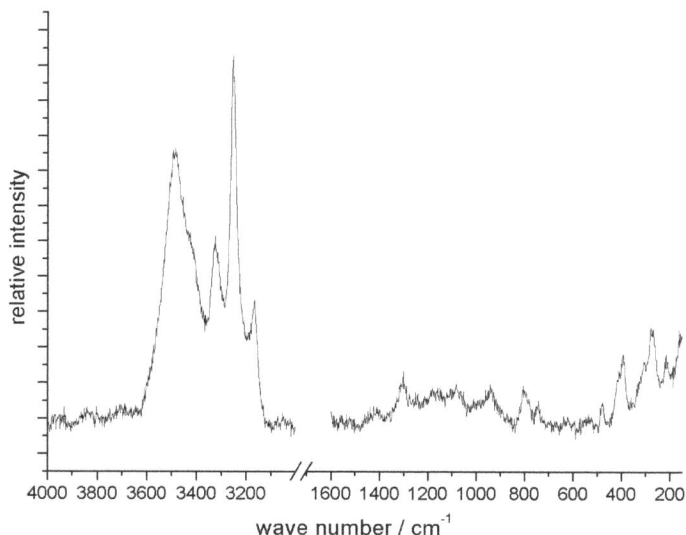

**Figure 4.** Raman spectrum of $[Zn_2(NH_3)_2(NH_2)_3]Cl$.

The second range is due to the symmetric (1302 cm$^{-1}$) and asymmetric (1621 cm$^{-1}$) scissoring modes, which are not resolved for the different crystallographic sites. These vibrations also appear in the spectra of $[Zn(NH_3)_4]Br_2$ (1607 and 1246 cm$^{-1}$ in IR [4]) and $[Zn(NH_3)_4]I_2$ (1615, 1600, 1256 and 1242 cm$^{-1}$ [21]). In comparison to the ammoniates of zinc fluoride, these signals are shifted to lower wave numbers, again as a result of reduced hydrogen bond strength. In the third region (1031–737 cm$^{-1}$), rocking and wagging modes are present. The rocking vibrations are not present in the spectrum of free ammonia molecules, but are known to increase in intensity with increasing covalent character in complexes [22]. The last range (476–134 cm$^{-1}$) is due to Zn–N skeleton vibrations.

The absence of signals above 3500 cm$^{-1}$ indicates the absence of significant impurities of OH$^-$ and $H_2O$, which should provoke a prominent mode at around 3700 cm$^{-1}$ [23].

## 3. Materials and Methods

### 3.1. Synthesis

All ammonothermal syntheses were carried out in custom-built autoclaves from austenitic nickel-chromium-based alloy 718 with an inner volume of 97 mL and equipped with a high-pressure valve, a filling line, a rupture disc and a pressure transmitter (HBM P2VA1/5000 bar), which enables continuous pressure monitoring during synthesis [24]. A platinum liner was introduced to minimize contact of the solution with the autoclave wall and thus reduce corrosion. A tubular furnace (LOBA, HTM Reetz GmbH, Berlin, Germany) in vertical position was used to heat the autoclave bodies. Due to unheated installations at the top of the vessel, a natural temperature gradient with a temperature difference of about 100 K developed, resulting in suitable convection for a chemical material transport. Additionally, the applied furnace temperature, which is referred to in the synthesis description, is about 150 K higher than the average reaction temperature [25].

Colorless transparent plate-like crystals $[Zn_2(NH_3)_2(NH_2)_3]Cl$ with maximum size of several hundred μm were obtained from Zn and $NH_4Cl$ (molar ratio 10:1) under ammonothermal conditions (873 K, 97 MPa for 6 h, in 38 h cooling to RT) in the cold zone of the autoclave. In a similar reaction,

[Zn(NH$_3$)$_2$(NH$_2$)]Br was synthesized from Zn and NH$_4$Br (molar ratio 1:1.7; 773 K, 230 MPa for 12 h, 72 h cooling duration), and colorless crystals were collected from the cold zone. Both compounds are sensitive to moist air and immediately lose their transparency on contact. In the hot zone of the autoclaves, there was always well crystallized Zn(NH$_2$)$_2$ present along with small amounts of unreacted Zn.

Due to the hygroscopic nature of the reactants and products, all manipulations were carried out inside an argon filled glovebox ($p$(O$_2$) < 0.1 ppm). A known amount of ammonia was condensed into the autoclave with use of a tensieudiometer [26] for the simultaneous pressure and temperature measurement and a dry ice ethanol cooling bath ($T$ = 198 K). After successful synthesis, the excess ammonia was vented and the autoclave was subsequently evacuated.

### 3.2. Diffraction Data Collection and Structure Refinements

Selected single crystals were isolated under Ar, sealed in glass capillaries and immediately mounted on a Bruker–Nonius Kappa-CCD diffractometer (Billerica, MA, USA). The chloride was measured at room temperature, while successful intensity data collection on the bromide was only possible if directly cooled to 100 K and subsequently measured (Mo-$K\alpha$ radiation). Structure solution and refinements were carried out using the program package *SHELXL97* [27]. All atom positions except hydrogen were determined by direct methods and anisotropically refined. The hydrogen positions attached to amide groups of the complex amidozincate ions were refined using the riding model, constraining the interatomic N–H distances to 89(2) pm. The isotropic displacement factors $U_{iso}$ were restrained to 1.2/1.5 times the $U_{iso}$ of the nitrogen atom to which they are attached to. For absorption correction, a linear scaling was applied using the absorption correction coefficients 12.93 mm$^{-1}$ ([Zn$_2$(NH$_3$)$_2$(NH$_2$)$_3$]Cl) and 11.17 mm$^{-1}$ ([Zn(NH$_3$)$_2$(NH$_2$)]Br) [28].

[Zn(NH$_3$)$_2$(NH$_2$)]Br suffers from twinning by partial merohedry via a two-fold rotation around [100] in a cell setting in $P112_1/a$. The partial merohedry leads to overlap of reflections ($hkl$) with $h = 2n$ with those reflections ($h\bar{k}\bar{l}$) of the second twin domain. Reflections with $h = 2n + 1$ are unaffected, but additional reflections ($h\frac{1}{2}kl$) with $h = 2n + 1$ appear. In the standard setting, $P12_1/n1$, the twinning occurs via a rotation around the face diagonal [101]. Assignment of the reflections to the domains was carried out with the program TWINXLI [29]. Due to application of the twin law, the reliability factors significantly drop from $R_1$ = 0.1136, $wR_2$ = 0.227, GooF = 1.685 (all reflections) to $R_1$ = 0.0445, $wR_2$ = 0.0864, GooF = 1.047 and the residual electron density from $\rho_{max}$ = 13.29 $\times$ 10$^{-6}$ pm$^{-3}$, $\rho_{min}$ = $-1.51$ $\times$ 10$^{-6}$ pm$^{-3}$ to $\rho_{max}$ = 1.23 $\times$ 10$^{-6}$ pm$^{-3}$, $\rho_{min}$ = $-1.13$ $\times$ 10$^{-6}$ pm$^{-3}$. Tables A1 and A2 additionally give the final anisotropic displacement parameters. Further details of the crystal structure investigation are available from the Fachinformationszentrum Karlsruhe, 76344 Eggenstein-Leopoldshafen, Germany (Fax: +49-7247-808-666; E-mail: crysdata@fiz-karlsruhe.de, http://www.fiz-karlsruhe.de/request_for_deposited_data.html#c665) on quoting the depository numbers CSD-432188 ([Zn$_2$(NH$_3$)$_2$(NH$_2$)$_3$]Cl) and CSD-432189 ([Zn(NH$_3$)$_2$(NH$_2$)]Br).

### 3.3. Vibrational Spectroscopy

The solid state Raman spectrum of ([Zn$_2$(NH$_3$)$_2$(NH$_2$)$_3$]Cl was measured with a Horiba XploRa Raman spectrometer (Kyoto, Japan) coupled with a confocal polarization microscope (Olympus BX51, Tokio, Japan) employing a 638 nm solid state laser at a single crystal sealed in a glass capillary.

## 4. Conclusions

Zinc metal easily dissolves in ammonia, particularly in ammonoacidic or -basic solutions under ammonothermal conditions. From these solutions at increased temperatures, zinc amide crystallizes. However, quasi-ternary ammoniates of zinc amide halides can also be obtained. With increasing temperatures, condensation of the cationic ions in diamminetriamidodizinc chloride [Zn$_2$(NH$_3$)$_2$(NH$_2$)$_3$]Cl and diamminemonoamidozinc bromide [Zn(NH$_3$)$_2$(NH$_2$)]Br occurs via amide ions towards two- and three-dimensional moieties, which can be viewed as sections of the crystal structure of binary zinc nitride, $Zn_3N_2$. The formation of amides in solutions containing halide ions, hence nominally ammonoacidic conditions, might be understood by the large excess of Zn, providing an ammonobasic buffer solution $Zn(NH_2)_2/ZnX_2$ in ammonia. Such investigations may pave the way to ammonothermal zinc nitride synthesis and crystal growth, a potential novel nitride semiconductor material for which currently no technique for bulk crystal production is known.

**Acknowledgments:** We thank Falk Lissner for collecting the diffraction intensity data. We are much obliged to Werner Massa for providing the program TWINXLI. Furthermore, we thank our cooperation partners of the interdisciplinary research group Ammono-FOR 1600 for the good cooperation and the Deutsche Forschungsgemeinschaft for generous funding.

**Author Contributions:** Nicolas S. A. Alt and Eberhard Schlücker designed the experimental autoclaves; Theresia M. M. Richter performed the experiments; Theresia M. M. Richter, Sabine Strobel and Rainer Niewa analyzed the data and wrote the paper.

## Appendix

**Table A1.** Anisotropic displacement parameters $U_{ij}$ ($10^4$ $pm^2$) for $[Zn_2(NH_3)_2(NH_2)_3]Cl$.

| Atom | $U_{11}$ | $U_{22}$ | $U_{33}$ | $U_{23}$ | $U_{13}$ | $U_{12}$ |
|---|---|---|---|---|---|---|
| Zn | 0.0195(4) | 0.0198(5) | 0.0312(5) | −0.0017(4) | 0.0006(4) | 0.0009(4) |
| Cl | 0.023(1) | 0.026(1) | 0.025(1) | 0 | 0 | 0 |
| N(1) | 0.014(3) | 0.023(3) | 0.047(4) | −0.008(4) | 0 | 0 |
| N(2) | 0.037(3) | 0.039(3) | 0.029(3) | 0 | 0.005(3) | 0 |
| N(3) | 0.022(4) | 0.027(5) | 0.031(5) | 0 | 0 | 0 |

**Table A2.** Anisotropic displacement parameters $U_{ij}$ ($10^4$ $pm^2$) for $[Zn_2(NH_3)_2(NH_2)]Br$.

| Atom | $U_{11}$ | $U_{22}$ | $U_{33}$ | $U_{23}$ | $U_{13}$ | $U_{12}$ |
|---|---|---|---|---|---|---|
| Br | 0.028(3) | 0.0153(4) | 0.021(3) | 0 | −0.003(2) | 0 |
| Zn | 0.019(3) | 0.0126(5) | 0.017(3) | 0 | 0 | 0.002(2) |
| N(1) | 0.03(1) | 0.02(1) | 0.01(1) | 0 | 0 | 0 |
| N(2) | 0.03(1) | 0 | 0.03(1) | −0.02(1) | 0 | 0 |
| N(3) | 0.019(9) | 0.011(4) | 0.015(9) | 0 | 0 | 0 |

## References

1.  Ehrentraut, D.; Meissner, E.; Bockowski, M. *Technology of Gallium Nitride Crystal Growth*; Springer Materials Sciences: Berlin/Heidelberg, Germany, 2010.

2.  Richter, T.M.M.; Le Tonquesse, S.; Alt, N.S.A.; Schlücker, E.; Niewa, R. Trigonal-Bipyramidal Coordination in First Ammoniates of $ZnF_2$: $ZnF_2(NH_3)_3$ and $ZnF_2(NH_3)_2$. *Inorg. Chem.* **2016**, *55*, 2488–2498. [CrossRef] [PubMed]

3.  MacGillavry, C.; Bijvoet, J. The crystal structure of $Zn(NH_3)_2Cl_2$ and $Zn(NH_3)_2Br_2$. *Z. Kristallogr.* **1936**, *94*, 249–255.

4.  Eßmann, R. Influence of coordination on N–H⋯X⁻ hydrogen bonds. 1. $[Zn(NH_3)_4]Br_2$ and $[Zn(NH_3)_4]I_2$. *J. Mol. Struct.* **1995**, *356*, 201–206. [CrossRef]

5.  Kraus, F.; Korber, N. Hydrogen bonds in potassium amide-ammonia(1/2), $KNH_2·2NH_3$. *Z. Anorg. Allg. Chem.* **2005**, *631*, 1032–1034. [CrossRef]

6.  Zachwieja, U.; Jacobs, H. Tri-μ-amido-bis-[triamminechromium(III)]iodide, $[Cr_2(NH_2)_3(NH_3)_6]I_3$, ein neuer "hexagonaler Perowskit". *Z. Kristallogr.* **1993**, *206*, 247–254.

7.  Ketchum, D.; Schimek, G.; Pennington, W.; Kolis, J. Synthesis of new Group III fluoride-ammonia adducts in supercritical ammonia: Structures of $AlF_3(NH_3)_2$ and $InF_2(NH_2)(NH_3)$. *Inorg. Chim. Act.* **1999**, *294*, 200–206. [CrossRef]

8.  Jander, J.; Doetsch, V.; Jander, G. *Anorganische und Allgemeine Chemie in Flüssigem Ammoniak*; Akademischer Verlag: Berlin, Germany, 1966.

9.  Binnewies, M.; Glaum, R.; Schmidt, M.; Schmidt, P. *Chemical Vapor Transport Reactions*; De Gruyter: Berlin, Germany, 2011.

10. Hertweck, B.; Steigerwald, T.G.; Alt, N.S.A.; Schluecker, E. Different corrosion behaviour of autoclaves made of nickel base alloy 718 in ammonobasic and ammonoacidic environments. *J. Supercrit. Fluids* **2014**, *95*, 158–166. [CrossRef]

11. Juza, R. Amides of the Alkali and the Alkaline earth Metals. *Angew. Chem. Int. Ed.* **1964**, *3*, 471–481. [CrossRef]

12. Liebau, F. *Structural Chemistry of Silicates*; Springer: Berlin, Germany, 1985.

13. Ishikawa, D.N.; Téllez, S.C.A. Infrared and Raman spectra of $Zn(NH_3)_2Br_2$ with $^{15}N$ and $^2H$ isotopic substitution. *Vibr. Spectrosc.* **1994**, *8*, 87–95. [CrossRef]

14. Nakamoto, K. *Infrared and Raman Spectra of Inorganic and Coordination Compounds*, 4th ed.; Wiley: New York, NY, USA, 1997.

15. Schmidt, K.; Hauswirth, W.; Müller, A. Vibrational Spectra of Nitrogen-15-substituted Hexa-amminenickel(II) Chloride, Hexa-amminecobalt(III) Chloride, and Tetra-amminezinc(II) Iodide. *Dalton Trans.* **1975**, *21*, 2199–2201. [CrossRef]

16. Acevedo, R.; Díaz, G. Normal coordinate analysis for the $M(NH_3)_4{}^{2+}$ complex ions in $D_{4h}$ and $T_d$ symmetries. Simplified molecular models. *Spectrosc. Lett.* **1986**, *19*, 653–668. [CrossRef]

17. Gauglitz, G.; Vo-Dinh, T. *Handbook of Spectroscopy*; Wiley-VCH: Weinheim, Germany, 2003.

18. Lutz, H.D. Structure and strengt hof hydrogen bonds in inorganic solids. *J. Mol. Struct.* **2003**, *646*, 227–236. [CrossRef]

19. Novak, A. Hydrogen Bonding in Solids. Correlation of Spectroscopic and Crystallographic Data. *Struct. Bond.* **1973**, *18*, 177–216.

20. Svatos, G.F.; Curran, C.; Quagliano, J.V. Infrared Absorption Spectra of Inorganic Coördination Complexes. V. The N–H Stretching Vibration in Coördination compounds. *J. Am. Chem. Soc.* **1955**, *77*, 6159–6163. [CrossRef]

21. Schmidt, K.H.; Müller, A. Vibrational spectra and force constants of pure ammine complexes. *Coord. Chem. Rev.* **1976**, *19*, 41–97. [CrossRef]

22. Mizushima, S.; Svatos, G.F.; Quagliano, J.V.; Curran, C. A Comparison of the Deformation Vibration Frequencies of the Methyl Group With Those of the Coordinated Ammonia Molecule. *Anal. Chem.* **1955**, *27*, 325.

23. Weidlein, J.; Müller, U.; Dehnicke, K. *Schwingungsspektroskopie: Eine Einführung*; Band 2, Thieme: Stuttgart, Germany, 1988.

24. Alt, N.S.A.; Meissner, E.; Schlücker, E.; Frey, L. In situ monitoring technologies for ammonothermal reactors. *Phys. Status Solidi C* **2012**, *9*, 436–439. [CrossRef]

25. Zhang, S. Intermediates during the Formation of GaN under Ammonothermal Conditions. Ph.D. Thesis, University of Stuttgart, Stuttgart, Germany, 2014.

26. Hüttig, G.F. Apparatur zur gleichzeitigen Druck- und Raummessung von Gasen. (Tensi-Eudiometer). *Z. Anorg. Allg. Chem.* **1920**, *114*, 161–173. [CrossRef]

27. Sheldrick, G.M. *SHELXL-97*, Program for the Refinement of Crystal Structure; University of Göttingen: Göttingen, Germany, 1997.

28. Otwinowski, Z.; Minor, W. DENZO and SCALEPACK. In *International Tables for Crystallography F*; Springer: Dordrecht, The Netherlands, 2001; pp. 226–235.

29. Hahn, F.; Massa, W. *TWINXL3.1*, Programm zur Aufbereitung von Datensätzen verzwillingter Kristalle; University of Marburg: Marburg, Germany, 2000.

# 10

# *P*-Fluorous Phosphines as Electron-Poor/Fluorous Hybrid Functional Ligands for Precious Metal Catalysts: Synthesis of Rh(I), Ir(I), Pt(II), and Au(I) Complexes Bearing *P*-Fluorous Phosphine Ligands

Shin-ichi Kawaguchi [1,*], Yuta Saga [2,3], Yuki Sato [2], Yoshiaki Minamida [2], Akihiro Nomoto [2] and Akiya Ogawa [2,*]

[1]   Center for Education and Research in Agricultural Innovation, Faculty of Agriculture, Saga University, 152-1 Shonan-cho Karatsu, Saga 847-0021, Japan
[2]   Department of Applied Chemistry, Graduate School of Engineering, Osaka Prefecture University, 1-1 Gakuen-cho, Nakaku, Sakai, Osaka 599-8531, Japan; y.saga@katayamakagaku.co.jp (Yut.S.); su108027@edu.osakafu-u.ac.jp (Yuk.S.); yomi191919@yahoo.co.jp (Y.M.); nomoto@chem.osakafu-u.ac.jp (A.N.)
[3]   Katayama Chemical Industries Co., Ltd., 26-22, 3-Chome, Higasinaniwa-cho, Amagasaki, Hyogo 660-0892, Japan
*   Correspondence: skawa@cc.saga-u.ac.jp (S.K.); ogawa@chem.osakafu-u.ac.jp (A.O.)

Academic Editor: Lee J. Higham

**Abstract:** *P*-Fluorous phosphine ($R_2PR_f$), in which the perfluoroalkyl group is directly bonded to the phosphorus atom, is a promising ligand because it has a hybrid functionality, i.e., electron-poor and fluorous ligands. However, examples of *P*-fluorous phosphine–metal complexes are still rare, most probably because the *P*-fluorous group is believed to decrease the coordination ability of the phosphines dramatically. In contrast, however, we have succeeded in synthesizing a series of *P*-fluorous phosphine–coordinated metal complexes such as rhodium, iridium, platinum, and gold. Furthermore, the electronic properties of $R_2P^nC_{10}F_{21}$ are investigated by X-ray analysis of $PtCl_2(Ph_2P^nC_{10}F_{21})_2$ and the infrared CO stretching frequency of $RhCl(CO)(R_2P^nC_{10}F_{21})_2$. $IrCl(CO)(Ph_2P^nC_{10}F_{21})_2$- and $AuCl(R_2P^nC_{10}F_{21})$-catalyzed reactions are also demonstrated.

**Keywords:** phosphine-metal complex; fluorous; electron-poor phosphine ligand; precious metal complex

## 1. Introduction

*P*-Fluorous phosphine ($R_2PR_f$), in which the perfluoroalkyl group is directly bonded to the phosphorus atom, is a hybrid functional phosphine ligand having both "electron-poor" [1] and "fluorous" [2–8] characteristics. Since strongly electron-withdrawing ligands are known to promote reductive elimination steps in catalytic reactions, the former property may be exploited, not only to optimize known reactions, but also to develop new catalytic reactions [9–12]. As to the latter property, the use of a fluorous biphasic system (FBS) may make it possible to recover catalysts and ligands easily and to reuse them for catalytic reactions. We recently developed a catalytic reaction using *P*-fluorous phosphines as ligands, i.e., a palladium-catalyzed cross-coupling reaction between acid chlorides and terminal alkynes, and have demonstrated the recyclability of the catalyst and the ligand [13]. Namely, the poor electron density of the *P*-fluorous phosphine ligands induced the palladium-catalyzed cross-coupling reaction, even under copper-free conditions, and the fluorous affinity of *P*-fluorous phosphines enabled the reuse of their Pd-complexes by using an FBS.

Our group recently developed three types of convenient synthetic methods of $P$-fluorous phosphine ligands ($R_2PR_f$) (see, Scheme 1): method A, the photoinduced direct displacement of $R_3P$ with $R_fI$ under radical conditions [13]; method B, the photoinduced $S_H2$ reaction of $(Ph_2P)_2$ with $R_fI$ [14]; method C, the reductive substitution reaction of diphenyl(2,4,6-trimethylbenzoyl)phosphine oxide (TMDPO, known as a representative radical initiator for polymerization) with $R_fI$ under light [15].

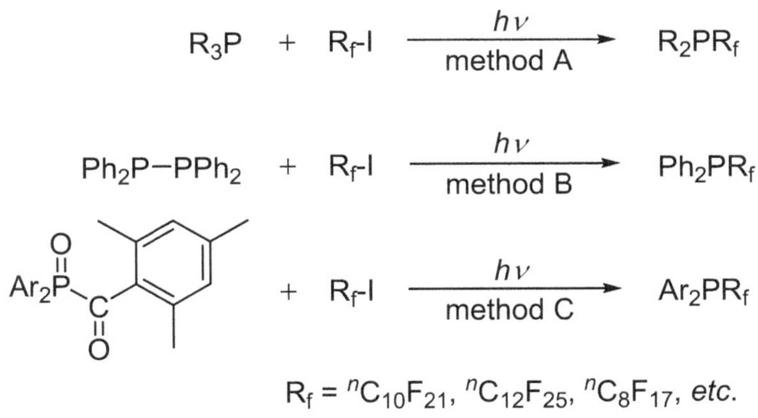

$$R_3P \ + \ R_f\text{-}I \ \xrightarrow[\text{method A}]{h\nu} \ R_2PR_f$$

$$Ph_2P\text{--}PPh_2 \ + \ R_f\text{-}I \ \xrightarrow[\text{method B}]{h\nu} \ Ph_2PR_f$$

$$+ \ R_f\text{-}I \ \xrightarrow[\text{method C}]{h\nu} \ Ar_2PR_f$$

$$R_f = {}^nC_{10}F_{21}, \ {}^nC_{12}F_{25}, \ {}^nC_8F_{17}, \ etc.$$

Scheme 1. Synthetic methods of $P$-fluorous phosphines.

To synthesize a diphenyl-substituted $P$-fluorous phosphine ligand ($Ph_2P^nC_{10}F_{21}$ **1a**), method C is the best because method C can be applied to the gram-scale synthesis of **1a** (Equation (1)) [15]. In the case of the synthesis of the dialkyl-substituted $P$-fluorous phosphine ligand ($^nBu_2P^nC_{10}F_{21}$ **1b**), method A is the most suitable because the gram scale of **1b** can be obtained by this method (Equation (2)).

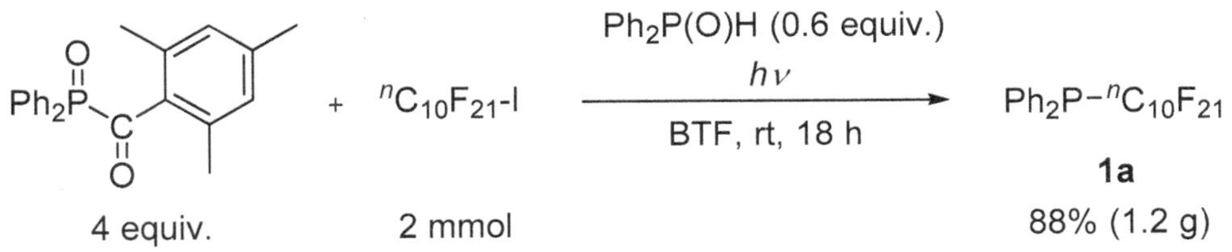

$$(1)$$

$$(2)$$

Our recent success in palladium-catalyzed cross-coupling using $P$-fluorous phosphines [13] and the establishment of their synthetic methods [13–15] strongly suggest their promising usability as functional ligands in many transition metal–catalyzed reactions. However, examples of $P$-fluorous phosphine–metal complexes are still rare in the literature [16–20], and are limited to metal-$R_2PR_f$ complexes bearing a short-chain perfluoroalkyl (light fluorous) group, which cannot be separated by a FBS due to the low content of fluorine atoms. To encourage the use of $P$-fluorous phosphine ligands in various catalytic reactions, we investigated their reactions with representative transition metal catalysts to prepare the corresponding $P$-fluorous phosphine–coordinated metal catalysts. In this paper, we report the synthesis of $P$-fluorous phosphine–coordinated rhodium, iridium, platinum, and gold complexes, and also describe their electronic properties, structural features, and catalytic activities.

## 2. Results and Discussion

First, we investigated the coordination of $R_2PR_f$ to Pt. When a mixture of $Ph_2P^nC_{10}F_{21}$ (**1a**) and $PtCl_2(CH_3CN)_2$ was stirred in $CHCl_3$ for two days, the $P$-perfluoroalkylated phosphine–platinum complex, $PtCl_2(Ph_2P^nC_{10}F_{21})_2$ (**2a**), was obtained successfully (Equation (3)). The $^{31}P$ NMR spectrum showed a triplet with satellites ($J_{P-F}$ = 55.8 Hz, $J_{P-Pt}$ = 3878 Hz). This indicated the successful complexation between the $P$-fluorous phosphine–platinum complex. When $^nBu_2P^nC_{10}F_{21}$ (**1b**) was used, instead of **1a**, for complexation with platinum under similar conditions, the $P$-perfluoroalkylated phosphine–platinum complex $trans$-$PtCl_2(^nBu_2P^nC_{10}F_{21})_2$ (**2b**) was also obtained in good yield (Equation (3)).

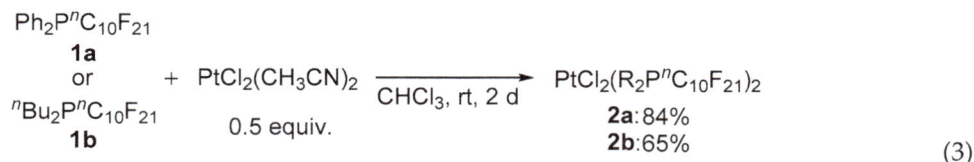

$$
\begin{array}{c}
Ph_2P^nC_{10}F_{21} \\
\textbf{1a} \\
\text{or} \\
^nBu_2P^nC_{10}F_{21} \\
\textbf{1b}
\end{array}
+ \quad
\begin{array}{c}
PtCl_2(CH_3CN)_2 \\
\text{0.5 equiv.}
\end{array}
\xrightarrow{CHCl_3,\ rt,\ 2\ d}
\begin{array}{c}
PtCl_2(R_2P^nC_{10}F_{21})_2 \\
\textbf{2a}{:}84\% \\
\textbf{2b}{:}65\%
\end{array}
\tag{3}
$$

We fortunately obtained a single crystal of complex **2a** by recrystallization from $CHCl_3$, which was suitable for X-ray analysis. The Oak Ridge Thermal Ellipsoid Plot (ORTEP) representation of **2a** is shown in Figure 1. The P–Pt–P bond angle is 178.66°. This indicates that the platinum complex $PtCl_2(Ph_2P^nC_{10}F_{21})_2$ has the structure of the $trans$ form. The Pt–P bond lengths in $trans$-$PtCl_2(Ph_2P^nC_{10}F_{21})_2$ are 2.3027(6) and 2.3039(6) Å. These are shorter than the corresponding bond lengths in $trans$-$PtCl_2(PPh_3)_2$ (2.319(3) and 2.316(3) Å) [21]. The P–$CF_2$ bond lengths in $trans$-$PtCl_2(Ph_2P^nC_{10}F_{21})_2$ are 1.913(3) and 1.909(3). They are longer than the P–$CF_2$ bond length in $Ph_2PC_2F_5$ (1.891(3) Å) [18]. The four P–C($ipso$) lengths in $trans$-$PtCl_2(Ph_2P^nC_{10}F_{21})_2$ are 1.810(4), 1.813(3), 1.810(4), and 1.815(3) Å. These are shorter than the P–C($ipso$) bond lengths in $Ph_2PC_2F_5$ (1.832(3) and 1.835(3) Å) [22]. On the other hand, the average Pt–P–C($ipso$) angle in $trans$-$PtCl_2(Ph_2P^nC_{10}F_{21})_2$ is 115.3°. It is similar to the average Pt–P–C($ipso$) angle in $trans$-$PtCl_2(PPh_3)_2$ (113.8°). The ORTEP representation also shows two perfluoroalkyl chains aligned in a parallel direction. The nearest distance between the two F atoms of each perfluoroalkyl chain is 2.734 Å, which is slightly shorter than the sum of the van der Waals radii of two F atoms (2.94 Å). Moreover, the packing diagram of $trans$-$PtCl_2(Ph_2P^nC_{10}F_{21})_2$ shows that the long perfluoroalkyl chains are assembled into a fluorous layer (Figure 2) [23–25]. In the case of the reported light fluorous phosphine-metal complexes, such a parallel direction of the fluorous group was not observed [18,19].

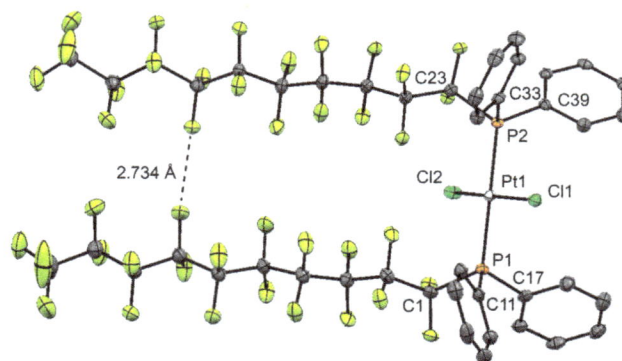

**Figure 1.** ORTEP representation (thermal ellipsoids at 50%) of $PtCl_2(Ph_2P^nC_{10}F_{21})_2$. Space group: $P$-1 (#2), $Z$ = 2, $R_1$ = 0.0312, $wR_2$ = 0.0886, selected bond lengths (Å) and angles (°): Pt(1)–Cl(1) = 2.332(10), Pt(1)–P1(1) = 2.303(6), P(1)–C(1) = 1.913(3), P(1)–C(11) = 1.810(4), Cl(1)–Pt(1)–Cl(2) = 172.79(3), Cl(1)–Pt(1)–P(1) = 89.10(3), P(1)–Pt(1)–P(2) = 178.66(4), Pt(1)–P(1)–C(1) = 117.15(8), Pt(1)–P(1)–C(11) = 116.41(9), Pt(1)–P(1)–C(17) = 116.41(9), C(1)–P(1)–C(11) = 98.16(14).

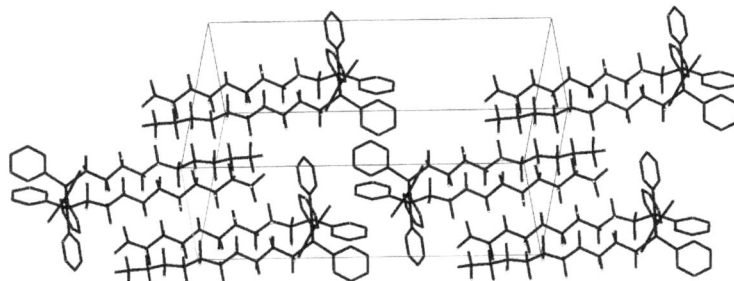

**Figure 2.** Packing diagram of $PtCl_2(Ph_2P^nC_{10}F_{21})_2$.

The infrared CO stretching frequency ($\nu_{(CO)}$) of the $RhCl(CO)(PR_3)_2$ complex provides information on the electronic properties of the phosphines [26]. Therefore, we investigated the reaction of a $P$-perfluoroalkylated phosphine with $[RhCl(CO)_2]_2$ for the synthesis of $RhCl(CO)(R_2PR_f)_2$. When the reaction of **1a** or **1b** with $[RhCl(CO)_2]_2$ was examined according to a reported method [9], $RhCl(CO)(Ph_2P^nC_{10}F_{21})_2$ (**3a**) and $RhCl(CO)(^nBu_2P^nC_{10}F_{21})_2$ (**3b**) were obtained in good yields, respectively (Equation (4)). Coupling in the $^{31}P$ NMR spectrum was assigned to the interaction between $^{31}P$ and $^{103}Rh$ ($J_{P-Rh}$ = 143.8 Hz), confirming complexation between the $P$-perfluoroalkylated phosphine and rhodium. Additionally, the symmetry of the $^{31}P$ NMR spectrum indicates that the steric configuration is *trans*.

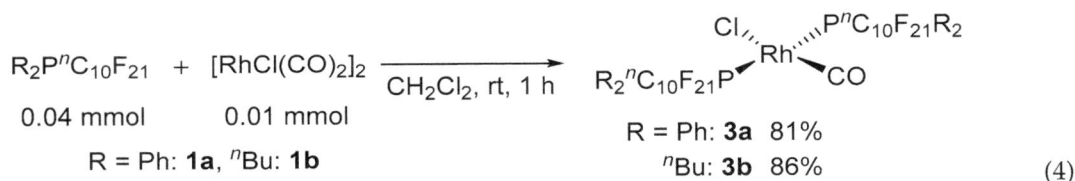

$$R_2P^nC_{10}F_{21} \quad + \quad [RhCl(CO)_2]_2 \xrightarrow{CH_2Cl_2,\ rt,\ 1\ h}$$

0.04 mmol      0.01 mmol

R = Ph: **1a**, $^n$Bu: **1b**

R = Ph: **3a**  81%
$^n$Bu: **3b**  86%  (4)

Using the same method, a series of $RhCl(CO)(PR_3)_2$ complexes were synthesized and their infrared CO stretching frequencies ($\nu_{(CO)}$) were measured to evaluate the electronic properties of $R_2PR_f$ (Table 1). Large $\nu_{(CO)}$ values indicate poor electron-donating abilities of the phosphine ligand. The $\nu_{(CO)}$ of $RhCl(CO)(Ph_2P^nC_{10}F_{21})_2$ and $RhCl(CO)[P(C_6F_5)_3]_2$ are almost the same (2010, 2008 cm$^{-1}$), and therefore the electron-donating ability of $Ph_2P^nC_{10}F_{21}$ is as poor as that of $P(C_6F_5)_3$. The $\nu_{(CO)}$ of $RhCl(CO)(^nBu_2P^nC_{10}F_{21})_2$ is 1987 cm$^{-1}$, which is similar to that of $RhCl(CO)[P(4-CF_3C_6H_4)_3]_2$ (1990 cm$^{-1}$). Thus, the electron-donating ability of $^nBu_2P^nC_{10}F_{21}$ is similar to that of $P(4-CF_3C_6H_4)_3$.

**Table 1.** The $\nu_{(CO)}$ data of $RhCl(CO)(PR_3)_2$.

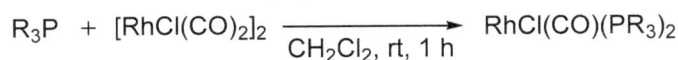

$$R_3P \quad + \quad [RhCl(CO)_2]_2 \xrightarrow{CH_2Cl_2,\ rt,\ 1\ h} RhCl(CO)(PR_3)_2$$

| Entry | $RhCl(CO)(PR_3)_2$ | $\nu_{(CO)}$ (cm$^{-1}$) |
|---|---|---|
| 1 | $RhCl(CO)(Ph_2P^nC_{10}F_{21})_2$ | 2010 |
| 2 | $RhCl(CO)(^nBu_2P^nC_{10}F_{21})_2$ | 1987 |
| 3 | $RhCl(CO)(PPh_3)_2$ | 1967 |
| 4 | $RhCl(CO)(P(4-CF_3C_6H_4)_3)_2$ | 1990 |
| 5 | $RhCl(CO)(P(C_6F_5)_3)_2$ | 2008 |

An iridium-phosphine complex (iridium is a congener of rhodium) was also investigated. Vaska-type complexes [27], such as $IrCl(CO)(PR_3)_2$, are well-known iridium-phosphine complexes. We examined the preparation of a Vaska-type complex of $P$-perfluoroalkylated phosphine from $IrCl_3 \cdot 3H_2O$ and an excess amount of $Ph_2P^nC_{10}F_{21}$ via a reduction process [28,29]. As a result, $IrCl(CO)(Ph_2P^nC_{10}F_{21})_2$ (**4a**) was successfully obtained (Equation (5)). The $\nu_{(CO)}$ of **4a** was shifted to a higher frequency (1990 cm$^{-1}$) compared with that of the Vaska complex $IrCl(CO)(PPh_3)_2$ (1944 cm$^{-1}$) [27], as well as that of the rhodium complex. The symmetry of the $^{31}P$ NMR spectrum ($\delta_P$ 45.0 ppm, t, $J_{P-F}$ = 34.6 Hz) indicates that the steric configuration of **4a** is *trans*.

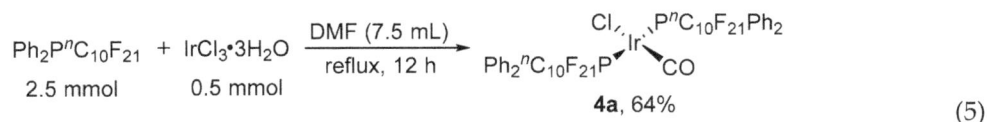

$$Ph_2P^nC_{10}F_{21} + IrCl_3 \cdot 3H_2O \xrightarrow[\text{reflux, 12 h}]{\text{DMF (7.5 mL)}} \quad \substack{Cl \\ \diagdown \\ Ph_2{}^nC_{10}F_{21}P} Ir \substack{P^nC_{10}F_{21}Ph_2 \\ \diagup \\ CO}$$

2.5 mmol　　0.5 mmol　　　　　　　　　　　　　　　　　**4a**, 64%　　　　　　　　(5)

We next investigated the application of **4a** as a catalyst. Several iridium-catalyzed hydrosilylation reactions of alkynes have been reported [30–33]. We therefore attempted alkyne hydrosilylation using **4a** as a catalyst. When a mixture of 1-octyne and triethylsilane was heated in the presence of 1 mol % of **4a**, vinylsilane derivatives as hydrosilylation products were obtained in 81% yield ($E/Z$ = 46/54) (Equation (6)). The result clearly indicates that the iridium complex **4a** can catalyze the hydrosilylation of alkynes as well as $IrCl(CO)(PPh_3)_2$.

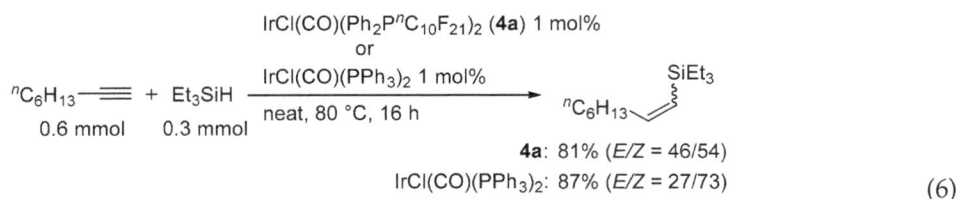

$$^nC_6H_{13}{-}{\equiv} + Et_3SiH \xrightarrow[\text{neat, 80 °C, 16 h}]{\substack{IrCl(CO)(Ph_2P^nC_{10}F_{21})_2 \text{ (4a) 1 mol\%} \\ \text{or} \\ IrCl(CO)(PPh_3)_2 \text{ 1 mol\%}}} \quad {}^nC_6H_{13}\diagdown\!\diagup^{SiEt_3}$$

0.6 mmol　　0.3 mmol

**4a**: 81% ($E/Z$ = 46/54)
$IrCl(CO)(PPh_3)_2$: 87% ($E/Z$ = 27/73)　　(6)

Finally, we investigated a *P*-perfluoroalkylated phosphine–gold(I) complex. $AuCl(SMe_2)$ was selected as the starting gold complex [34]. When an equimolar amount of $R_2P^nC_{10}F_{21}$ (**1a** and **1b**) and $AuCl(SMe_2)$ was stirred at room temperature, the desired complexes, $AuCl(R_2P^nC_{10}F_{21})$ (**5a** and **5b**), were obtained in quantitative yields, respectively (Equation (7)). The $^{19}F$ NMR and $^{31}P$ NMR analyses confirmed the complexation of **1a** with gold(I). The $^{19}F$ NMR signal of $P-CF_2-CF_2-$ appeared at −107.8 ppm ($CDCl_3$) as a doublet of triplets ($^2J_{F-P}$ = 64.1 Hz, $^3J_{F-F}$ = 14.2 Hz), which was shifted downfield to 0.9 ppm compared with the free $P-CF_2-CF_2-$ of $Ph_2P^nC_{10}F_{21}$, due to metal complexation. The $^{31}P$ NMR signal appeared at 40.1 ppm as a triplet of triplets ($^2J_{P-F}$ = 65.0 Hz, $^3J_{P-F}$ = 12.9 Hz) in $CH_2Cl_2$. HRMS (FAB) analysis further confirmed the complexation: the found value 900.9848 (calculated value for $C_{22}H_{10}F_{21}PAu$ [M − Cl]$^+$: 900.9850) indicated the presence of the $Au(Ph_2P^nC_{10}F_{21})$ moiety.

$$R_2P^nC_{10}F_{21} + AuCl(SMe_2) \xrightarrow[\text{CH}_2\text{Cl}_2, \text{ rt, 1 h}]{} AuCl(R_2P^nC_{10}F_{21})$$

0.1 mmol　　0.1 mmol

R = Ph: **1a**, nBu: **1b**

R = Ph: **5a**　98%
$^nBu$: **5b**　99%　　　　(7)

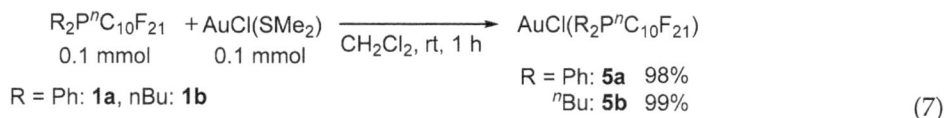

A catalytic hydroalkoxylation of an alkene was demonstrated using gold complex **5**, according to the literature [35]. In the presence of a catalytic amount of **5a** and AgOTf, the addition reaction of 2-chloroethanol to 1-octene took place to give 2-(2'-chloroethoxy)octane in a good yield (Equation (8)). In the case of $AuCl(^nBu_2P^nC_{10}F_{21})$, the desired adduct was also obtained in a good yield. The synthesized gold(I) complex, $AuCl(Ph_2P^nC_{10}F_{21})$ and $AuCl(^nBu_2P^nC_{10}F_{21})$, was found to exhibit more excellent catalytic activity with the ethanol addition to 1-octene compared with $AuClPPh_3$.

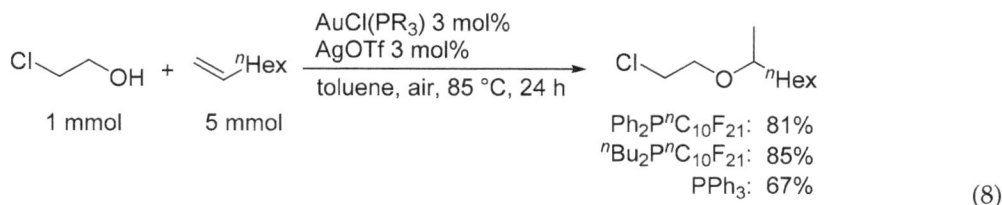

$$Cl\diagdown\!\diagup^{OH} + \diagup\!\diagdown^{nHex} \xrightarrow[\text{toluene, air, 85 °C, 24 h}]{\substack{AuCl(PR_3) \text{ 3 mol\%} \\ AgOTf \text{ 3 mol\%}}} Cl\diagdown\!\diagup^{O}\diagup\!\diagdown^{nHex}$$

1 mmol　　　5 mmol

$Ph_2P^nC_{10}F_{21}$: 81%
$^nBu_2P^nC_{10}F_{21}$: 85%
$PPh_3$: 67%　　　　(8)

It is assumed that metal-$R_2PR_f$ complexes have fluorous affinities because they contain perfluoroalkyl groups. Therefore, the fluorous affinities of these metal-$Ph_2P^nC_{10}F_{21}$ complexes were investigated; namely, the solubility of **2a–5a** in a fluorous solvent (FC-72, perfluorohexane) was measured at 25 °C. The results are as follows: $PtCl_2(Ph_2P^nC_{10}F_{21})_2$, 0.32 g/L; $RhCl(CO)(Ph_2P^nC_{10}F_{21})_2$, 0.40 g/L; $IrCl(CO)(Ph_2P^nC_{10}F_{21})_2$, 0.13 g/L; $AuCl(Ph_2P^nC_{10}F_{21})$, did not dissolve. Because the fluorine content of $AuCl(Ph_2P^nC_{10}F_{21})$ is low (42.6%), it did not dissolve in FC-72. These results show that these metal-$Ph_2P^nC_{10}F_{21}$ complexes, i.e., $PtCl_2(R_2PR_f)_2$, $RhCl(CO)(R_2PR_f)_2$, and $IrCl(CO)(R_2PR_f)_2$, have fluorous affinities, and therefore it is possible to extract them by using a fluorous solvent and an appropriate organic solvent.

## 3. Materials and Methods

### 3.1. General Comments

$Ph_2P^nC_{10}F_{21}$ (**1a**) [15] and $^nBu_2P^nC_{10}F_{21}$ (**1b**) [13] were synthesized according to the literature. Other materials were obtained from commercial suppliers and used without purification before use.

### 3.2. Synthesis of P-Perfluoroalkylated Phosphine Complex with Pt(II)

Under inert atmosphere, $Ph_2P^nC_{10}F_{21}$ (21.1 mg, 0.03 mmol), $PtCl_2(CH_3CN)_2$ (5.2 mg, 0.015 mmol), and $CHCl_3$ (0.6 mL) were added to a sealed test tube. After standing for two days, white precipitate was formed and filtered. Then enough pure $trans$-$PtCl_2(Ph_2P^nC_{10}F_{21})_2$ was obtained in 84% yield.

*trans-Dichlorobis((perfluorodecyl)diphenylphosphine)palladium(II)* (**2a**): white solid; melting point (mp) 188–190 °C; $^1H$ NMR (396 MHz, $CD_2Cl_2$): δ 7.52–7.55 (m, 8H), 7.59–7.62 (m, 4H), 7.94 (dd, $J_{H–H}$ = 7.3 Hz, $J_{H–P}$ = 12.7 Hz, 8H); $^{31}P$ NMR (160 MHz, $CD_2Cl_2$): δ 25.8 (t with satellites, $J_{P–F}$ = 55.8 Hz, $J_{P–Pt}$ = 3878 Hz); $^{19}F$ NMR (373 MHz, $CD_2Cl_2$): δ −126.2 (4F), −122.8 (4F), −122.0 (8F), −121.8 (8F), −121.2 (4F), −114.3 (4F), −101.8 (d, $J_{F–P}$ = 57.0 Hz, 4F), −81.0 (6F); HRMS (FAB) Calcd. for $C_{44}H_{20}Cl_2F_{42}P_2Pt$ [M]$^+$: 1671.9373, Found: 1671.9396.

Under inert atmosphere, $^nBu_2P^nC_{10}F_{21}$ (132.8 mg, 0.2 mmol), $PtCl_2(PhCN)_2$ (47.2 mg, 0.1 mmol), and $CHCl_3$ (0.6 mL) were added to a sealed test tube. After standing for two days, white precipitate was formed and then, by filtration, enough pure $trans$-$PtCl_2(^nBu_2P^nC_{10}F_{21})_2$ was obtained in 65% yield.

*trans-Dichlorobis(dibutyl(perfluorodecyl)phosphine)palladium(II)* (**2b**): white solid; mp 71–72 °C; $^1H$ NMR (396 MHz, $CDCl_3$): δ 0.97 (t, $J_{H–H}$ = 7.3 Hz, 12H), 1.44–1.54 (m, 12H), 1.70–1.83 (m, 4H), 2.18–2.30 (m, 4H), 2.36–2.46 (m, 4H); $^{31}P$ NMR (160 MHz, $CDCl_3$): δ 24.6 (quint with satellites, $J_{P–F}$ = 25.8 Hz, $J_{P–Pt}$ = 2720 Hz); $^{19}F$ NMR (373 MHz, $CDCl_3$): δ −126.2 (4F), −122.8 (4F), −122.0 (4F), −121.8 (12F), −121.5 (4F), −116.4 (4F), −109.1 (dt, $J_{F–P}$ = 57.0 Hz, $J$ = 28.5 Hz, 4F), −80.9 (t, $J_{F–F}$ =11.4 Hz, 6F); HRMS (FAB) Calcd. for $C_{36}H_{36}Cl_2F_{42}P_2{}^{194}Pt$ [M]$^+$: 1592.0625, Found: 1592.0603. The copies of $^1H$ NMR, $^{19}F$ NMR, and $^{31}P$ NMR spectra of platinum complex (**2a,b**) are shown in Supplementary Materials.

### 3.3. Synthesis of P-Perfluoroalkylated Phosphine Complex with Rh(I)

Rhodium complexes ($RhCl(CO)(PR_3)_2$) were synthesized according to the reported method [9]. Phosphine (0.044 mmol), $[RhCl(CO)_2]_2$ (4.3 mg, 0.011 mmol), and dichloromethane (1.0 mL) were added to a 20 mL two-necked round-bottomed flask under argon. The solution was stirred at room temperature for 1 h. The solution was filtrated, and concentrated under reduced pressure. The resulting solid was purified by recrystallization (solvent: $CHCl_3$).

*trans-carbonylchlorobis((perfluorodecyl)diphenylphosphine)rhodium(I)* (**3a**): yellow solid; mp 190–192 °C (decomposition); $^1H$ NMR (400 MHz, $CDCl_3$): δ 7.44 (dd, $J_{H–H}$ = 7.3 Hz, $J_{H–P}$ = 8.0 Hz, 8H), 7.52 (t, $J$ = 7.3 Hz, 4H), 7.93 (q, $J$ = 6.3 Hz, 8H); $^{31}P$ NMR (162 MHz, $CDCl_3$): δ 52.1 (dquint, $J_{P–Rh}$ = 143.8 Hz, $J_{P–F}$ = 33.0 Hz); $^{19}F$ NMR (376 MHz, $CDCl_3$): δ −125.9 (4F), −122.5 (4F), −121.7 (4F), −121.5 (12F), −121.4 (4F), −113.7 (4F), −104.8 (m, 4F), −80.7 (6F); IR (KBr) 3067, 2010, 1481, 1439, 1373, 1339, 1246, 1207, 1153, 1099, 745, 691, 648 cm$^{-1}$; HRMS (ESI) Calcd. for $C_{45}H_{20}ClF_{42}NaOP_2Rh$ [M + Na]$^+$:

1596.8960, Found: 1596.8960; Anal. Calcd. for $C_{45}H_{20}ClF_{42}OP_2Rh$: C, 34.09; H, 1.55%, Found: C, 34.32; H, 1.28%.

*trans-carbonylchlorobis(dibutyl(perfluorodecyl)phosphine)rhodium(I)* (**3b**): yellow solid; mp 49–50 °C; $^1H$ NMR (400 MHz, $CDCl_3$): δ 0.95 (t, $J_{H-H}$ = 7.7 Hz, 12H), 1.48 (sextet, $J_{H-H}$ = 7.3 Hz, 8H), 1.61–1.78 (m, 8H), 2.18–2.37 (m, 8H); $^{31}P$ NMR (162 MHz, $CDCl_3$): δ 46.3 (dquint, $J_{P-Rh}$ = 133.3 Hz, $J_{P-F}$ = 30.1 Hz); $^{19}F$ NMR (376 MHz, $CDCl_3$): δ −126.3 (4F), −122.9 (4F), −122.0 (4F), −121.8 (12F), −121.5 (4F), −116.5 (4F), −109.0 (m, 4F), −81.0 (t, $J_{F-F}$ = 11.4 Hz, 6F); IR (KBr) 2960, 1987, 1372, 1210, 1151, 1111, 973, 852 cm$^{-1}$; HRMS (FAB) Calcd. for $C_{37}H_{36}ClF_{42}NaOP_2Rh$ [M + Na]$^+$: 1517.0212, Found: 1517.0245. The copies of $^1H$ NMR, $^{19}F$ NMR, and $^{31}P$ NMR spectra of rhodium complex (**3a,b**) are shown in Supplementary Materials.

RhCl(CO)[P($C_6F_5$)$_3$]$_2$ [9], RhCl(CO)[P(4-$CF_3C_6H_4$)$_3$]$_2$ [25], and RhCl(CO)[P($C_6H_5$)$_3$]$_2$ [9] were reported in the literature, respectively.

### 3.4. Synthesis of P-Perfluoroalkylated Phosphine Complex with Ir(I)

IrCl(CO)(Ph$_2$P$^n$C$_{10}$F$_{21}$)$_2$ was successfully synthesized according to the reported method [28,29]. Under inert atmosphere, Ph$_2$P$^n$C$_{10}$F$_{21}$ (1.76 g, 2.5 mmol), IrCl$_3$·3H$_2$O (176 mg, 0.5 mmol), and DMF (7.5 mL) were added to a 30 mL two necked flask. The mixture was refluxed for 12 h and the hot solution was filtered. The filtrate was recrystallized from MeOH and *trans*-IrCl(CO)(Ph$_2$P$^n$C$_{10}$F$_{21}$)$_2$ was obtained in 64% yield.

*trans-carbonylchlorobis((perfluorodecyl)diphenylphosphine)iridium(I)* (**4a**): yellow solid; mp 181–183 °C; $^1H$ NMR (400 MHz, $CDCl_3$): δ 7.44–7.56 (m, 12H), 7.93 (dd, $J_{H-H}$ = 6.4, $J_{P-H}$ = 6.4 Hz, 8H); $^{31}P$ NMR (162 MHz, $CDCl_3$): δ 45.0 (t, $J_{P-F}$ = 34.6 Hz); $^{19}F$ NMR (376 MHz, $CDCl_3$): δ −126.0 (4F), −122.6 (4F), −121.8 (8F), −121.6 (8F), −121.2 (4F), −113.9 (4F), −104.5 (d, $J_{F-P}$ = 34.6 Hz, 4F), −80.7 (6F); IR (KBr) 1990, 1242, 1207, 1153 cm$^{-1}$; HRMS (ESI) Calcd. for $C_{45}H_{20}ClF_{42}OP_2Ir$ [M]$^+$: 1663.9636, Found: 1663.9631. The copies of $^1H$ NMR, $^{19}F$ NMR, and $^{31}P$ NMR spectra of iridium complex (**4a**) are shown in Supplementary Materials.

### 3.5. Hydrosilylation Reaction Catalyzed by Iridium Complex (4a)

Under inert atmosphere, **4a** (5.0 mg, 0.003 mmol), Et$_3$SiH (34.9 mg, 0.3 mmol), and 1-octyne (66.1 mg, 0.6 mmol) were added to a sealed test tube. The solution was heated at 80 °C for 16 h. The production of hydrosilylation adducts [36] was confirmed by $^1H$ NMR using 1,4-dioxane as internal standard.

### 3.6. Synthesis of P-Perfluoroalkylated Phosphine Complex with Au(I)

Under inert atmosphere, Ph$_2$P$^n$C$_{10}$F$_{21}$ (145 mg, 0.2 mmol) or $^n$Bu$_2$P$^n$C$_{10}$F$_{21}$ (132.8 mg, 0.2 mmol), AuCl(SMe$_2$) (59 mg, 0.2 mmol), and CH$_2$Cl$_2$ (5 mL) were added to a sealed test tube. The solution was stirred for 1 h. After the solvent was removed in vacuo, solid was obtained as enough pure AuCl(Ph$_2$P$^n$C$_{10}$F$_{21}$) in 98% yield or AuCl($^n$Bu$_2$P$^n$C$_{10}$F$_{21}$) in 99% yield.

*Chloro((perfluorodecyl)diphenylphosphine)gold(I)* (**5a**): white solid; mp 130–131 °C; $^1H$ NMR (400 MHz, $CDCl_3$): δ 7.56–7.62 (m, 4H), 7.66–7.72 (m, 2H), 7.95 (dd, $J_{H-H}$ = 7.8, $J_{P-H}$ = 13.7 Hz, 4H); $^{31}P$ NMR (162 MHz, $CH_2Cl_2$): δ 40.1 (tt, $^2J_{P-F}$ = 65.0 Hz, $^3J_{P-F}$ = 12.9 Hz); $^{19}F$ NMR (376 MHz, $CDCl_3$): δ −126.0 (2F), −122.6 (2F), −121.8 (2F), −121.6 (6F), −121.2 (2F), −114.8 (2F), −104.5 (dt, $J_{F-P}$ = 64.1 Hz, $J_{F-F}$ = 14.2 Hz, 2F), −80.7 (t, $J_{F-F}$ =10.1 Hz, 3F); HRMS (FAB) Calcd. for $C_{22}H_{10}F_{21}PAu$ [M − Cl]$^+$: 900.9850, Found: 900.9848.

*Chloro(dibutyl(perfluorodecyl)phosphine)gold(I)* (**5b**): reddish purple solid; mp 62–63 °C; $^1H$ NMR (400 MHz, $CDCl_3$): δ 1.52 (sextet, $J_{H-H}$ = 7.3 Hz, 4H), 1.59–1.78 (m, 4H), 1.98−2.21 (m, 4H); $^{31}P$ NMR (162 MHz, $CH_2Cl_2$): δ 40.1 (t, $J_{P-F}$ = 65.0 Hz); $^{19}F$ NMR (376 MHz, $CDCl_3$): δ −126.0 (2F), −122.6 (2F), −121.7 (2F), −121.5 (6F), −121.1 (2F), −116.5 (2F), −110.9 (dt, $J_{F-P}$ = 62.6 Hz, $J_{F-F}$ = 17.1 Hz, 2F), −80.6 (3F); HRMS (FAB) Calcd. for $C_{18}H_{18}AuClF_{21}NaP$ [M + Na]$^+$: 919.0063,

Found: 919.0031. The copies of $^1$H NMR, $^{19}$F NMR, and $^{31}$P NMR spectra of gold complex (**5a,b**) are shown in Supplementary Materials.

### 3.7. Au Complex-Catalyzed Addition Reaction of 2-Chloroethanol to 1-Octene

The addition reaction was conducted according to the reported method [35]. Under inert atmosphere, **5** (0.03 mmol), AgOTf (7.7 mg, 0.03 mmol), 2-chloroethanol (8.0 mg, 1 mmol), 1-octene (561.2 mg, 5 mmol), and toluene (1 mL) were added to a three-necked flask with a condenser. The solution was heated at 85 °C for 24 h.

## 4. Conclusions

We demonstrated the synthesis of *P*-fluorous phosphine–coordinated metal complexes, $PtCl_2(R_2P^nC_{10}F_{21})_2$, $RhCl(CO)(R_2P^nC_{10}F_{21})_2$, $IrCl(CO)(Ph_2P^nC_{10}F_{21})_2$, and $AuCl(R_2P^nC_{10}F_{21})$. The structure of $PtCl_2(Ph_2P^nC_{10}F_{21})_2$ was revealed and discussed. The catalytic activities of the iridium and gold complexes for some synthetic reactions were also shown. The promising *P*-fluorous phosphine–transition metal complexes will be used to catalyze novel reactions in future work.

**Acknowledgments:** This research was supported by a Grant-in-Aid for Exploratory Research (26620149, Akiya Ogawa, 26860168, Shin-ichi Kawaguchi), from the Ministry of Education, Culture, Sports, Science and Technology, Japan.

**Author Contributions:** Shin-ichi Kawaguchi and Akiya Ogawa conceived and designed the experiments; Shin-ichi Kawaguchi, Yuta Saga, Yuki Sato and Yoshiaki Minamida performed the experiments; Akihiro Nomoto analyzed the data; Shin-ichi Kawaguchi and Akiya Ogawa wrote the paper.

## References

1. Banger, K.K.; Brisdon, A.K.; Herbert, C.J.; Ghaba, H.A.; Tidmarsh, I.S. Fluoroalkenyl, fluoroalkynyl and fluoroalkyl phosphines. *J. Fluor. Chem.* **2009**, *130*, 1117–1129. [CrossRef]

2. Gládysz, J.A.; Curran, D.P.; Horváth, I.T. *Handbook of Fluorous Chemistry*; Wiley-VCH: Weinheim, Germany, 2004.

3. Betzemeier, B.; Knochel, P. Palladium-catalyzed cross-coupling of organozinc bromides with aryl iodides in perfluorinated solvents. *Angew. Chem. Int. Ed.* **1997**, *36*, 2623–2624. [CrossRef]

4. Horváth, I.T. Fluorous biphase chemistry. *Acc. Chem. Res.* **1998**, *31*, 641–650. [CrossRef]

5. Rábai, J.; Szabó, D.; Borbás, E.K.; Kövesi, I.; Kövsedi, I.; Csámpai, A.; Gömöry, A.; Pashinnik, V.E.; Shermolovich, Y.G. Practice of fluorous biphase chemistry: Convenient synthesis of novel fluorophilic ethers via a mitsunobu reaction. *J. Fluor. Chem.* **2002**, *114*, 199–207. [CrossRef]

6. Moineau, J.; Pozzi, G.; Quici, S.; Sinou, D. Palladium-catalyzed Heck reaction in perfluorinated solvents. *Tetrahedron Lett.* **1999**, *40*, 7683–7686. [CrossRef]

7. Tzschucke, C.C.; Markert, C.; Glatz, H.; Bannwarth, W. Fluorous biphasic catalysis without perfluorinated solvents: Application to Pd-mediated Suzuki and Sonogashira couplings. *Angew. Chem. Int. Ed.* **2002**, *41*, 4500–4503. [CrossRef]

8. Horváth, I.T.; Rábai, J. Facile catalyst separation without water—Fluorous biphase hydroformylation of olefins. *Science* **1994**, *266*, 72–75. [CrossRef] [PubMed]

9. Korenaga, T.; Ko, A.; Uotani, K.; Tanaka, Y.; Sakai, T. Synthesis and application of 2,6-bis(trifluoromethyl)-4-pyridyl phosphanes: The most electron-poor aryl phosphanes with moderate bulkiness. *Angew. Chem. Int. Ed.* **2011**, *50*, 10703–10707. [CrossRef] [PubMed]

10. Korenaga, T.; Abe, K.; Ko, A.; Maenishi, R.; Sakai, T. Ligand electronic effect on reductive elimination of biphenyl from *cis*-[Pt(Ph)$_2$(diphosphine)] complexes bearing electron-poor diphosphine: Correlation study between experimental and theoretical results. *Organometallics* **2010**, *29*, 4025–4035. [CrossRef]

11.  Korenaga, T.; Osaki, K.; Maenishi, R.; Sakai, T. Electron-poor chiral diphosphine ligands: High performance for Rh-catalyzed asymmetric 1,4-addition of arylboronic acids at room temperature. *Org. Lett.* **2009**, *11*, 2325–2328. [CrossRef] [PubMed]

12.  Tian, P.; Dong, H.-Q.; Lin, G.-Q. Rhodium-catalyzed asymmetric arylation. *ACS Catal.* **2012**, *2*, 95–119. [CrossRef]

13.  Kawaguchi, S-i.; Minamida, Y.; Okuda, T.; Sato, Y.; Saeki, T.; Yoshimura, A.; Nomoto, A.; Ogawa, A. Photoinduced synthesis of *P*-perfluoroalkylated phosphines from triarylphosphines and their application in the copper-free cross-coupling of acid chlorides and terminal alkynes. *Adv. Synth. Catal.* **2015**, *357*, 2509–2519. [CrossRef]

14.  Kawaguchi, S-i.; Minamida, Y.; Ohe, T.; Nomoto, A.; Sonoda, M.; Ogawa, A. Synthesis and properties of perfluoroalkyl phosphine ligands: Photoinduced reaction of diphosphines with perfluoroalkyl iodides. *Angew. Chem. Int. Ed.* **2013**, *52*, 1748–1752. [CrossRef] [PubMed]

15.  Sato, Y.; Kawaguchi, S-i.; Ogawa, A. Photoinduced reductive perfluoroalkylation of phosphine oxides: Synthesis of *P*-perfluoroalkylated phosphines using TMDPO and perfluoroalkyl iodides. *Chem. Commun.* **2015**, *51*, 10385–10388. [CrossRef] [PubMed]

16.  Schnabel, R.C.; Roddick, D.M. (Fluoroalkyl)phosphine complexes of rhodium and iridium—Synthesis and reactivity properties of [(dfepe)Ir($\mu$-Cl)]$_2$. *Inorg. Chem.* **1993**, *32*, 1513–1518. [CrossRef]

17.  Banger, K.K.; Banham, R.P.; Brisdon, A.K.; Cross, W.I.; Damant, G.; Parsons, S.; Pritchard, R.G.; Sousa-Pedrares, A. Synthesis and coordination chemistry of perfluorovinyl phosphine derivatives. Single crystal structures of PPh(CF=CF$_2$)$_2$, *cis*-[PtCl$_2$\{PPh$_2$(CF=CF$_2$)\}$_2$] and [\{AuCl[PPh$_2$(CF=CF$_2$)]\}$_2$]. *J. Chem. Soc. Dalton Trans.* **1999**, 427–434. [CrossRef]

18.  Palcic, J.D.; Kapoor, P.N.; Roddick, D.M.; Peters, R.G. Perfluoroalkylphosphine coordination chemistry of platinum: Synthesis of (C$_2$F$_5$)$_2$PPh and (C$_2$F$_5$)PPh$_2$ complexes of platinum(II). *Dalton Trans.* **2004**, 1644–1647. [CrossRef] [PubMed]

19.  Lewis-Alleyne, L.C.; Murphy-Jolly, M.B.; Le Goff, X.F.; Caffyn, A.J.M. Synthesis and complexation of heptafluoroisopropyldiphenylphosphine. *Dalton Trans.* **2010**, *39*, 1198–1200. [CrossRef] [PubMed]

20.  Barnes, N.A.; Brisdon, A.K.; William Brown, F.R.; Cross, W.I.; Crossley, I.R.; Fish, C.; Herbert, C.J.; Pritchard, R.G.; Warren, J.E. Synthesis of gold(I) fluoroalkyl and fluoroalkenyl-substituted phosphine complexes and factors affecting their crystal packing. *Dalton Trans.* **2011**, *40*, 1743–1750. [CrossRef] [PubMed]

21.  Johansson, M.H.; Otto, S. *trans*-Dichlorobis(triphenylphosphine-*P*)platinum(II). *Acta Crystallogr. Sect. C* **2000**, *56*, e12–e15. [CrossRef]

22.  Clarke, M.L.; Ellis, D.; Mason, K.L.; Orpen, A.G.; Pringle, P.G.; Wingad, R.L.; Zaher, D.A.; Baker, R.T. The electron-poor phosphines P\{C$_6$H$_3$(CF$_3$)$_2$-3,5\}$_3$ and P(C$_6$F$_5$)$_3$ do not mimic phosphites as ligands for hydroformylation. A comparison of the coordination chemistry of P\{C$_6$H$_3$(CF$_3$)$_2$-3,5\}$_3$ and P(C$_6$F$_5$)$_3$ and the unexpectedly low hydroformylation activity of their rhodium complexes. *Dalton Trans.* **2005**, 1294–1300. [CrossRef]

23.  Yajima, T.; Tabuchi, E.; Nogami, E.; Yamagishi, A.; Sato, H. Perfluorinated gelators for solidifying fluorous solvents: Effects of chain length and molecular chirality. *RSC Adv.* **2015**, *5*, 80542–80547. [CrossRef]

24.  Skalická, V.; Rybáčková, M.; Skalický, M.; Kvíčalová, M.; Cvačka, J.; Březinová, A.; Čejka, J.; Kvíčala, J. Polyfluoroalkylated tripyrazolylmethane ligands: Synthesis and complexes. *J. Fluor. Chem.* **2011**, *132*, 434–440. [CrossRef]

25.  Haar, C.M.; Huang, J.; Nolan, S.P.; Petersen, J.L. Synthetic, thermochemical, and catalytic studies involving novel R$_2$P(OR$_f$) [R = alkyl or aryl; R$_f$ = CH$_2$CH$_2$(CF$_2$)$_5$CF$_3$] ligands. *Organometallics* **1998**, *17*, 5018–5024. [CrossRef]

26.  Roodt, A.; Otto, S.; Steyl, G. Structure and solution behaviour of rhodium(I) Vaska-type complexes for correlation of steric and electronic properties of tertiary phosphine ligands. *Coord. Chem. Rev.* **2003**, *245*, 121–137. [CrossRef]

27.  Vaska, L.; DiLuzio, J.W. Carbonyl and hydrido-carbonyl complexes of iridium by reaction with alcohols. Hydrido complexes by reaction with acid. *J. Am. Chem. Soc.* **1961**, *83*, 2784–2785. [CrossRef]

28.  Rappoli, B.J.; Churchill, M.R.; Janik, T.S.; Rees, W.M.; Atwood, J.D. Crystal structure of the quasitetrahedral iridium(I) complex, Ir(COCH$_2$CMe$_3$)[P(*p*-tolyl)$_3$]$_2$[C$_2$(CO$_2$Me)$_2$]. An intermediate in cyclotrimerization of activated alkynes by 16-electron alkyl complexes of iridium, *trans*-RIr(CO)L$_2$ [R = Me, CH$_2$CMe$_3$; L = PPh$_3$, P(*p*-tolyl)$_3$]. *J. Am. Chem. Soc.* **1987**, *109*, 5145–5149.

29. Vrieze, K.; Collman, J.P.; Sears, C.T.; Kubota, M.; Davison, A.; Shawl, E.T. *Trans*-chlorocarbonylbis (tri-phenylphosphine)iridium. In *Inorganic Synthesis*; John Wiley & Sons, Inc.: Hoboken, NJ, USA, 1968; pp. 101–104.

30. Tanke, R.S.; Crabtree, R.H. Unusual activity and selectivity in alkyne hydrosilylation with an iridium catalyst stabilized by an oxygen-donor ligand. *J. Am. Chem. Soc.* **1990**, *112*, 7984–7989. [CrossRef]

31. Apple, D.C.; Brady, K.A.; Chance, J.M.; Heard, N.E.; Nile, T.A. Iridium complexes as hydrosilylation catalysts. *J. Mol. Catal.* **1985**, *29*, 55–64. [CrossRef]

32. Hesp, K.D.; Wechsler, D.; Cipot, J.; Myers, A.; McDonald, R.; Ferguson, M.J.; Schatte, G.; Stradiotto, M. Exploring the utility of neutral rhodium and iridium $\kappa^2$-*P,O* and $\kappa^2$-*P(S),O* complexes as catalysts for alkene hydrogenation and hydrosilylation. *Organometallics* **2007**, *26*, 5430–5437. [CrossRef]

33. Murai, T.; Nagaya, E.; Shibahara, F.; Maruyama, T.; Nakazawa, H. Rhodium(I) and iridium(I) imidazo[1,5-*a*]pyridine-1-ylalkylalkoxy complexes: Synthesis, characterization and application as catalysts for hydrosilylation of alkynes. *J. Organomet. Chem.* **2015**, *794*, 76–80. [CrossRef]

34. Shin, S. Tris-(pentafluorophenyl)phosphine gold(I) complexes as new highly efficient catalysts for the oxycarbonylation of homopropargyl carbonates. *Bull. Korean Chem. Soc.* **2005**, *26*, 1925–1926. [CrossRef]

35. Hirai, T.; Hamasaki, A.; Nakamura, A.; Tokunaga, M. Enhancement of reaction efficiency by functionalized alcohols on gold(I)-catalyzed intermolecular hydroalkoxylation of unactivated olefins. *Org. Lett.* **2009**, *11*, 5510–5513. [CrossRef]

36. Takeuchi, R.; Tanouchi, N. Solvent-controlled stereoselectivity in the hydrosilylation of alk-1-ynes catalysed by rhodium complexes. *J. Chem. Soc. Perkin Trans. 1* **1994**, 2909–2913. [CrossRef]

# 11

# Synthesis, Crystal Structure, Polymorphism, and Magnetism of $Eu(CN_3H_4)_2$ and First Evidence of $EuC(NH)_3$

**Arno L. Görne** [1], **Janine George** [1], **Jan van Leusen** [1] and **Richard Dronskowski** [1,2,*]

[1] Institute of Inorganic Chemistry, RWTH Aachen University, Landoltweg 1, 52056 Aachen, Germany; arno.goerne@ac.rwth-aachen.de (A.L.G.); janine.george@ac.rwth-aachen.de (J.G.); jan.vanleusen@ac.rwth-aachen.de (J.v.L.)

[2] Jülich-Aachen Research Alliance, JARA-HPC, RWTH Aachen University, 52056 Aachen, Germany

* Correspondence: drons@HAL9000.ac.rwth-aachen.de

Academic Editor: Steve Liddle

**Abstract:** We report the first magnetically coupled guanidinate, $\alpha$-$Eu(CN_3H_4)_2$ (monoclinic, $P2_1$, $a$ = 5.8494(3) Å, $b$ = 14.0007(8) Å, $c$ = 8.4887(4) Å, $\beta$ = 91.075(6)°, $V$ = 695.07(6) Å$^3$, $Z$ = 4). Its synthesis, polymorphism, crystal structure, and properties are complemented and supported by density-functional theory (DFT) calculations. The $\alpha$-, $\beta$- and $\gamma$-polymorphs of $Eu(CN_3H_4)_2$ differ in powder XRD, while the $\gamma$-phase transforms into the $\beta$-form over time. In $\alpha$-$Eu(CN_3H_4)_2$, Eu is octahedrally coordinated and sits in one-dimensional chains; the guanidinate anions show a hydrogen-bonding network. The different guanidinate anions are theoretically predicted to adopt *syn*-, *anti*- and all-*trans*-conformations. Magnetic measurements evidence ferromagnetic interactions, presumably along the Eu chains. Finally, $EuC(NH)_3$ (isostructural to $SrC(NH)_3$ and $YbC(NH)_3$, hexagonal, $P6_3/m$, $a$ = 5.1634(7) Å, $c$ = 7.1993(9) Å, $V$ = 166.23(4) Å$^3$, $Z$ = 2) is introduced as a possible ferromagnet.

**Keywords:** europium; guanidinate; DFT; IR; liquid ammonia

---

## 1. Introduction

At the beginning of the 21st century, rare-earth metals are critical materials in high-technology applications [1]. Within the recent decades, several technological innovations disrupted the rare-earth market [2], in turn stimulating the scientific quest for future materials. One vibrant field is the study of $Eu^{2+}$ compounds whose complex crystal structures are coupled with application-relevant properties including, to name only some recent examples, luminescence [3–6], field-induced reversal of the magnetoresistive effect [7], and complex magnetism [8,9]. The most renowned magnetic compounds are the europium chalcogenides that are considered ideal 3D Heisenberg systems [10]. While EuO is a ferromagnet with a Curie temperature of 69.3 K [11–13], EuS is also a ferromagnet but with a far lower Curie temperature of 18.7 K, showing weak, secondary antiferromagnetic interactions [13].

Our interest lies in nitrogen-based materials. For $Eu^{2+}$, there are a number of simple amides, thiocyanates, and carbodiimides such as $Eu(NH_2)_2$ [14,15], $Eu(NCS)_2$ [16], and EuNCN [17], but also a growing number of more exotic and intriguing examples including $Eu_2Si_5N_8$ [18,19], $Eu_3[NBN]_2$ [20], $Eu_2Cl_2NCN$ [21], and $EuSi_2O_2N_2$ [22]. Low-dimensional magnetic properties have been reported in $Eu^{2+}$ coordination polymers with 2,2'-bipyridime showing 1D ferromagnetic interactions [23] and in $LiEu_2(NCN)I_3$ and $LiEu_4(NCN)_3I_3$ [24], also with low-dimensional ferromagnetic ordering and possibly conflicting antiferromagnetic interactions at very low temperatures.

Here, we present the first europium guanidinates, inorganic salts derived from the molecule guanidine $CN_3H_5$ [25,26]. Our group has already pioneered the deprotonation of this strongly basic molecule (Figure 1), demonstrated by the preparation of the alkali-metal guanidinates [27–29]. Progressing from there, we recently reported *doubly* deprotonated guanidinates—in $SrC(NH)_3$ and $YbC(NH)_3$—and the first magnetic guanidinate, $Yb(CN_3H_4)_3$, a non-Curie–Weiss paramagnet [30,31]. In the following, we present the first magnetically coupled guanidinate, $\alpha$-$Eu(CN_3H_4)_2$, with probable 1D ferromagnetic order. We detail its synthesis, polymorphism, crystal structure, and properties, complemented and supported by density-functional theory (DFT) calculations. Also, we present a preliminary report on $EuC(NH)_3$ and a first indication of its magnetic properties.

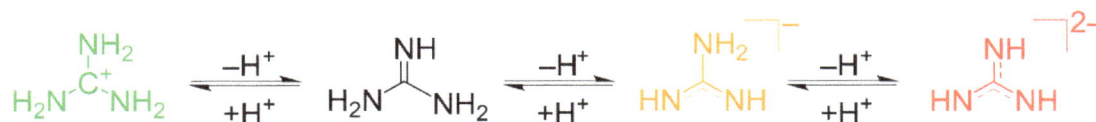

**Figure 1.** Protonation and deprotonation of guanidine from guanidinium on the left to the doubly deprotonated guanidinate on the right.

## 2. Results and Discussion

### 2.1. Polymorphism of Eu(CN3H4)2

Depending on the synthetic conditions, three polymorphs of $Eu(CN_3H_4)_2$ could be prepared, which we call $\alpha$, $\beta$, and $\gamma$. The polymorphs show different powder X-ray diffraction (PXRD) patterns (Figure 2a). For the $\alpha$-phase, the crystal structure was solved (see below). $\alpha$-$Eu(CN_3H_4)_2$ was prepared at temperatures around 65 °C, the $\beta$-phase at a lower 50 °C, and the $\gamma$-phase exclusively around room temperature. Under these conditions, $Eu^{2+}$ is the stable oxidation state, and $Eu^{3+}$ would only form at temperatures starting around 300 °C [32]. The oxidation state was also corroborated by the magnetic measurements (see Section 2.3). Interestingly, the $\gamma$-phase spontaneously transforms to the $\beta$-phase over several weeks (Figure 2b), so $\gamma$-$Eu(CN_3H_4)_2$ must be a metastable phase.

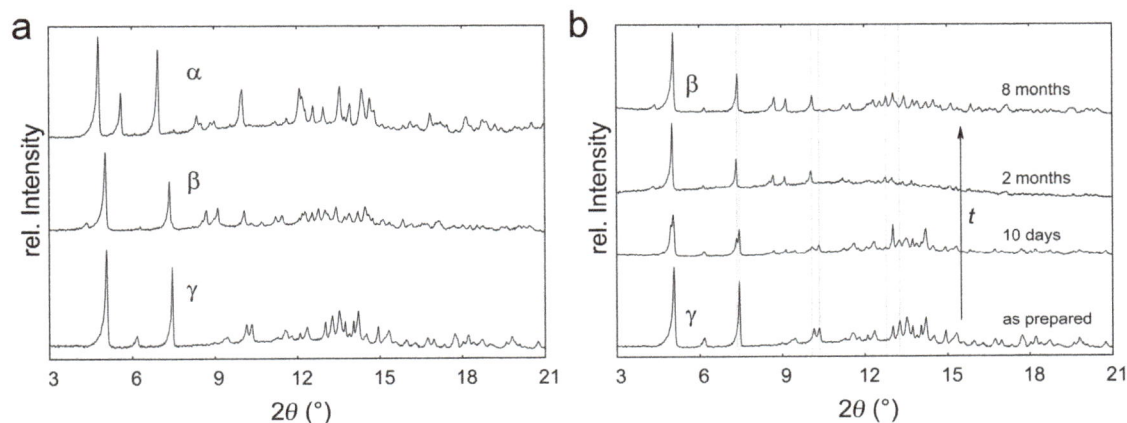

**Figure 2.** Diffraction patterns of the different polymorphs of $Eu(CN_3H_4)_2$ (**a**) and the transformation from the $\gamma$-phase to the $\beta$-phase over time (**b**).

IR measurements indicate a strong similarity between the $\alpha$- and $\beta$-phase (Figure 3), not surprisingly so because the spectrum is dominated by the $CN_3H_4^-$ anion vibrations [31]. In addition, there is no trace of an IR contribution of the $C(NH)_3^{2-}$ unit, which supports the proposed composition. The IR spectrum of $\alpha$-$Eu(CN_3H_4)_2$ was also calculated by DFT from the density of phonon states and the Born effective charges as explained in references [33] and [34]. All observed signals were reproduced, while the differing intensity of the simulated signals could be caused by a thermal effect (calculated 0 K vs experimental 300 K). To identify the vibrations, the IR-active phonons at the $\Gamma$-point were visualized (Table 1).

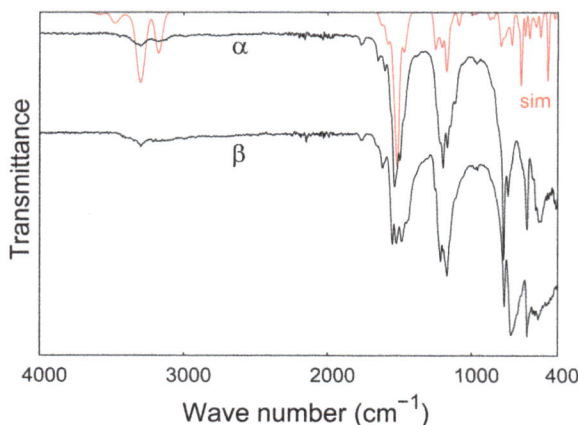

**Figure 3.** IR measurement of the β- (with an offset) and α-polymorphs of $Eu(CN_3H_4)_2$ and simulation of the latter.

**Table 1.** Assignment of vibrational bands of α-$Eu(CN_3H_4)_2$ as obtained with density-functional theory (DFT).

| Vibration | α-$Eu(CN_3H_4)_2$ (cm$^{-1}$) |
|---|---|
| $\nu_{s,\,as}$(N–H) | 3299, 3172 |
| $\delta_{sciss}$(N–H) | 1652, 1602 |
| $\nu_s$, $\delta_{sciss}$(C–N); $\delta_{sciss}$(N–H) | 1533, 1495 |
| $\delta_{rock}$(N–H) | 1197–1167 |
| $\delta_{wagg}$(N–H) | 1147–1114 |
| $\nu_{breath}$(CN$_3$) | 965 |
| $\delta_{twist}$(N–H) | 776 |
| C-inversion by CN$_3$ plane | 746 |
| $\delta_{rock}$(C–N), $\delta_{rock}$(N–H) | 614 |
| $\delta_{sciss}$(C–N), $\delta_{sciss}$(N–H) | 532–514 |

Preliminary thermogravimetric analysis (TGA) measurements of α- and β-$Eu(CN_3H_4)_2$ show a two-step decay. The first step around 155 °C corresponds to the loss of two equivalents of ammonia, typical for guanidinates [30,31,35], to arrive at the hydrogen cyanamide $Eu(NCNH)_2$. This phase has not been prepared before. The second step around 250 °C does not plateau in the measurement range up to 350 °C. This step could be the transformation of europium hydrogen cyanamide to the carbodiimide by releasing $H_2NCN$, as observed, for example, in the transition-metal hydrogen cyanamides of Fe, Co, and Ni [36,37]. EuNCN could not be prepared as a single-phase material in the reported synthesis at 1300 K [17]. Thus, the guanidinates appear as interesting precursor materials for new (hydrogen) cyanamides.

*2.2. Refinement and Crystal Structure of α-Eu(CN₃H₄)₂*

α-$Eu(CN_3H_4)_2$ crystallizes in the acentric, monoclinic space group $P2_1$ with $a = 5.8494(3)$ Å, $b = 14.0007(8)$ Å, $c = 8.4887(4)$ Å, $\beta = 91.075(6)°$, $V = 695.07(6)$ Å$^3$, and $Z = 4$ (Table 2). The asymmetric unit consists of two Eu atoms and four independent guanidinate units. The large number of parameters, the limited number of reflections, and the domination of the X-ray scattering by the heavy Eu atoms required a number of restraints and constraints to obtain a reasonable structure: while the Eu atoms were refined anisotropically, the C and N atoms of each guanidinate unit were constrained to a single $U_{iso}$ value. Also, the C–N bond lengths, N–C–N angles, and CN$_3$ torsion angles were restrained to sensible values (as obtained from similar guanidinates in the literature). The obtained structural model fits the PXRD measurement well (Figure 4). Different C–N bond lengths allowed for a distinction between amine and imine groups, while the assignment was confirmed by the DFT calculations. In addition, we detected a minor side phase of EuO, likely formed during handling of Eu metal under argon. Such an impurity can often be seen in the literature [38]. The Rietveld refinement estimates the amount of EuO as 1.3(2) wt %.

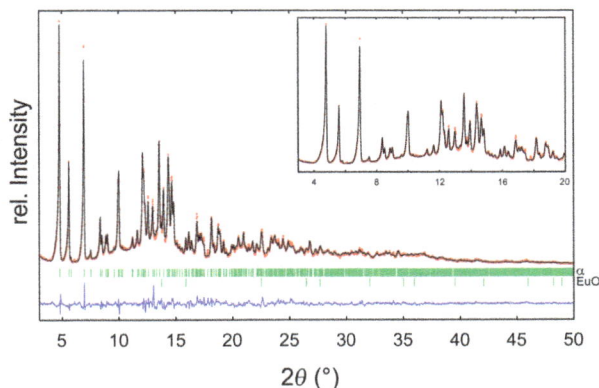

**Figure 4.** Rietveld refinement of $\alpha$-Eu(CN$_3$H$_4$)$_2$ with a minor side phase of EuO measured with Mo $K\alpha_1$ radiation.

**Table 2.** Crystallographic data and refinement details for EuC(NH)$_3$ and $\alpha$-Eu(CN$_3$H$_4$)$_2$.

| Formula | $\alpha$-Eu(CN$_3$H$_4$)$_2$ | EuC(NH)$_3$ |
|---|---|---|
| Formula weight (g·mol$^{-1}$) | 268.09 | 209.02 |
| Crystal system | Monoclinic | Hexagonal |
| Space group | $P2_1$ (Nr. 4) | $P6_3/m$ (Nr. 176) |
| Temperature (K) | 298 | 298 |
| $a$ (Å) | 5.8494(3) | 5.1634(7) |
| $b$ (Å) | 14.0007(8) | $= a$ |
| $c$ (Å) | 8.4887(4) | 7.1993(9) |
| $\beta$ (°) | 91.075(6) | 90 |
| $V$ (Å$^3$) | 695.07(6) | 166.22(4) |
| $Z$ | 4 | 2 |
| Cryst. density (g·cm$^{-3}$) | 2.4848(2) | 4.1176(9) |
| Radiation | Mo $K\alpha_1$ | Mo $K\alpha_1$ |
| No. reflections | 870 | 108 |
| No. restraints/constraints | 24/4 | 0/1 |
| No. refined parameters | 71 | 14 |
| $R_\mathrm{p}$, $wR_\mathrm{p}{}^a$ | 3.7/4.9 | 5.1/7.3 |
| $R_\mathrm{Bragg}$, $R_F{}^b$ | 11.3/6.8 | 18.7/11.7 |

$$^a\ R_\mathrm{p} = \frac{\sum|y(\mathrm{obs})-y(\mathrm{calc})|}{\sum y(\mathrm{obs})} \times 100;\ wR_\mathrm{p} = \sqrt{\frac{\sum w(y(\mathrm{obs})-y(\mathrm{calc}))^2}{\sum wy(\mathrm{obs})^2}} \times 100$$

$$^b\ R_\mathrm{Bragg} = \frac{\sum|I_{obs}-I_{calc}|}{\sum|I_{obs}|} \times 100;\ R_F = \frac{\sum|F_{obs}|-|F_{calc}|}{\sum F_{obs}} \times 100$$

Neutron diffraction experiments—to improve the structural model and to localize the hydrogen atoms—were not feasible owing to the high neutron absorption of Eu. Synchrotron PXRD measurements degraded the sample visibly and led to a color change from yellow to black. The obtained diffractograms were of poor quality, something which was also observed for long measurements of $\beta$- and $\gamma$-Eu(CN$_3$H$_4$)$_2$ with our in-house X-ray diffractometer. For that reason, the crystal structures of $\beta$- and $\gamma$-Eu(CN$_3$H$_4$)$_2$ could not be determined, very unfortunately.

Thus, we used our structural model of $\alpha$-Eu(CN$_3$H$_4$)$_2$ as the starting point for DFT calculations, the ultima ratio in this difficult case. First, we began with full structural optimizations and tests for dynamic stability to evaluate the plausibility of our structural model. To that end, the hydrogen positions were varied such as to find the energetically most stable structure for further phonon calculations. We will come back to the variation of the hydrogen positions below. The resulting structure shows only minor differences from the experimental structure, and it is also dynamically stable (i.e., its density of phonon states does not contain significant imaginary modes). This is no surprise because dispersion-corrected DFT calculations have been shown to be a powerful tool to validate experimental molecular crystal structures [39]. The good agreement between experimental and fully optimized structure by DFT and the dynamic stability strongly support the plausibility of our experimental structure model.

In $\alpha$-Eu(CN$_3$H$_4$)$_2$, the Eu atoms are coordinated 6-fold in distorted octahedra (Figure 5, Table 3). The octahedra are condensed to edge-sharing chains along the $b$-axis with short Eu–Eu distances of 3.66 and 3.77 Å, each bridged by two imine N atoms of different guanidinate units. Another guanidinate unit connects the corners of two octahedra via its N–C–N core, tilting the octahedra towards each other. This motif of two Eu octahedra, corresponding to the asymmetric unit, is repeated in a zigzag fashion along the chain, propagated by the $2_1$ screw axis.

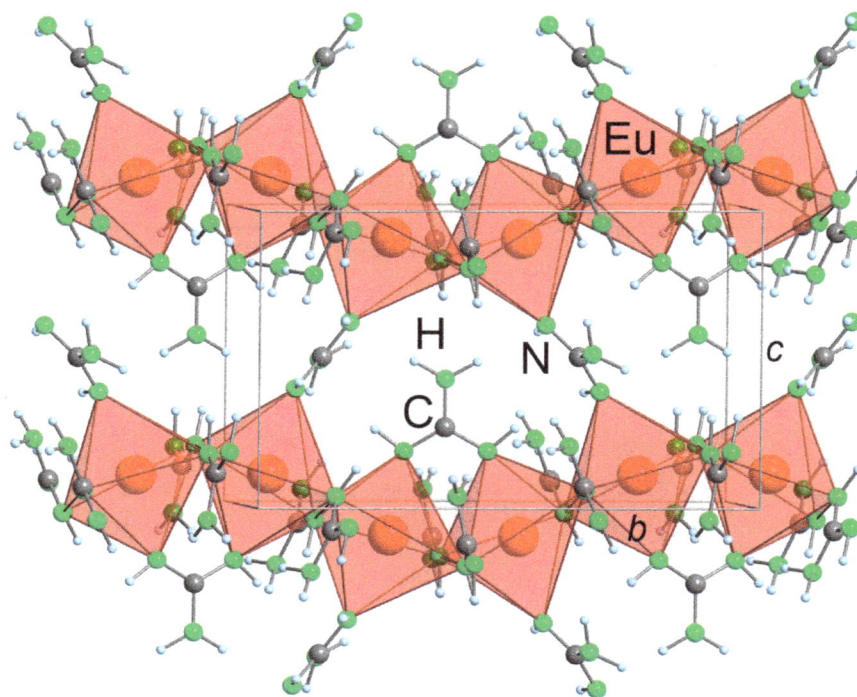

**Figure 5.** Crystal structure of $\alpha$-Eu(CN$_3$H$_4$)$_2$ shown along the $a$-axis as optimized by DFT.

Let us return to the first guanidinate units, bridging the Eu chains by an N imine atom. The very same guanidinate units interconnect the chains along the $a$-axis, this time with the other imine N atom over their N–C–N body. Here, the Eu–Eu distance is equal to the lattice parameter $a = 5.85$ Å. Finally, along the $c$-axis, the chains are connected via the last corner of each Eu octahedron over the N–C–N body of a guanidinate with a Eu–Eu distance of 7.55 Å.

The guanidinate units are connected with each other in a hydrogen-bonding network. While other functional groups have relatively short N–H$\cdots$N contacts, only those from amine to imine groups should be considered to be hydrogen-*bonded* [40].

An unusual feature of $\alpha$-Eu(CN$_3$H$_4$)$_2$ is the conformation of the imine hydrogen atoms of the guanidinate units (Figure 6). In the gas phase, the most stable conformation was calculated as the *syn*-conformation [35], adopted in the solid state in KCN$_3$H$_4$, RbCN$_3$H$_4$, and CsCN$_3$H$_4$ [27,28,41]. The energetically less favorable *anti*-conformation is adopted in LiCN$_3$H$_4$ and NaCN$_3$H$_4$, owing presumably to improved packing and hydrogen-bonding [28,29]. In $\alpha$-Eu(CN$_3$H$_4$)$_2$, both the *syn*- and *anti*-conformation are adopted by guanidinate units. Furthermore, one unit also adopts an all-*trans*-conformation that has never been observed before for a guanidinate. While this conformation is unfavorable in the gas phase by 40 kJ·mol$^{-1}$ [35], it is the predicted conformation for a hypothetical Li$^+$CN$_3$H$_4^-$ *ion pair* in the gas phase [29]. *Substituted* guanidinates can also adopt this conformation [42,43]. In the case of $\alpha$-Eu(CN$_3$H$_4$)$_2$, this conformation is taken by the guanidinate unit connecting the motif of the two tilted Eu octahedra, and this conformational change seemingly allows for a better coordination by the imine groups.

**Table 3.** Atomic positions of $\alpha$-Eu(CN$_3$H$_4$)$_2$ in space group $P2_1$ (all on Wyckoff position $2a$) as determined from DFT.

| Atom | $x$ | $y$ | $z$ |
|------|-----|-----|-----|
| Eu1 | 0.5358 | 0.7086 | 0.0995 |
| Eu2 | 0.5091 | 0.4477 | 0.1043 |
| C1 | 0.4778 | 0.5939 | −0.2622 |
| C2 | 0.0096 | 0.5851 | 0.1186 |
| C3 | 0.9894 | 0.8521 | 0.0723 |
| C4 | 0.4115 | 0.8357 | 0.5014 |
| N1 | 0.4261 | 0.6722 | −0.1801 |
| N2 | 0.5249 | 0.5123 | −0.1822 |
| N3 | 0.4775 | 0.5942 | −0.4249 |
| N4 | 0.2256 | 0.5851 | 0.1783 |
| N5 | −0.1886 | 0.5721 | 0.1963 |
| N6 | −0.0171 | 0.6030 | −0.0391 |
| N7 | 0.8237 | 0.8338 | −0.0357 |
| N8 | 1.2141 | 0.8350 | 0.0569 |
| N9 | 0.9070 | 0.8881 | 0.2132 |
| N10 | 0.5037 | 0.7872 | 0.3822 |
| N11 | 0.5109 | 0.8953 | 0.6061 |
| N12 | 0.1747 | 0.8252 | 0.5170 |
| H1 | 0.3877 | 0.7278 | −0.2540 |
| H2 | 0.5761 | 0.4617 | −0.2613 |
| H3 | 0.4914 | 0.6572 | −0.4856 |
| H4 | 0.5196 | 0.533 | −0.4836 |
| H5 | 0.2196 | 0.5831 | 0.2991 |
| H6 | −0.1546 | 0.5563 | 0.3121 |
| H7 | 0.1267 | 0.6207 | −0.1006 |
| H8 | −0.1558 | 0.5741 | −0.0972 |
| H9 | 0.8935 | 0.7988 | −0.1291 |
| H10 | 1.2891 | 0.8532 | 0.1636 |
| H11 | 0.7685 | 0.9334 | 0.2022 |
| H12 | 1.0256 | 0.9096 | 0.2948 |
| H13 | 0.6727 | 0.8069 | 0.3758 |
| H14 | 0.6814 | 0.8996 | 0.5808 |
| H15 | 0.1090 | 0.7649 | 0.4674 |
| H16 | 0.1149 | 0.8394 | 0.6260 |

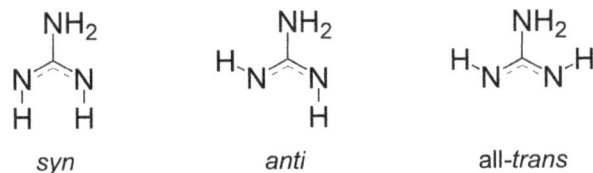

**Figure 6.** Possible conformations for the singly deprotonated guanidinate CN$_3$H$_4^-$.

To further computationally test the calculated hydrogen positions and conformations, we optimized four additional cells with *anti*- or *syn*-conformation, replacing the all-*trans*-conformation. All results were energetically significantly less favorable by at least 12 kJ·mol$^{-1}$ per formula unit of

$Eu(CN_3H_4)_2$. Thus, we consider the DFT prediction to be reliable, but eagerly wait for further experimental corroboration.

## 2.3. Magnetism of α-Eu(CN₃H₄)₂

Magnetic measurements of α-Eu(CN₃H₄)₂ were conducted with different applied magnetic fields (Figure 7). The field-dependence of the effective magnetic moment $\mu_{eff}$ at room temperature reveals a small ferromagnetic impurity: tiny fragments of the steel autoclaves contaminating the sample, described in reference [31]. The maximum at ca. 70 K further reveals traces of EuO within the sample, consistent with the Rietveld analysis. Hence, the data for $T > 20$ K were corrected as detailed in the Methods section and reference [10].

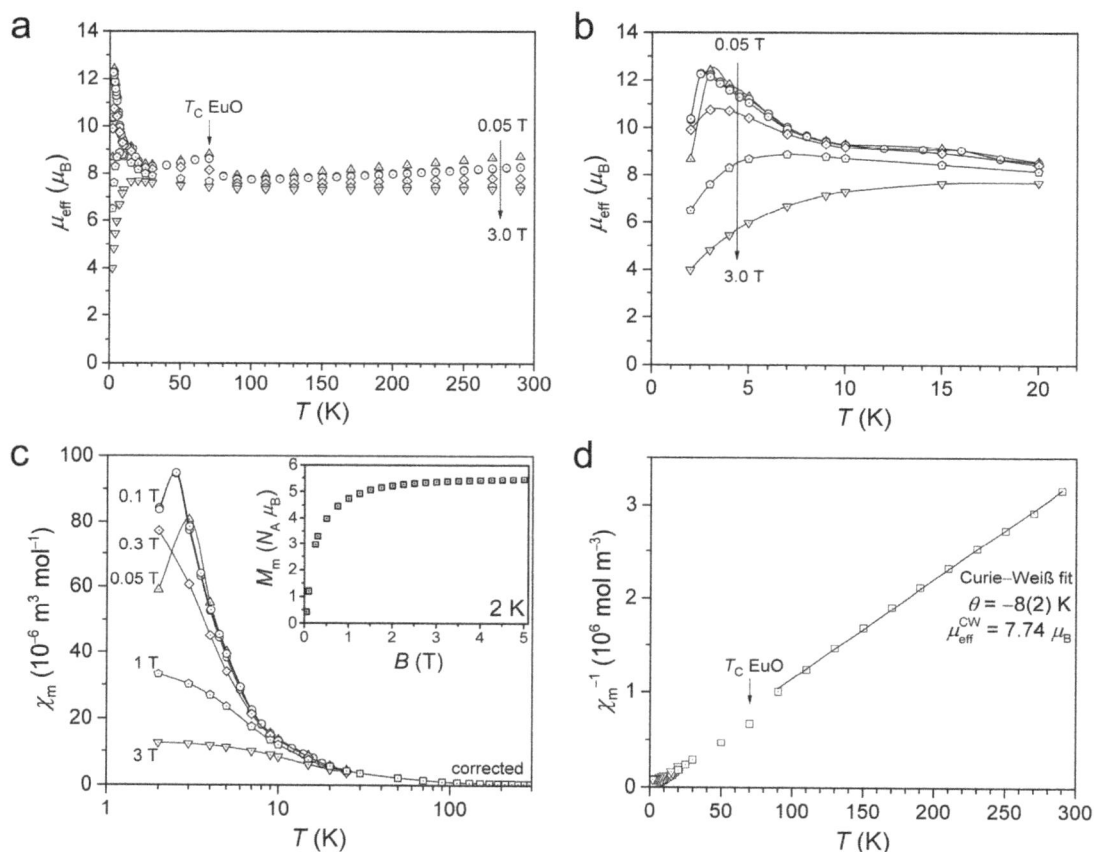

**Figure 7.** Effective magnetic moment (**a**) vs temperature for α-Eu(CN₃H₄)₂ at different applied fields; effective magnetic moment in more detail (**b**) revealing ferromagnetic exchange interactions (lines in (**b**) and (**c**) as guide to the eye only). Molar, corrected magnetic susceptibility (**c**) vs temperature for different applied fields; inset: magnetization versus applied field at 2.0 K. Corrected inverse molar susceptibility (**d**) vs temperature, and fit to the Curie–Weiss law.

At room temperature, the $\mu_{eff}$ value of noninteracting $Eu^{2+}$ ($4f^7$, $g_J = 2$, $J = 7/2$) ions is expected to be close to 7.91 $\mu_B$, lowered by noticeable spin-orbit coupling contributions from the spin-only value of 7.94 $\mu_B$ [44,45]. A Curie–Weiss fit of the corrected data ($T > 90$ K) gives an effective magnetic moment of 7.74 $\mu_B$ and a slightly negative Curie temperature with −8(2) K. Although both values imply antiferromagnetic exchange interactions between the $Eu^{2+}$ ions, they should be looked at as artifacts of the correction method since exchange interactions between lanthanide centers are usually very small ($-2J < 2$ cm$^{-1}$).

The field-dependent molar magnetization $M_m$ curve at 2.0 K hints at a saturation value of $M_{m,sat} \leq 6\,N_A\,\mu_B$, significantly lower than the expected saturation value of $7.0\,N_A\,\mu_B$ for noninteracting $Eu^{2+}$ ($4f^7$) ions. The ratio of these saturation magnetizations is different from the squared ratio of the effective moments $\mu_{eff}$ at room temperature; that is, there is not a common factor that could scale both values to reach the expectation. Thus, we can conclude the existence of exchange interactions within the compound, while their nature is ambiguous. The field-dependent maxima in the $\mu_{eff}$ vs $T$ data in the temperature range 2–6 K indicate ferromagnetic exchange interactions.

It should be noted that $Eu^{3+}$ has a distinctly different magnetic behavior [10]: its magnetic susceptibility as a function of temperature is almost constant and hence does *not* exhibit Curie–Weiss behavior. Its high-temperature $\mu_{eff}$ value is expected to be only 3.5 $\mu_B$, while $\mu_{eff}$ shows a strong temperature dependence. This is all in stark contrast to the magnetic measurements of $\alpha$-Eu(CN$_3$H$_4$)$_2$, thus conclusively showing that Eu is in the oxidation state +2.

In summary, the low-temperature data indicate ferromagnetic exchange interactions, most likely due to one-dimensional, weak interactions along the $Eu^{2+}$ chains of the crystal structure. Different, minor antiferromagnetic exchange pathways may additionally characterize the compound as indicated by the negative Curie temperature; they are, however, subject to speculation due to the uncertainties arising from the necessary correction for ferromagnetic impurities at $T > 20$ K.

### 2.4. Introduction of EuC(NH)$_3$

Finally, we want to give a preliminary account of EuC(NH)$_3$. This compound is isostructural to SrC(NH)$_3$ [30] and YbC(NH)$_3$ [31] and crystallizes in the hexagonal space group $P6_3/m$ with $a = 5.1634(7)$ Å, $c = 7.1993(9)$ Å, $V = 166.23(4)$ Å$^3$, and $Z = 2$ (Figure 8; Table 2). As for YbC(NH)$_3$, DFT calculations were used to locate the hydrogen atoms, a method validated in reference [46] (Table 4). So far, EuC(NH)$_3$ was only obtained together with an unidentified side phase.

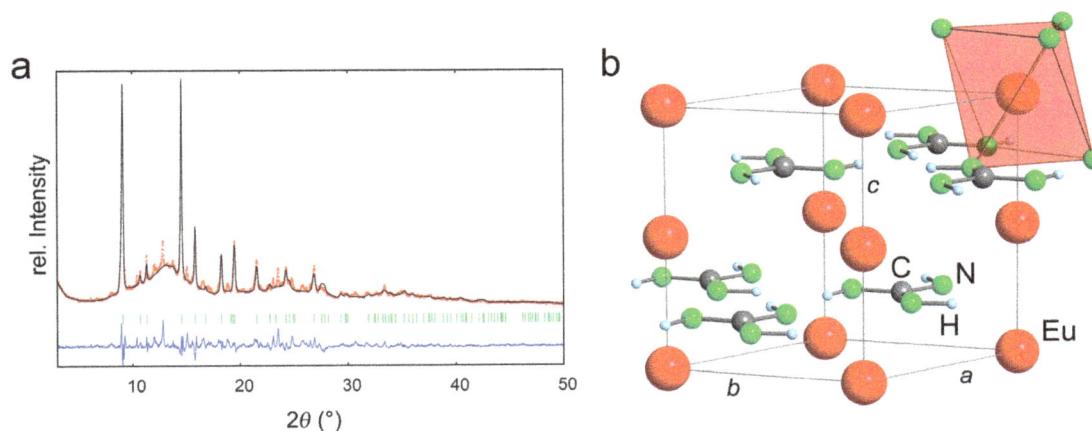

**Figure 8.** Rietveld refinement of EuC(NH)$_3$ (**a**), showing several unidentified reflections. The crystal structure of EuC(NH)$_3$ (**b**).

**Table 4.** Atomic positions and displacement parameters of EuC(NH)$_3$ in space group $P6_3/m$. Hydrogen position determined from DFT.

| Atom | Wyckoff position | $x$ | $y$ | $z$ | $U_{iso}$ or $U_{eq}$ (Å$^2$) |
|---|---|---|---|---|---|
| Eu | 2$b$ | 0 | 0 | 0 | 0.058(1) |
| C | 2$c$ | 1/3 | 2/3 | 1/4 | 0.005(7) |
| N | 6$h$ | 0.065(4) | 0.422(3) | 1/4 | 0.005(7) |
| H (DFT) | 6$h$ | −0.092 | 0.486 | 1/4 | – |

Compared to both $SrC(NH)_3$ [30] and $YbC(NH)_3$ [31], $EuC(NH)_3$ shows a shorter $a$- and a longer $c$-axis, while the volume falls in-between the two. The C–N bonds are found to be somewhat short at 1.328(1) Å, close to a double bond although the bond order should be $1^1/_3$. As for the isostructural compounds, no hydrogen-bonding is expected.

The first magnetic measurements evidence ferromagnetic exchange interactions at low temperatures indicated by the occurrence of maxima in the $\mu_{\rm eff}$ vs $T$ curve (Figure 9), but without phase-pure samples, this result is only tentative. In particular, the $\mu_{\rm eff}$ vs $T$ curve for $T > 25$ K reveals, as for $Eu(CN_3H_4)_2$, ferromagnetic impurities that can be assigned to EuO and autoclave material. Furthermore, the unidentified side phase could be magnetic and contribute to the measured susceptibility.

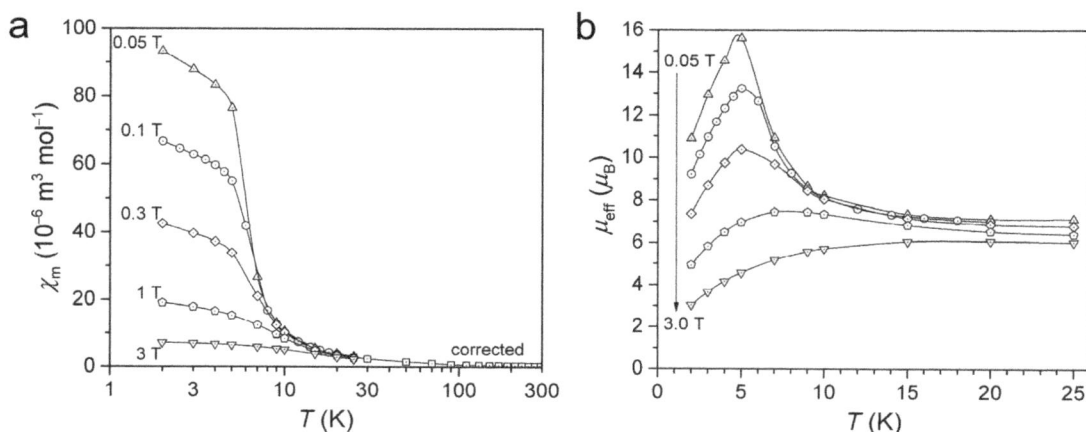

**Figure 9.** Magnetic susceptibility of $EuC(NH)_3$ vs temperature (**a**) and effective magnetic moment vs temperature (**b**), assuming molar weight of a pure $EuC(NH)_3$ sample (lines to guide the eye only).

## 3. Materials and Methods

### 3.1. Syntheses

The highly moisture-sensitive compounds were handled in an argon-filled glove box (MBRAUN, Garching, Germany) to prevent degradation. Reactants were used as obtained from the manufacturers mentioned below.

Guanidine $CN_3H_5$ was prepared in a one-pot synthesis in liquid ammonia in steel autoclaves as described in [31]. The autoclaves were constructed from stainless steel 1.4571 and a copper ring as a sealing gasket with a reaction volume of about 75 $cm^3$. A detailed description of the autoclaves can be found in [47].

$Eu(CN_3H_4)_2$ crystallizes in three different polymorphs, depending on the reaction temperature, as differentiated by PXRD. In all cases, stoichiometric reactants (0.2–1 mmol of Eu metal; Smart Elements, Vienna, Austria, 99.99%) were weighed in steel autoclaves and 15 $cm^3$ of dried, *solid* ammonia (Linde, Pullach, Germany, 99.999%, without further purification) were added. For the $\alpha$-polymorph, compounds of highest crystallinity were obtained by heating for 5–10 days to 65 °C. $\beta$-$Eu(CN_3H_4)_2$ could be prepared by shorter reaction times of 2–5 days at only 50 °C, but higher quality and crystallinity could be obtained after the $\gamma$-polymorph converted to $\beta$-$Eu(CN_3H_4)_2$ (see below). $\gamma$-$Eu(CN_3H_4)_2$ was synthesized by reacting for 4–8 days at room temperature, sometimes yielding almost amorphous samples. During storage under argon, $\gamma$-$Eu(CN_3H_4)_2$ converted into $\beta$-$Eu(CN_3H_4)_2$ over the course of weeks, as evidenced from PXRD. All products showed traces of EuO (about 1 wt % from PXRD), most likely formed during handling of Eu metal in the glove box. Yields were 70%–80%. All $Eu(CN_3H_4)_2$ compounds were of a bright-yellow color.

$EuC(NH)_3$ was obtained from stoichiometric reactants (0.3–1 mmol of Eu) in steel autoclaves with 5–20 $cm^3$ of dried, solid ammonia. Reaction times were 4–14 days at 50–70 °C to yield 60%–80% of an orange powder. In some cases, however, only amorphous samples were obtained or $EuC(NH)_3$ was a product when $Eu(CN_3H_4)_2$ was targeted at suboptimal reaction conditions. Unfortunately, no phase-pure products could be achieved, but only mixtures with an unidentified side-phase (volume fraction estimated from PXRD 10%–20%).

### 3.2. Powder X-Ray Diffraction

For PXRD, the samples were sealed in 0.3 mm glass capillaries and measured with a STADI MP diffractometer (STOE, Darmstadt, Germany) with monochromatic Mo $K\alpha_1$ radiation and a Mythen detector. The measurement ranges were 3°–75° in $2\theta$ with a step size of 0.015° for both $EuC(NH)_3$ and $\alpha$-$Eu(CN_3H_4)_2$ and limited scans of 3°–21° in $2\theta$ for $\gamma$- and $\beta$-$Eu(CN_3H_4)_2$. High-resolution synchrotron powder-diffraction data were collected using beamline 11-BM at the Advanced Photon Source (APS), Argonne National Laboratory using an average wavelength of 0.414170 Å. During the measurement, darkening of the samples was observed and the diffractograms were different from those obtained from our in-house diffractometer. Apparently, the samples decomposed under X-ray radiation, even more so when exposed to intense synchrotron radiation. For $\gamma$- and $\beta$-$Eu(CN_3H_4)_2$, similar loss in crystallinity was observed for long measurements with our in-house diffractometer, so limited scans were used for identification purposes.

The crystal structure of $\alpha$-$Eu(CN_3H_4)_2$ was solved by charge-flipping with SUPERFLIP [48] as implemented in the Jana2006 suite [49] and further refined with the suite. To obtain a sensible structural model, the $U_{iso}$ of the C and N atoms of each guanidinate unit were constrained. Also, the C–N distances, N–C–N angles, and $CN_3$ torsion angles were restrained to established values for the guanidinate unit as obtained from neutron-diffraction measurements [31,41]. Finally, a secondary phase of EuO [50] was added in the refinement, reaching a weight-percentage of 1.3(2) wt %.

The DFT-optimized structure of $\alpha$-$Eu(CN_3H_4)_2$ in the experimental lattice parameters also describes the PXRD pattern well. In this Rietveld refinement, only the profile parameters—a single $U_{iso}$ parameter for the Eu atoms, and another for all C and N atoms—were refined. These thermal displacement parameters were deposited with the calculated atomic positions as a CIF file.

For $EuC(NH)_3$, Rietveld refinements were performed with the Jana2006 suite using the reported $SrC(NH)_3$ structure type [30] as the starting model. The hydrogen atoms were located by DFT calculations (see below). All CIF data may also be obtained from FIZ Karlsruhe, 76344 Eggenstein-Leopoldshafen, Germany (fax: (+49)-7247-808-666; e-mail: crysdata@fiz-karlsruhe.de), on quoting the depository numbers CSD-432390 for $\alpha$-$Eu(CN_3H_4)_2$ and CSD-432391 for $EuC(NH)_3$.

### 3.3. DFT Calculations

DFT calculations were computed at the PBE+D3(BJ)/PAW level [51–55] as implemented in VASP [56–59]. The cutoff energy for the plane-wave expansion was 500 eV; the $k$-meshes used for the calculations were sufficiently large. Phonon calculations did not use spin-polarization, and finite displacements of 0.01 Å were applied. The supercells for the phonon calculations of $\alpha$-$Eu(CN_3H_4)_2$ and $EuC(NH)_3$ were $3 \times 1 \times 2$ and $4 \times 4 \times 3$, respectively. To arrive at hydrogen positions fitting our $EuC(NH)_3$ experimental results at 300 K, we started out from the latter and placed the hydrogen atoms as observed for $SrC(NH)_3$ [30] and $YbC(NH)_3$ [31]. Then, we selectively optimized the hydrogen positions, leaving lattice vectors and all other positions fixed. The resulting hydrogen positions are expected to be qualitatively comparable to those from neutron-diffraction experiments [46].

### 3.4. Magnetometry

Magnetic properties of both $Eu(CN_3H_4)_2$ and $EuC(NH)_3$ were measured with a superconducting quantum interference device (SQUID) magnetometer (MPMS-5XL, Quantum Design Inc., San Diego, CA, USA). Each polycrystalline sample was compacted and immobilized into cylindrical polytetrafluoroethylene (PTFE) capsules. Measurements included field- and temperature-dependent molar magnetic susceptibilities (0.05–5.0 T, 2–290 K) and determination of the molar magnetization as a function of the applied field at 2 K. At applied fields of 0.1 T, the magnetic susceptibility was measured in field cooled (FC) and zero-field cooled (ZFC) mode, showing no significant difference. The data were corrected for diamagnetic contributions of sample holder and compound (Pascal's constants, $\chi_{dia} = -1.68 \times 10^{-9} \; m^3 \cdot mol^{-1}$, $Eu(CN_3H_4)_2$ and $-1.31 \times 10^{-9} \; m^3 \cdot mol^{-1}$, $EuC(NH)_3$).

Field-dependent measurements of the molar magnetic susceptibility $\chi_g$ allowed to correct for a small ferromagnetic impurity by applying the formula below for each temperature ($T > 20$ K) [10].

$$\chi_g(H) = \chi_g(\infty) + \frac{\sigma^s}{H}$$

For this formula, the magnetization must be a linear function of the field. Therefore, we included data of fields of up to 0.3 T to rule out errors caused by a saturation of $\alpha$-$Eu(CN_3H_4)_2$ or $EuC(NH)_3$. Extrapolations to infinitely high fields yield the corrected values $\chi_m(\infty)$ through multiplying $\chi_g(\infty)$ by the molar mass of $\alpha$-$Eu(CN_3H_4)_2$ or $EuC(NH)_3$.

### 3.5. IR Spectroscopy

An ALPHA FT-IR-spectrometer (Bruker, Billerica, MA, USA) placed in an argon-filled glove box and equipped with an ATR Platinum Diamond sample holder with a measurement range of $4000-400 \; cm^{-1}$ was employed to measure the IR spectrum of $Eu(CN_3H_4)_2$. The results were compared to a DFT-based calculation of an IR spectrum of $\alpha$-$Eu(CN_3H_4)_2$. The frequencies and eigenvectors at the $\Gamma$-point were derived by a finite displacement approach as implemented PHONOPY [60], and the Born effective charge tensor was calculated by density-functional perturbation theory as implemented in VASP ("LEPSILON=.TRUE."). The IR intensities were derived from these values as described in references [33] and [34]. Also, a Gaussian broadening was applied to the spectrum.

## 4. Conclusions

In summary, we synthesized the $\alpha$-, $\beta$- and $\gamma$-polymorphs of $Eu(CN_3H_4)_2$ and identified them by PXRD. The $\gamma$-phase transforms into the $\beta$-form over time. The IR spectra are dominated by the anion and are interpretable with the help of DFT calculations. Preliminary TGA measurements show that the guanidinates could be precursors for the preparation of (hydrogen) cyanamides. The crystal structure of $\alpha$-$Eu(CN_3H_4)_2$ was solved by PXRD, and DFT was used to optimize the structural model. In $\alpha$-$Eu(CN_3H_4)_2$, Eu is coordinated in double zigzag chains that are connected by the hydrogen-bonded guanidinate anions. The $CN_3H_4^-$ anions are predicted to adopt the *syn-*, *anti-*, and all-*trans*-conformations. The all-*trans*-conformation is found for the first time in a guanidinate. Magnetic measurements show paramagnetism at high temperatures and ferromagnetic exchange interactions below 6 K, presumably in one dimension along the Eu chains. Finally, $EuC(NH)_3$, isostructural to $SrC(NH)_3$ [30] and $YbC(NH)_3$ [31], is introduced as a possible low-temperature ferromagnet.

**Acknowledgments:** We would like to thank Paul Müller for PXRD measurements, Brigitte Jansen for TGA measurements and Christina Houben for SQUID magnetometry measurements. Use of the Advanced Photon Source at Argonne National Laboratory was supported by the U.S. Department of Energy, Office of Science, Office of Basic Energy Sciences, under Contract No. DE-AC02-06CH11357. The financial support by Deutsche Forschungsgemeinschaft and the Fonds der Chemischen Industrie (scholarship to Janine George) is gratefully acknowledged. The DFT calculations were performed with computing resources thankfully granted by JARA-HPC from RWTH Aachen University under project JARA0069.

**Author Contributions:** Richard Dronskowski initiated the research; Arno L. Görne conceived and designed the experiments, performed them and analyzed the data; Janine George performed the DFT calculations; Jan van Leusen and Arno L. Görne analyzed the magnetic data; all authors contributed to the writing of the paper.

## References

1.  Abraham, D.S. The Next Resource Shortage? *The New York Times*, 20 November 2015.

2.  Natural Environment Research Council. *Rare Earth Elements*; British Geological Survey: Nottingham, UK, 2011.

3.  Zurawski, A.; Mai, M.; Baumann, D.; Feldmann, C.; Müller-Buschbaum, K. Homoleptic imidazolate frameworks $^3_\infty[Sr_{1-x}Eu_x(Im)_2]$-hybrid materials with efficient and tuneable luminescence. *Chem. Commun.* **2011**, *47*, 496–498. [CrossRef] [PubMed]

4.  Rybak, J.-C.; Hailmann, M.; Matthes, P.R.; Zurawski, A.; Nitsch, J.; Steffen, A.; Heck, J.G.; Feldmann, C.; Götzendörfer, S.; Meinhardt, J.; et al. Metal–Organic Framework Luminescence in the Yellow Gap by Codoping of the Homoleptic Imidazolate $^3_\infty[Ba(Im)_2]$ with Divalent Europium. *J. Am. Chem. Soc.* **2013**, *135*, 6896–6902. [CrossRef] [PubMed]

5.  Pust, P.; Weiler, V.; Hecht, C.; Tücks, A.; Wochnik, A.S.; Henß, A.-K.; Wiechert, D.; Scheu, C.; Schmidt, P.J.; Schnick, W. Narrow-band red-emitting $Sr[LiAl_3N_4]:Eu^{2+}$ as a next-generation LED-phosphor material. *Nat. Mater.* **2014**, *13*, 891–896. [CrossRef] [PubMed]

6.  Kuda-Wedagedara, A.N.W.; Wang, C.; Martin, P.D.; Allen, M.J. Aqueous $Eu^{II}$-Containing Complex with Bright Yellow Luminescence. *J. Am. Chem. Soc.* **2015**, *137*, 4960–4963. [CrossRef] [PubMed]

7.  Slabon, A.; Mensing, C.; Kubata, C.; Cuervo-Reyes, E.; Nesper, R. Field-Induced Inversion of the Magnetoresistive Effect in the Zintl Phase $Eu_{5+x}Mg_{18-x}Si_{13}$ ($x$ = 2.2). *Angew. Chem. Int. Ed.* **2013**, *52*, 2122–2125. [CrossRef] [PubMed]

8.  Rushchanskii, K.Z.; Kamba, S.; Goian, V.; Vaněk, P.; Savinov, M.; Prokleška, J.; Nuzhnyy, D.; Knížek, K.; Laufek, F.; Eckel, S.; et al. A multiferroic material to search for the permanent electric dipole moment of the electron. *Nat. Mater.* **2010**, *9*, 649–654. [CrossRef] [PubMed]

9.  Niehaus, O.; Ryan, D.H.; Flacau, R.; Lemoine, P.; Chernyshov, D.; Svitlyk, V.; Cuervo-Reyes, E.; Slabon, A.; Nesper, R.; Schellenberg, I.; et al. Complex physical properties of EuMgSi—A complementary study by neutron powder diffraction and $^{151}$Eu Mossbauer spectroscopy. *J. Mater. Chem. C* **2015**, *3*, 7203–7215. [CrossRef]

10. Lueken, H. *Magnetochemie: Eine Einführung in Theorie und Anwendung*; Teubner Verlag: Stuttgart, Leipzig, 1999.

11. Matthias, B.T.; Bozorth, R.M.; Van Vleck, J.H. Ferromagnetic Interaction in EuO. *Phys. Rev. Lett.* **1961**, *7*, 160–161. [CrossRef]

12. Kornblit, A.; Ahlers, G.; Buehler, E. Heat capacity of $RbMnF_3$ and EuO near the magnetic phase transitions. *Phys. Lett. A* **1973**, *43*, 531–532. [CrossRef]

13. Wachter, P. Europium chalcogenides: EuO, EuS, EuSe and EuTe. In *Handbook on the Physics and Chemistry of Rare Earths*; Elsevier: Amsterdam, The Netherlands, 1979; Volume 2, pp. 507–574.

14. Juza, R.; Hadenfeldt, C. Darstellung und Eigenschaften von Europium(II)-amid. *Naturwissenschaften* **1968**, *55*, 229. [CrossRef]

15. Hulliger, F. Ferromagnetism of europium amide $Eu(NH_2)_2$. *Solid State Commun.* **1970**, *8*, 1477–1478. [CrossRef]

16. Wickleder, C. $M(SCN)_2$ ($M$ = Eu, Sr, Ba): Kristallstruktur, thermisches Verhalten, Schwingungsspektroskopie. *Z. Anorg. Allg. Chem.* **2001**, *627*, 1693–1698. [CrossRef]

17. Reckeweg, O.; DiSalvo, F.J. $EuCN_2$—The First, but Not Quite Unexpected Ternary Rare Earth Metal Cyanamide. *Z. Anorg. Allg. Chem.* **2003**, *629*, 177–179. [CrossRef]

18. Huppertz, H.; Schnick, W. $Eu_2Si_5N_8$ and $EuYbSi_4N_7$. The First Nitridosilicates with a Divalent Rare Earth Metal. *Acta Crystallogr. C* **1997**, *53*, 1751–1753. [CrossRef]

19. Höppe, H.A.; Trill, H.; Mosel, B.D.; Eckert, H.; Kotzyba, G.; Pöttgen, R.; Schnick, W. Hyperfine interactions in the 13 K ferromagnet $Eu_2Si_5N_8$. *J. Phys. Chem. Solids* **2002**, *63*, 853–859. [CrossRef]

20. Carrillo-Cabrera, W.; Somer, M.; Peters, K.; Schnering, H.G.V. Crystal structure of trieuropium bis(dinitridoborate), $Eu_3[BN_2]_2$. *Z. Kristallogr. New Cryst. Struct.* **2001**, *216*, 43–44. [CrossRef]

21. Liao, W.; Dronskowski, R. Carbodiimides with Extended Structures by an Azide-Cyanide Route: Synthesis and Crystal Structure of $M_2Cl_2NCN$ (M = Eu and Sr). *Z. Anorg. Allg. Chem.* **2005**, *631*, 496–498. [CrossRef]

22. Stadler, F.; Oeckler, O.; Höppe, H.A.; Möller, M.H.; Pöttgen, R.; Mosel, B.D.; Schmidt, P.; Duppel, V.; Simon, A.; Schnick, W. Crystal Structure, Physical Properties and HRTEM Investigation of the New Oxonitridosilicate $EuSi_2O_2N_2$. *Chem. Eur. J.* **2006**, *12*, 6984–6990. [CrossRef] [PubMed]

23. Zucchi, G.; Thuéry, P.; Rivière, E.; Ephritikhine, M. Europium(II) compounds: Simple synthesis of a molecular complex in water and coordination polymers with 2,2'-bipyrimidine-mediated ferromagnetic interactions. *Chem. Commun.* **2010**, *46*, 9143–9145. [CrossRef] [PubMed]

24. Liao, W.; Hu, C.; Kremer, R.K.; Dronskowski, R. Formation of Complex Three- and One-Dimensional Interpenetrating Networks within Carbodiimide Chemistry: $NCN^{2-}$-Coordinated Rare-Earth-Metal Tetrahedra and Condensed Alkali-Metal Iodide Octahedra in Two Novel Lithium Europium Carbodiimide Iodides, $LiEu_2(NCN)I_3$ and $LiEu_4(NCN)_3I_3$. *Inorg. Chem.* **2004**, *43*, 5884–5890. [PubMed]

25. Yamada, T.; Liu, X.; Englert, U.; Yamane, H.; Dronskowski, R. Solid-State Structure of Free Base Guanidine Achieved at Last. *Chem. Eur. J.* **2009**, *15*, 5651–5655. [CrossRef] [PubMed]

26. Sawinski, P.K.; Meven, M.; Englert, U.; Dronskowski, R. Single-Crystal Neutron Diffraction Study on Guanidine, $CN_3H_5$. *Cryst. Growth Des.* **2013**, *13*, 1730–1735. [CrossRef]

27. Hoepfner, V.; Dronskowski, R. $RbCN_3H_4$: The First Structurally Characterized Salt of a New Class of Guanidinate Compounds. *Inorg. Chem.* **2011**, *50*, 3799–3803. [CrossRef] [PubMed]

28. Sawinski, P.K.; Dronskowski, R. Solvothermal Synthesis, Crystal Growth, and Structure Determination of Sodium and Potassium Guanidinate. *Inorg. Chem.* **2012**, *51*, 7425–7430. [CrossRef] [PubMed]

29. Sawinski, P.K.; Deringer, V.L.; Dronskowski, R. Completing a family: $LiCN_3H_4$, the lightest alkali metal guanidinate. *Dalton Trans.* **2013**, *42*, 15080–15087. [CrossRef] [PubMed]

30. Missong, R.; George, J.; Houben, A.; Hoelzel, M.; Dronskowski, R. Synthesis, Structure, and Properties of $SrC(NH)_3$, a Nitrogen-Based Carbonate Analogue with the Trinacria Motif. *Angew. Chem. Int. Ed.* **2015**, *54*, 12171–12175. [CrossRef] [PubMed]

31. Görne, A.L.; George, J.; van Leusen, J.; Dück, G.; Jacobs, P.; Chogondahalli Muniraju, N.K.; Dronskowski, R. Ammonothermal Synthesis, Crystal Structure, and Properties of the Ytterbium(II) and Ytterbium(III) Amides and the First Two Rare-Earth-Metal Guanidinates, $YbC(NH)_3$ and $Yb(CN_3H_4)_3$. *Inorg. Chem.* **2016**, *55*, 6161–6168. [CrossRef] [PubMed]

32. Jacobs, H.; Fink, U. Untersuchung des Systems Kalium/Europium/Ammoniak. *Z. Anorg. Allg. Chem.* **1978**, *438*, 151–159. [CrossRef]

33. Karhánek, D.; Bučko, T.; Hafner, J. A density-functional study of the adsorption of methane-thiol on the (111) surfaces of the Ni-group metals: II. Vibrational spectroscopy. *J. Phys. Condens. Matter* **2010**, *22*, 265006. [CrossRef] [PubMed]

34. Baroni, S.; de Gironcoli, S.; Dal Corso, A.; Giannozzi, P. Phonons and related crystal properties from density-functional perturbation theory. *Rev. Mod. Phys.* **2001**, *73*, 515–562. [CrossRef]

35. Hoepfner, V. Synthese und quantenchemische Untersuchung von Alkalimetallguanidinaten. Dissertation, RWTH Aachen University, Aachen, Germany, 2012.

36. Krott, M.; Liu, X.; Fokwa, B.P.T.; Speldrich, M.; Lueken, H.; Dronskowski, R. Synthesis, Crystal–Structure Determination and Magnetic Properties of Two New Transition-Metal Carbodiimides: CoNCN and NiNCN. *Inorg. Chem.* **2007**, *46*, 2204–2207. [CrossRef] [PubMed]

37. Liu, X.; Stork, L.; Speldrich, M.; Lueken, H.; Dronskowski, R. FeNCN and $Fe(NCNH)_2$: Synthesis, Structure, and Magnetic Properties of a Nitrogen-Based Pseudo-oxide and -hydroxide of Divalent Iron. *Chem. Eur. J.* **2009**, *15*, 1558–1561. [CrossRef] [PubMed]

38. Pöttgen, R.; Johrendt, D. Equiatomic Intermetallic Europium Compounds: Syntheses, Crystal Chemistry, Chemical Bonding, and Physical Properties. *Chem. Mater.* **2000**, *12*, 875–897. [CrossRef]

39. Van de Streek, J.; Neumann, M.A. Validation of experimental molecular crystal structures with dispersion-corrected density functional theory calculations. *Acta Crystallogr. B* **2010**, *66*, 544–558. [CrossRef] [PubMed]

40. Hoepfner, V.; Deringer, V.L.; Dronskowski, R. Hydrogen-Bonding Networks from First-Principles: Exploring the Guanidine Crystal. *J. Phys. Chem. A* **2012**, *116*, 4551–4559. [CrossRef] [PubMed]

41. Hoepfner, V.; Jacobs, P.; Sawinski, P.K.; Houben, A.; Reim, J.; Dronskowski, R. RbCN$_3$H$_4$ and CsCN$_3$H$_4$: A Neutron Powder and Single-Crystal X-ray Diffraction Study. *Z. Anorg. Allg. Chem.* **2013**, *639*, 1232–1236. [CrossRef]

42. Bailey, P.J.; Blake, A.J.; Kryszczuk, M.; Parsons, S.; Reed, D. The first triazatrimethylenemethane dianion: crystal structure of dilithio-triphenylguanidine Li$_2$[C(NPh)$_3$] as its tetrahydrofuran solvate. *J. Chem. Soc. Chem. Commun.* **1995**, 1647–1648. [CrossRef]

43. Bailey, P.J.; Mitchell, L.A.; Parsons, S. Guanidine anions as chelating ligands; syntheses and crystal structures of [Rh(η-C$_5$Me$_5$){η$^2$-(NPh)$_2$CNHPh}Cl] and [Ru(η-MeC$_6$H$_4$Pr$^i$-$p$)-{η$^2$-(NPh)$_2$CNHPh}Cl]. *J. Chem. Soc. Dalton Trans.* **1996**, 2839–2841. [CrossRef]

44. Shuskus, A.J. Electron Spin Resonance of Gd$^{3+}$ and Eu$^{2+}$ in Single Crystals of CaO. *Phys. Rev.* **1962**, *127*, 2022–2024. [CrossRef]

45. Baker, J.M.; Williams, F.I.B. Electron Nuclear Double Resonance of the Divalent Europium Ion. *Proc. R. Soc. A* **1962**, *267*, 283–294. [CrossRef]

46. Deringer, V.L.; Hoepfner, V.; Dronskowski, R. Accurate Hydrogen Positions in Organic Crystals: Assessing a Quantum-Chemical Aide. *Cryst. Growth Des.* **2011**, *12*, 1014–1021. [CrossRef]

47. Missong, R. Synthese und Charakterisierung von Strontium- und Bariumguanidinat. Dissertation, RWTH Aachen University, Aachen, Germany, 2016.

48. Palatinus, L.; Chapuis, G. SUPERFLIP—A computer program for the solution of crystal structures by charge flipping in arbitrary dimensions. *J. Appl. Cryst.* **2007**, *40*, 786–790. [CrossRef]

49. Petříček, V.; Dušek, M.; Palatinus, L. Crystallographic Computing System JANA2006: General features. *Z. Kristallogr.* **2014**, *229*, 345–352. [CrossRef]

50. McWhan, D.B.; Souers, P.C.; Jura, G. Magnetic and Structural Properties of Europium Metal and Europium Monoxide at High Pressure. *Phys. Rev.* **1966**, *143*, 385–389. [CrossRef]

51. Perdew, J.P.; Burke, K.; Ernzerhof, M. Generalized Gradient Approximation Made Simple. *Phys. Rev. Lett.* **1996**, *77*, 3865–3868. [CrossRef] [PubMed]

52. Grimme, S.; Antony, J.; Ehrlich, S.; Krieg, H. A consistent and accurate ab initio parametrization of density functional dispersion correction (DFT-D) for the 94 elements H–Pu. *J. Chem. Phys.* **2010**, *132*, 154104. [CrossRef] [PubMed]

53. Grimme, S.; Ehrlich, S.; Goerigk, L. Effect of the damping function in dispersion corrected density functional theory. *J. Comput. Chem.* **2011**, *32*, 1456–1465. [CrossRef] [PubMed]

54. Blöchl, P.E. Projector augmented-wave method. *Phys. Rev. B* **1994**, *50*, 17953–17979. [CrossRef]

55. Kresse, G.; Joubert, D. From ultrasoft pseudopotentials to the projector augmented-wave method. *Phys. Rev. B* **1999**, *59*, 1758–1775. [CrossRef]

56. Kresse, G.; Hafner, J. *Ab initio* molecular dynamics for liquid metals. *Phys. Rev. B* **1993**, *47*, 558–561. [CrossRef]

57. Kresse, G.; Hafner, J. *Ab initio* molecular-dynamics simulation of the liquid-metal–amorphous-semiconductor transition in germanium. *Phys. Rev. B* **1994**, *49*, 14251–14269. [CrossRef]

58. Kresse, G.; Furthmüller, J. Efficient iterative schemes for *ab initio* total-energy calculations using a plane-wave basis set. *Phys. Rev. B* **1996**, *54*, 11169–11186. [CrossRef]

59. Kresse, G.; Furthmüller, J. Efficiency of ab-initio total energy calculations for metals and semiconductors using a plane-wave basis set. *Comput. Mater. Sci.* **1996**, *6*, 15–50. [CrossRef]

60. Togo, A.; Tanaka, I. First principles phonon calculations in materials science. *Scr. Mater.* **2015**, *108*, 1–5. [CrossRef]

# 12

# Structural Study of Mismatched Disila-Crown Ether Complexes

**Kirsten Reuter, Fabian Dankert, Carsten Donsbach and Carsten von Hänisch** *

Fachbereich Chemie and Wissenschaftliches Zentrum für Materialwissenschaften (WZMW), Philipps-Universität Marburg, Hans-Meerwein Straße 4, D-35032 Marburg, Germany; kirsten.reuter@staff.uni-marburg.de (K.R.); Dankert@students.uni-marburg.de (F.D.); donsbach@students.uni-marburg.de (C.D.)
* Correspondence: haenisch@chemie.uni-marburg.de

Academic Editor: Matthias Westerhausen

**Abstract:** Mismatched complexes of the alkali metals cations $Li^+$ and $Na^+$ were synthesized from 1,2-disila[18]crown-6 (**1** and **2**) and of $K^+$ from 1,2,4,5-tetrasila[18]crown-6 (**4**). In these alkali metal complexes, not all crown ether O atoms participate in the coordination, which depicts the coordination ability of the C-, Si/C-, and Si-bonded O atoms. Furthermore, the inverse case—the coordination of the large $Ba^{2+}$ ion by the relatively small ligand 1,2-disila[15]crown-5—was investigated, yielding the dinuclear complex **5**. This structure represents a first outlook on sandwich complexes based on hybrid crown ethers.

**Keywords:** hybrid crown ether; siloxane; disilane; mismatch complex; host–guest chemistry

---

## 1. Introduction

The nature of the Si–O bond has been intensively studied over the past six decades. In the 1960s especially, the large valence angle in disiloxanes and the unusual short Si–O bond length, e.g., in $O(SiH_2Me_2)_2$, were issued in numerous publications [1,2]. The low basicity of siloxanes was originally attributed to an electron-withdrawing tendency of the silyl groups of the type p(O)→d(Si) [3–5]. This approach was later discarded in favor of hyperconjugation interactions between p(O)→σ*(Si–C) [6–8]. Alternatively, in an opposed model based on calculations of the electron density function, the Si–O bond was described as essentially ionic due to the high difference in electronegativity between Si and O [9,10]. Careful theoretical studies on the basicity of $O(SiH_2Me_2)_2$ and $OEt_2$ revealed that the lower electrostatic attraction in siloxanes results from the repulsion between the positively charged Si atoms and Lewis acids [11]. This proceeding has recently been extended on cyclosiloxanes [12], which were previously described as pseudo crown ethers or inorganic crown ethers [13–15]. However, the structural analogy to organic crown ethers is poor, since siloxanes feature O atoms linked by $-SiMe_2-$ rather than $-CH_2CH_2-$. Additionally, organic ring-contracted crown ethers exhibit an eminently reduced coordination ability, as has been shown in the referencing of [17]crown-6, in which only one $-CH_2CH_2-$ unit was replaced by $-CH_2-$ [16,17]. Consequently, higher comparability between organic crown ethers and cyclosiloxanes can be provided by extension of the $-SiMe_2-$ unit to $-SiMe_2SiMe_2-$. Recent studies of hybrid [12]crown-4 featuring one or two disilane fragments in a residuary organic crown ether framework revealed an increasing coordination ability towards $Li^+$ in the series C–O–C < C–O–Si < Si–O–Si (Scheme 1) [18,19].

Another deviation between the hitherto discussed cyclosiloxanes and organic crown ethers concerns the substituents at Si and C. Up to date, neither cyclosiloxanes with H-substituents at the Si atoms nor permethylated crown ethers have been synthesized, which complicates a meaningful comparison of the two types of ligands. Calculation of the energy changes for crown ethers, cyclosiloxanes and hybrid crown ethers going from the free ligand geometries to complex geometries—determined as relaxation energy—revealed that $SiMe_2$ or $Si_2Me_4$ containing ligands require steadily more energy for adopting the complex geometry [12,18,19]. The complex stability is directly affected by the relaxation energy, which is in the case of the hybrid crown ethers compensated by the particularly high donor ability of the O atoms [18].

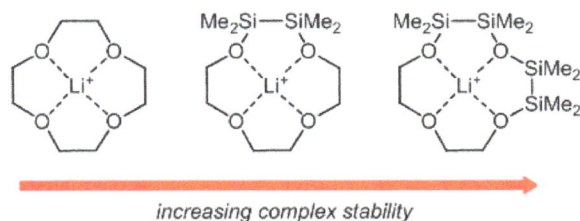

*increasing complex stability*

**Scheme 1.** Binding modes and relative binding affinities of $Li^+$ in [12]crown-4, 1,2-disila[12]crown-4, and 1,2,4,5-tetrasila[12]crown-4.

The hitherto described hybrid crown ethers exhibit up to three different types of O atoms—all C-, C/Si-, and all Si-bonded ones (Scheme 1). To experimentally explore the competition between the basicity of the inequivalent O atoms and the energy effort for reaching the ligand geometry in the complex, we performed complexation reactions using small alkali and alkaline earth metal ions and comparatively large ligands. As a result, the ligand exceeds with its ring diameter the ionic radius of the Lewis acid. Since particularly Si-based crown ethers show limited flexibility [11–15,18,19], we expected not all $O_{crown}$ atoms to participate in the coordination of the metal center [20–22]. The first mismatch structure of a hybrid crown ether was very recently published and is constituted of 1,2-disila[18]crown-6 and $Ca(OTf)_2$ (OTf = $^-OSO_2CF_3$) [23]. Therein, one of the C-bonded O atoms does not participate in the coordination of $Ca^{2+}$, showing the preference of the metal ion to be coordinated by the Si/C-linked O atoms. This preference depicts the coordination ability of the O atoms in partially Si-based crown ethers and is a matter of investigation in this work.

## 2. Results and Discussion

### 2.1. Mismatch Complexes Involving 1,2-Disila[18]crown-6 with $Li^+$ and $Na^+$

The hybrid ligand 1,2-disila[18]crown-6 was synthesized in a single step reaction from 1,2-dichlorodisilane and pentaethylene glycol (Scheme 2). Prior studies have shown that $Li^+$ matches well with 1,2-disila[12]crown-4 and $Na^+$ with 1,2-disila[15]crown-5 [18], so that the two cations together with 1,2-disila[18]crown-6 are supposed to fulfil the criteria of a mismatch. Reaction of 1,2-disila[18]crown-6 with lithium hexafluorophosphat in a 1:1 stoichiometry yielded a highly viscous oil. After freezing at $-196\,°C$ and subsequent storage at $-35\,°C$ for 3 days, Compound **1** crystallized in the space group $P2_1/c$ in the form of colorless planks. In the solid-state structure of Compound **1**, $Li^+$ is coordinated by five of the six crown ether O atoms (Figure 1). The non-coordinating completely carbon-bonded O atom O5 shows an atomic distance of 295.7(5) pm to the $Li^+$ cation. The $PF_6$ anion does not interact with the cation. The coordination polyhedron can be described as a distorted trigonal bipyramid (Figure 2). The three equatorial O atoms (O2, O4, O6) establish shorter bond lengths to the cation than the two axial O atoms. The shortest Li–O bond length has a value of 194.9(5) pm (Li1–O6), while the longest bond length measures 224.8(5) pm (Li1–O1). Compared to the hitherto known lithium complexes of hybrid sila-crown ethers, the Li1–O1 bond length is elongated, which may be the result of the strongly twisted ligand. Typically, the O atoms in sila-crown ethers complexes adopt an approximately planar conformation [13–15,18,19]. The disilane fragment in **1** is roughly coplanar to the thereon bonded O atoms O1 and O2, but the organic part of the ligand is strongly twisted and is wrapped around the metal center.

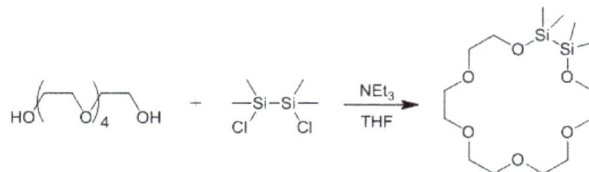

**Scheme 2.** Synthesis path for 1,2-disila[18]crown-6 [18].

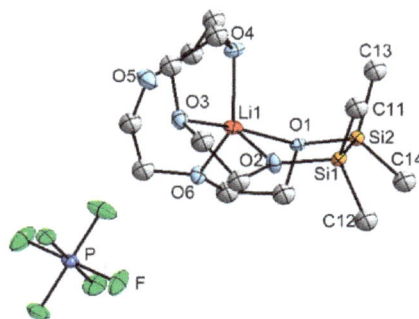

**Figure 1.** Molecular structure of [Li(1,2-disila[18]crown-6)]PF$_6$ (**1**) in the crystal. Thermal ellipsoids represent the 50% probability level. Hydrogen atoms are omitted for clarity. Selected bond lengths (pm) and angles (°): Si1–Si2: 235.1(1), Si1–O1: 168.9(2), Si2–O2: 167.9(2), Li1–O1: 224.8(5), Li1–O2: 200.4(5), Li1–O3: 212.1(5), Li1···O5: 295.7(5), Li1–O6: 194.9(5), O3–Li1–O1: 169.4(2), O2–Li1–O4: 115.9(2), O2–Li1–O6: 111.6(2), O4–Li1–O6: 132.5(2), O1–Li1–O4: 102.3(2), O1–Li1–O2: 88.9(2), O1–Li1–O6: 79.5(2), C11–Si2–Si1–C13: 9.6(1), C12–Si2–Si1–C14: 8.9(1).

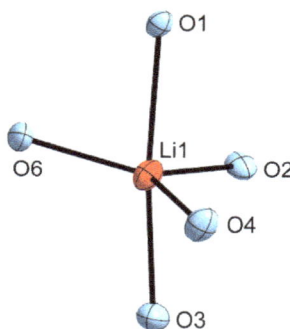

**Figure 2.** Trigonal bipyramidal coordination polyhedron of the lithium cation in [Li(1,2-disila[18]crown-6)]PF$_6$ (**1**).

Trigonal bipyramidal coordination is common for mismatched crown ether complexes of lithium [24,25], while in [12]crown-4 complexes the square-pyramid [26,27] or in sandwich complexes the square antiprism is the usual coordination polyhedron [28]. In prior studies on sila-crown ether complexes, it was already shown that the Me groups at the Si atoms take in a roughly eclipsed conformation [18,19,23]. In **1**, the Me groups adopt with dihedral angles of 9.6(1)° for C11–Si2–Si1–C13, and 8.9(1)° for C12–Si2–Si1–C14 the expected conformation of the complex. As a result, the attractive electrostatic interaction between the Si/C-bonded O atoms and the Li$^+$ cation must compensate for the required energy effort of the ecliptic arrangement. The $^{29}$Si{$^1$H}-NMR signal shifts from $\delta$ = 11.4 ppm in the free ligand to $\delta$ = 15.6 ppm in **1**, indicating a strong electrostatic interaction between Li$^+$ and O1 and O2. The strong shift also reflects the hard Lewis acidity in comparison to K$^+$, since in [K(1,2-disila[18]crown-6)PF$_6$] the respective $^{29}$Si{$^1$H}-NMR signal is at $\delta$ = 13.0 ppm [18].

By an analogous reaction of NaPF$_6$ with 1,2-disila[18]crown-6, single crystals in form of colorless blocks were obtained from dichloromethane/benzene (2:1). [Na(1,2-disila[18]crown-6)PF$_6$] (**2**) crystallizes in the triclinic space group $P\overline{1}$ as a monomeric contact ion pair (Figure 3). Na$^+$ is coordinated by five of the six crown ether O atoms and additionally by two F atoms of the $PF_6$ anion. The coordination sphere of $Na^+$ cannot be assigned to a hitherto described polyhedron as a result of its strong distortion. Compared to $K^+$, the ionic radius of $Na^+$ is still too small for the cavity diameter of 1,2-disila[18]crown-6. As a result, O1 is with a distance of 453.2(3) pm not participating in the coordination of the metal ion. This leads to a strong distortion of the ring system, as O1 is located significantly beneath the mean plane of the other crown ether O atoms. Additionally, the Me groups at the Si atoms show a staggered arrangement, which is the common structure in free hybrid crown ethers [18]. In the case of Compound **2**, the electrostatic attraction between O1 and $Na^+$ apparently does not compensate for the adoption of an ecliptic arrangement, so the cation is preferably coordinated by the C-bonded O atoms. The coordinating Si- and C-linked O atom O2 establishes a bond length of 238.5(3) pm to the metal, while the completely C-linked O atoms show values between 237.0(3) and 247.3(2) pm. The $^{29}Si\{^1H\}$-NMR signal of Compound **2** appears at $\delta = 14.3$ ppm and, according to the respective Lewis acidity of $Li^+$ and $Na^+$, is less low-field shifted compared to **1**.

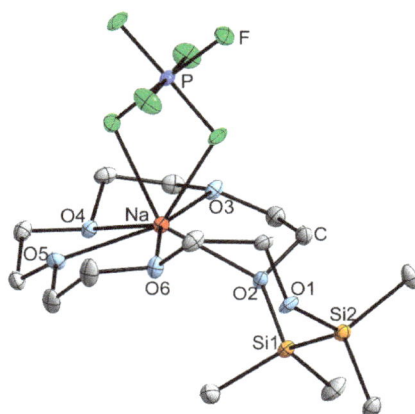

**Figure 3.** Molecular structure of $[Na(1,2\text{-disila}[18]crown\text{-}6)PF_6]$ (**2**) in the crystal. Thermal ellipsoids represent the 50% probability level. Hydrogen atoms are not displayed. Selected bond lengths (pm) and angles (°): Na···O1: 453.2(3), Na–O2: 238.5(3), Na–O3: 247.3(2), Na–O4: 244.9(3), Na–O5: 237.0(3), Si1–Si2: 235.8(1), Si2–O1: 166.2(3), Si1–O2: 167.5(2), Si2–O1–C10: 122.7(2), Si1–O2–C1: 121.7(2), C4–O4–C5: 113.5(2), C13–Si1–Si2–C11: 66.0(2), C14–Si1–Si2–C12: 65.6(2).

## 2.2. Determination of $\Delta E_{geom}$ in 1,2-Disila[18]crown-6 Complexes

As was shown in prior studies, hybrid crown ethers require steadily more energy than organic crown ethers for adopting the ligand structure within the complex [18,19,23]. The increase in relaxation energy was partially attributed to the sterically disfavored ecliptic arrangement of the Me groups bonded at the Si atoms. That was found to be the predominant conformation in the hybrid crown ether complex structures. The mismatched complexes **1** and **2** offer two different coordination modes of the Si/C-bonded O atoms: In the case of the Li-complex **1**, both Si-bonded O atoms participate in the coordination, while in the Na-complex **2**, only one of the Si/C-bonded O atoms binds to the metal center, which results in a staggered arrangement of the Me groups. It follows that 1,2-disila[18]crown-6 is expected to exhibit considerable different energy levels in the complex structures **1** and **2**. The energy difference $\Delta E_{geom}$ was determined by DFT calculations, implemented in Turbomole V7.0 [28], using the BP86 functional [29–32] and the def2-TZVP basis set with inclusion of dispersion interactions [33,34]. Accordingly, the energy of the ligand increases by 77.58 $kJ \cdot mol^{-1}$ for adopting the structure found within $[Li(1,2\text{-disila}[18]crown\text{-}6)]^+$ and by 29.24 $kJ \cdot mol^{-1}$ for $[Na(1,2\text{-disila}[18]crown\text{-}6)]^+$. The electrostatic attraction between the Si/C-bonded O atoms and $Na^+$ does not compensate for the ecliptic conformation of the Me groups. By contrast, $Li^+$ must exhibit a significantly increased electrostatic attraction to the hybrid-bonded O atoms. The mismatched hybrid crown ether complexes **1** and **2** therefore suggest that the cation exerts a major impact on the coordination modes of the ligand.

The optimized structure of the free ligand 1,2-disila[18]crown-6 shows, as expected, a staggered conformation of the methyl groups at the silicon atoms. The DFT calculated structures of the cations in Compounds **1** and **2** exhibit only very small differences in the structural parameter in comparison to the structures obtained by X-ray diffraction (see XYZ data in the ESI).

### 2.3. Mismatch Involving 1,2,4,5-Tetrasila[18]crown-6 and K+

The synthesis of hybrid crown ethers with a higher amount of disilane units was very recently described for 1,2,4,5-tetrasila[12]crown-4 [19]. In an analogous reaction of $O(Si_2Me_4Cl)_2$ with tetraethylene glycol, the ligand 1,2,4,5-tetrasila[18]crown-6 (**3**) was synthesized using high dilution of the agents to prevent polymerization (Scheme 3). Compound **3** is a highly viscous, colorless oil. Through the presence of two disilane units, the ring size is further increased in comparison to 1,2-disila[18]crown-6. In the $^{29}Si\{^1H\}$-NMR spectrum, Compound **3** shows two signals which can be assigned to the two types of Si atoms: The Si–O–Si entity appears at $\delta = 2.1$ ppm, the C–O–Si entity is low-field shifted and appears at $\delta = 11.0$ ppm.

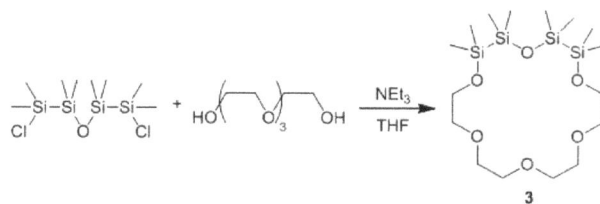

**Scheme 3.** Synthesis path of 1,2,4,5-tetrasila[18]crown-6 (**3**).

Treatment of **3** with $KPF_6$ yielded the corresponding, highly water sensitive complex [K(1,2,4,5-tetrasila[18]crown-6)$PF_6$] (**4**). Different to the hitherto known hybrid disila-crown ether complexes, **4** is directly after removal of the volatiles an oily compound, which crystallizes within 18 h at ambient temperature in form of colorless planks in the space group $P2_1/n$.

As observed in the Na+ complex **2**, Compound **4** is a monomeric contact ion pair (Figure 4). The cation is coordinated by five of the six crown ether O atoms and three F atoms of the anion, giving a coordination number of eight. The incorporation of two disilane units into the ring system leads to an increased ring diameter so that K+, which commonly matches perfectly with [18]crown-6, has a too small ionic radius for the ligand **3**. The inorganic part sticks out, showing an interatomic distance of 505.5(2) pm between the completely Si substituted O atom O2 and the metal ion. The Me groups at the Si atoms adopt an approximately staggered conformation with average dihedral angles of 84.2(2)° at Si1/Si2 and 59.8(2)° at Si3/Si4. Worth mentioning is the unusual orientation of Si4: In all hitherto known sila-crown ether complexes, the Si atoms bonded to coordinating O atoms are approximately arranged in plane with the crown ether O atoms [12–15,18,19,23]. In contrast to this, Si4 is considerably located beneath the mean plane of the coordinating O atoms. The Si/C-bonded O atoms O1 and O3 show O–K bond lengths of 283.8(2) and 279.4(2) pm, whereas the fully C-substituted O atoms O4–O6 establish average bond lengths of 273.7(2) pm. It can therefore be assumed that K+ is stronger coordinated by the carbon-based part of the hybrid crown ether **3**. Compared to [K(1,2-disila[18]crown-6)$PF_6$], which incorporates only one disilane unit and in which all crown ether atoms are participating in the coordination, the mean O–K bond lengths are in **4** considerably shorter [18]. This can be related to the coordination number of 8 in **4** compared to 9 in [K(1,2-disila[18]crown-6)$PF_6$]. The Si2–O2–Si3 bond angle is 143.8(1)°, this is a typical value for siloxanes [1,2]. Also the Si4–O3–C9 bond angle of 123.3(2)° is in the expected range [18,19]. Only the Si1–O1–C16 angle is with 117.8(1)° smaller than usually observed and is similar to that found in C–O–C bindings, e.g., C14–O6–C15 with 112.3(2)°.

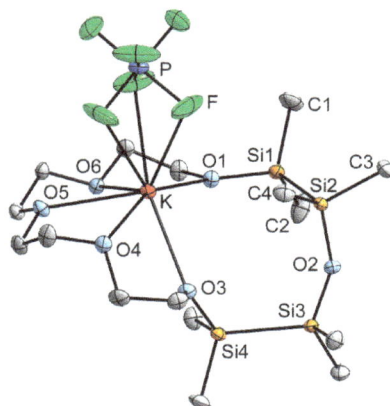

**Figure 4.** Molecular structure of [K(1,2,4,5-tetrasila[18]crown-6)PF$_6$] (**4**) in the crystal. Thermal ellipsoids represent the 50% probability level. Hydrogen atoms are not displayed. Selected bond lengths (pm) and angles (°): K–O1: 283.8(2), K···O2: 505.5(2), K–O3: 279.4(2), K–O4: 274.4(2), K–O5: 272.3(2), K–O6: 274.3(2), Si1–Si2: 234.9(1), Si1–O1: 166.5(2), Si2–O2: 164.5(2), Si3–O2: 165.2(2), Si4–O3: 166.7(2); Si4–O3–C9: 123.3(2), Si2–O2–Si3: 143.8(1), Si1–O1–C16: 117.8(1), C14–O6–C15: 112.3(2), C1–Si1–Si2–C4: 84.1(2), C2–Si1–Si2–C3: 84.2(1), C6–Si3–Si4–C8: 59.5(1), C5–Si3–Si4–C7: 60.1(1).

The reluctance of K$^+$ to interact with the Si-substituted O atoms was also observed in solution and can be deducted from the shifts in the $^{29}$Si{$^1$H}-NMR spectrum: The resonance signal of Si2/Si3 shows only a slight low-field shift to δ = 2.7 ppm (Δ(δ) = 0.6 ppm) and the signal of Si1/Si4 appears at δ = 11.9 ppm (Δ(δ) = 0.9 ppm). In comparison, the $^{29}$Si{$^1$H}-NMR signals of 1,2-disila[18]crown-6 shift from δ = 11.4 ppm in the free ligand to δ = 13.0 ppm in the potassium complex [18]. The small shift of the $^{29}$Si{$^1$H} signal indicates that also in solution O2 shows only minor interaction with the K$^+$ ion, owing to the high energy effort of Si$_2$Me$_4$ fragments to adopt the ecliptic geometry.

## 2.4. The Inverse Case: 1,2-Disila[15]crown-5 and Ba$^{2+}$

Beside experiments involving large ligands with comparatively small cations, we also investigated the inverse mismatch case, i.e., 1,2-disila[15]crown-5 with BaOTf$_2$ (OTf = $^-$OSO$_2$CF$_3$). Prior studies revealed that Ba$^{2+}$ perfectly matches with 1,2-disila[18]crown-6 and 1,2-disila-benzo[18]crown-6. In the corresponding complex, Ba$^{2+}$ is located in one plane with the coordinating O atoms and is saturated by two triflate groups, which are arranged upon and beneath the crown ether mean plane [23]. Reaction of 1,2-disila[15]crown-5 with BaOTf$_2$ in 1:1 stoichiometry yielded colorless blocks of [Ba(1,2-disila[15]crown-5)OTf$_2$]$_2$ (**5**) in the triclinic space group $P\bar{1}$. Different to the hitherto known sila-crown ether complexes, **5** forms a dinuclear complex (Figure 5). The four triflate anions act as bridges between the two metal centers and participate in the saturation of the coordination sphere with four O atoms, respectively. Furthermore, Ba$^{2+}$ is coordinated by the five crown ether O atoms, giving a coordination number of 9. The ion Ba(1) is located 156.8(2) pm above the calculated mean plane of the O$_{crown}$ atoms, which reflects the small ring diameter of 1,2-disila[15]crown-5 compared to the ionic radius of Ba$^{2+}$. The disilane units of the crown ethers show in opposite directions to each other as a result of the sterically demanding methyl groups. The typical approximately ecliptic arrangement of the methyl groups in sila-crown ethers complexes can also be found in Compound **5**. However, the dihedral angles have values of 26.1(3)° and 22.8(3)° and accordingly show stronger deviations from the ideal ecliptic arrangement compared to those found in other hybrid-crown ether complexes. The Si/C-bonded O atoms O1 and O5 establish bond lengths of 283.4(1) and 286.5(1) pm to the cation and are in a similar range with C-bonded O atoms, which show O–Ba bonds between 280.4(1) and 287.7(1) pm. Ba$^{2+}$ is furthermore strongly coordinated by the triflate O atoms since the bonding to Ba(1) has an average value of 275.7(4) pm. Another indication for the weak coordination of Ba$^{2+}$ by

1,2-disila[15]crown-5 was revealed by mass spectrometric analysis: Only [Na(1,2-disila[15]crown-5)]$^+$ was detected. Na$^+$ is a common impurity in mass spectrometers, so Ba$^{2+}$ was immediately replaced.

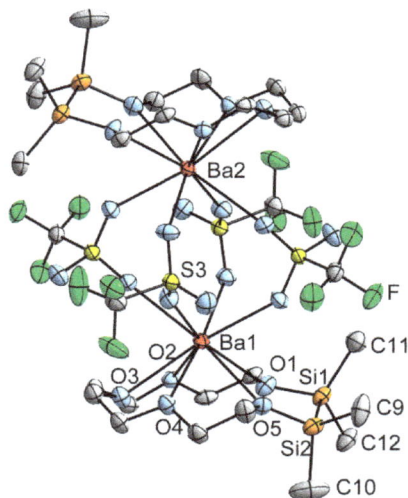

**Figure 5.** Molecular structure of [Ba(1,2disila[15]crown-5)OTf$_2$]$_2$ (**5**) in the crystal. Thermal ellipsoids represent the 50% probability level. Hydrogen atoms are not displayed. Selected bond lengths (pm) and angles (°): Ba1–O1: 283.4(1), Ba1–O2: 287.7(1), Ba1–O3: 280.4(1), Ba1–O4: 287.7(4), Ba1–O5: 286.5(1), Ba1–O$_{OTf}$: 269.4(1)–278.7(1), Si1–Si2: 234.6(2), Si1–O1: 167.2(4), 165.9(3), Ba1···Ba2: 527.1(2), Si1–O1–C1: 121.5(3), Si2–O5–C8: 121.6(3), C4–O3–C5: 114.9(4), C11–Si1–Si2–C9: 26.1(3), C12–Si1–Si2–C10: 22.8(3).

## 3. Materials and Methods

### 3.1. General Experimental Technique

All working procedures were conducted under exclusion of oxygen and moisture using Schlenk techniques under a nitrogen atmosphere. Solvents were dried and freshly distilled before use. Nuclear magnetic resonance (NMR) spectra were recorded with BRUKER Model AVANCE HD300, BRUKER Model DRX400, or BRUKER Model AVANCE500 spectrometers (Bruker Corporation, Rheinstetten, Germany) and were visualized with MestReNova [35]. Infrared (IR) spectra were recorded in attenuated total reflectance (ATR) mode on a BRUKER model ALPHA FT-IR. MS spectrometry was measured on a LTQ-FT (ESI, Thermo Fischer Scientific, Darmstadt, Germany) or on a JEOL AccuTOF-GC (LIFDI, JEOL, Freising, Germany). Elemental analysis data cannot be provided due to the presence of fluorine in the samples, which harm the elemental analysis devices. The ligands 1,2-disila[18]crown-6 and 1,2-disila[15]crown-5 [18] and O(Si$_2$Me$_4$Cl)$_2$ [19] were prepared by reported methods.

### 3.2. Computational Details

Calculations were performed with Turbomole V7.0 [28]. The resolution of identity (RI) approximation, dispersion corrections [29–32], and the conductor-like screening (COSMO) model [36] were applied, the latter with default settings. For all calculations the BP86 functional and def2-TZVP basis set [33,34] were chosen.

### 3.3. Crystal Structures

Data collection was performed on a Bruker D8 Quest or a Stoe IPDS II diffractometer at 100(2) K with Mo Kα radiation and graphite monochromatization. Structure solution was done by direct methods, refinement with full-matrix-least-squares against $F^2$ using shelxs-2014, shelxl-2014, shelxt-2014, and olex2 software (Table 1) [37,38]. The presentation of crystal structures was done with Diamond4.2.2 [39]. CCDC 1517535 (**1**), 1517536 (**2**), 1517537 (**4**), and 1517538 (**5**) contain the supplementary crystallographic data for this paper. These data can be obtained free of charge from The Cambridge Crystallographic Data Centre via www.ccdc.cam.ac.uk/data_request/cif.

**Table 1.** Crystal Structure Data.

| Empirical Formula | $C_{14}H_{30}Li_1O_6Si_2F_6P_1$ | $C_{14}H_{32}F_6Na_1O_6P_1Si_2$ | $C_{16}H_{40}F_6K_1O_6Si_4P_1$ | $C_{28}H_{55}Ba_2F_{12}O_{22}S_4Si_4$ |
|---|---|---|---|---|
| Formula weight (g·mol$^{-1}$) | 502.47 | 520.53 | 624.91 | 1487.00 |
| Crystal color, shape | colorless plank | colorless block | colorless plank | colorless block |
| Crystal size (mm) | 0.134 × 0.189 × 0.382 | 0.060 × 0.271 × 0.284 | 0.138 × 0.140 × 0.539 | 0.232 × 0.245 × 0.509 |
| Crystal system | monoclinic | triclinic | monoclinic | triclinic |
| Space group | $P2_1/c$ | $P\bar{1}$ | $P2_1/n$ | $P\bar{1}$ |
| Formula units | 4 | 2 | 4 | 2 |
| Temperature (K) | 100(2) | 100(2) | 100(2) | 100(2) |
| Unit cell dimensions | $a = 9.189(1)$ $b = 23.663(1)$ $c = 10.608(1)$ $\beta = 94.332(1)$ | $a = 8.512(1)$ $b = 11.468(1)$ $c = 13.996(1)$ $\alpha = 105.64(1)$ $\beta = 103.51(1)$ $\gamma = 103.20(1)$ | $a = 9.379(1)$ $b = 22.608(2)$ $c = 14.906(1)$ $\beta = 104.62(1)$ | $a = 11.269(2)$ $b = 12.458(3)$ $c = 20.817(4)$ $\alpha = 78.69(3)$ $\beta = 83.62(3)$ $\gamma = 78.63(3)$ |
| Cell volume (Å$^3$) | 2300.0(2) | 1215.83(17) | 3058.4(4) | 2801.6(11) |
| Pcalc (g/cm$^3$) | 1.451 | 1.422 | 1.357 | 1.763 |
| $\mu$ (Mo K$\alpha$) (mm$^{-1}$) | 0.298 | 0.301 | 0.446 | 1.739 |
| 2θ range | 2.384–25.299 | 2.578–25.237 | 2.289–25.319 | 1.695–26.373 |
| Reflections measured | 47204 | 13274 | 86701 | 24025 |
| Independent Reflections | 4181 [Rint = 0.0402] | 4422 [Rint = 0.0882] | 5566 [Rint = 0.0290] | 11417 [Rint = 0.0848] |
| $R1$ ($I > 2\sigma(I)$) | 0.0455 | 0.0535 | 0.0191 | 0.0435 |
| $wR2$ (all data) | 0.1103 | 0.1504 | 0.0360 | 0.1091 |
| GooF | 1.023 | 1.021 | 0.800 | 0.926 |
| Largest diff. peak and hole (e·Å$^{-3}$) | 1.02/−0.65 | 0.60/−0.56 | 0.60/−0.61 | 1.61/−2.25 |

## 3.4. Experimental Section

Li(1,2-disila[18]crown-6)]PF$_6$ (**1**): At ambient temperature, 159 mg (1.05 mmol, 1 equiv) of LiPF$_6$ was added to 370 mg (1.1 mmol, 1 equiv) of 1,2-disila[18]crown-6 in 15 mL of dichloromethane. The suspension was stirred for 18 h and was subsequently filtered. The solvent was removed under reduced pressure, and the residue was washed with *n*-pentane. The resulting colorless greasy solid was recrystallized with traces of dichloromethane after freezing at −196 °C and subsequently storage at −35 °C for 3 days. 45% (275 mg, 0.5 mmol) of **1** was obtained in form of colorless planks. $^1$H NMR (300 MHz, CD$_2$Cl$_2$): δ = 0.34 (s, 12H, C$H_3$), 3.71–3.73 (m, 4H, C$H_2$), 3.76 (s, 12H, C$H_2$), 3.81–3.82 ppm (m, 4H, C$H_2$); $^{13}$C{$^1$H} NMR (75 MHz, CD$_2$Cl$_2$): δ = −0.5 (s, CH$_3$), 61.7 (s, CH$_2$), 68.1 (s, CH$_2$), 68.2 (s, CH$_2$), 68.4 (s, CH$_2$), 71.4 ppm (s, CH$_2$); $^{29}$Si{$^1$H} NMR (CD$_2$Cl$_2$): δ = 15.6 ppm (s); $^7$Li NMR (194 MHz, CD$_2$Cl$_2$): δ = −0.9 ppm (s); $^{31}$P{$^1$H} NMR (117 MHz, CD$_2$Cl$_2$): δ = −144.0 ppm (h, $^1J_{PF}$ = 710 Hz); $^{19}$F NMR (283 MHz, CD$_2$Cl$_2$): δ = −73.7 ppm (d, $^1J_{PF}$ = 710 Hz). IR $\tilde{v}$ = 2962(w), 2885(vw), 1456(vw), 1410(vw), 1351(vw), 1258(m), 1057(s), 1011(s), 923(w), 789(vs), 701(w), 661(w), 635(w), 556(m), 466(m). MS (ESI$^+$): *m/z* 359.1893% [*M*]$^+$ − PF$_6$ (95).

[Na(1,2-disila[18]crown-6)PF$_6$] (**2**): At ambient temperature, 48 mg (0.3 mmol, 1 equiv) of NaPF$_6$ was added to 100 mg (0.28 mmol, 1 equiv) of 1,2-disila[18]crown-6 in 10 mL of dichloromethane. The suspension was stirred for 1 h, followed by filtration and removal of the solvent. The residue was washed twice with 10 mL of *n*-pentane and was dried in vacuo. Recrystallization from dichloromethane: benzene (2:1) at −35 °C yielded 44% (64 mg, 0.12 mmol) of **2** in form of colorless blocks after 1 day. $^1$H NMR (300 MHz, CD$_2$Cl$_2$): δ = 0.29 (s, 12H, C$H_3$), 3.61–3.63 (m, 4H, C$H_2$), 3.67 (s, 12H, C$H_2$), 3.80–3.83 ppm (m, 4H, C$H_2$); $^{13}$C{$^1$H} NMR (75 MHz, CD$_2$Cl$_2$): δ = −0.4 (s, CH$_3$), 62.8 (s, CH$_2$), 69.9 (s, CH$_2$), 72.0 (s, CH$_2$), 72.9 ppm (s, CH$_2$); $^{29}$Si{$^1$H} NMR (CD$_2$Cl$_2$): δ = 14.3 ppm (s); $^{31}$P{$^1$H} NMR (117 MHz, CD$_2$Cl$_2$): δ = −143.9 ppm (h, $^1J_{PF}$ = 710 Hz); $^{19}$F NMR (283 MHz, CD$_2$Cl$_2$): δ = −74.8 ppm (d, $^1J_{PF}$ = 710 Hz). IR $\tilde{v}$ = 2912(w), 2880(w), 1457(w), 1399(w), 1350(w), 1291(w), 1250(m), 1131(s), 1082(s), 1056(s), 955(s), 931(m), 834(vs), 816(vs), 794(s), 771(s), 740(m), 720(m), 635(m), 556(s), 504(w), 471(w). MS (ESI$^+$): $m/z$ 375.1634% $[M]^+$ − PF$_6$ (100).

1,2,4,5-Tetrasila[18]crown-6 (**3**): 0.7 mL (4.1 mmol, 1 equiv) of tetraethylene glycol and 1.1 mL (8.2 mmol, 2 equiv) of NEt$_3$ in 50 mL of THF was simultaneously, with 1.30 g (4.1 mmol, 1 equiv) of O(Si$_2$Me$_4$Cl)$_2$ in 50 mL of THF, dropped into a three-neck flask with 50 mL of stirred THF. The resulting white suspension was stirred for 12 h. Subsequently, the solvent was removed under reduced pressure, the product was extracted with 50 mL of $n$-pentane followed by filtration. The solvent was removed in vacuo, and 85% (1.5 g, 3.5 mmol) of **3** was obtained in form of a colorless oil. $^1$H NMR (300 MHz, CD$_2$Cl$_2$): δ = 0.20 (s, 12H, C$H_3$), 0.22 (s, 12H, C$H_3$), 3.54–3.56 (m, 4H, C$H_2$), 3.60 (s, 8H, C$H_2$), 3.72–3.76 ppm (m, 4H, C$H_2$); $^{13}$C{$^1$H} NMR (75 MHz, CD$_2$Cl$_2$): δ = −0.5 (s, CH$_3$), 2.9 (s, CH$_3$), 63.9 (s, CH$_2$), 71.2 (s, CH$_2$), 71.6 (s, CH$_2$), 73.1 (s, CH$_2$); $^{29}$Si{$^1$H} NMR (CD$_2$Cl$_2$): δ = 2.1 (s, $Si$O$Si$), 11.0 ppm (s, CO$Si$). IR $\tilde{v}$ = 2949(w), 2867(w), 1456(w), 1400(w), 1350(w), 1294(w), 1246(m), 1091(s), 1031(s), 947(m), 826(m), 797(s), 761(s), 682(m), 660(m), 635(m), 553(w), 546(w). MS (ESI$^+$): $m/z$ 441.1977% $[MH]^+$ (15).

[K(1,2,4,5-tetrasila[18]crown-6)PF$_6$] (**4**): 58 mg (0.32 mmol, 1 equiv) of KPF$_6$ was added to a stirred solution of 140 mg (0.32 mmol, 1 equiv) of 1,2,4,5-tetrasila[18]crown-6 in 15 mL of dichloromethane. The resulting suspension was stirred for 12 h at ambient temperature, followed by filtration. The solvent was removed in vacuo, and the product was obtained in form of a colorless, highly viscous oil. After 18 h at ambient temperature, colorless blocks were obtained, yielding 61% (120 mg, 0.2 mmol) of **4**. $^1$H NMR (300 MHz, CD$_2$Cl$_2$): δ = 0.08 (s, 12H, C$H_3$), 0.23 (s, 12H, C$H_3$), 3.54–3.57 (m, 4H, C$H_2$), 3.62 (s, 8H, C$H_2$), 3.73–3.77 ppm (m, 4H, C$H_2$); $^{13}$C{$^1$H} NMR (75 MHz, CD$_2$Cl$_2$): δ = −0.6 (s, CH$_3$), 2.8 (s, CH$_3$), 63.6 (s, CH$_2$), 71.0 (s, CH$_2$), 71.1 (s, CH$_2$), 73.1 ppm (s, CH$_2$); $^{29}$Si{$^1$H} NMR (CD$_2$Cl$_2$): δ = 2.7 (s, $Si$O$Si$), 11.9 ppm (s, CO$Si$); $^{31}$P{$^1$H} NMR (117 MHz, CD$_2$Cl$_2$): δ = −143.9 ppm (h, $^1J_{PF}$ = 710 Hz); $^{19}$F NMR (283 MHz, CD$_2$Cl$_2$): δ = −73.8 ppm (d, $^1J_{PF}$ = 710 Hz). IR $\tilde{v}$ = 2948(w), 2886(w), 1470(w), 1458(w), 1401(w), 1360(w), 1349(w), 1301(w), 1247(m), 1126(m), 1110(m), 1095(m), 1085(m), 1065(m), 1051(m), 1017(m), 945(m), 931(m), 916(m), 825(vs), 797(vs), 762(vs), 738(m), 719(w), 684(m), 659(m), 555(s), 441(w), 427(w), 414(w). MS (ESI$^+$): $m/z$ 479.1531% $[M]^+$ − PF$_6$ (100).

[Ba(1,2-disila[15]crown-5)OTf$_2$]$_2$ (**5**): 119 mg (0.27 mmol, 1 equiv) of BaOTf$_2$ was added to 84 mg (0.27 mmol, 1 equiv) of 1,2-disila[15]crown-5 in 15 mL of dichloromethane. The suspension was stirred for 18 h followed by filtration. The solvent was removed under reduced pressure, and the residue was washed twice with 15 mL of $n$-pentane. The product was recrystallized from dichloromethane and pentane (2:5). After 1 day at ambient temperature, colorless plates of **5** were obtained with 22% (87 mg, 0.06 mmol) yield. $^1$H NMR (300 MHz, CD$_2$Cl$_2$): δ = 0.37 (s, 24H, C$H_3$), 3.71–4.04 ppm (m, 32H, C$H_2$); $^{13}$C{$^1$H} NMR (75 MHz, CD$_2$Cl$_2$): δ = −0.8 (s, CH$_3$), 62.0 (s, CH$_2$), 69.0 (s, CH$_2$), 70.3 (s, CH$_2$), 72.6 (s, CH$_2$), 120.9 ppm (q, $^1J_{CF}$ = 322 Hz, CF$_3$); $^{19}$F NMR (283 MHz, CD$_2$Cl$_2$): δ = −79.4 ppm (s, CF$_3$); $^{29}$Si{$^1$H} NMR (CD$_2$Cl$_2$): δ = 17.8 ppm (s); IR $\tilde{v}$ = 2952(w), 2869(w), 1468(w), 1358(w), 1263(s), 1228(s), 1171(s), 1156(s), 1121(m), 1084(s), 1061(s), 1030(s), 948(s), 917(m), 867(m), 838(s), 793(s), 770(s), 728(s), 631(s), 575(s), 515(s), 454(w), 416(w); MS (LIFDI$^+$): $m/z$ 331.136% [1,2-disila[15]crown-5+Na]$^+$ (100).

## 4. Conclusions

In this work, the competing coordination ability of C-, Si/C-, and fully Si-bonded O atoms was studied. 1,2-disila[18]crown-6 as well as 1,2,4,5-tetrasila[18]crown-6 turned out to be suitable ligands, since the presence of Si$_2$ units further increases the ring diameter in comparison to the organic crown ether [18]crown-6. Single crystals of [Li(1,2-disila[18]crown-6)]PF$_6$ (**1**) and of [Na(1,2-disila[18]crown-6)PF$_6$] (**2**) were obtained and revealed a divergent coordination of the cation. In **1**, the highly flexible ligand completely saturates the coordination sphere of Li$^+$, while the PF$_6$ anion does not show any interaction with the cation. The Li$^+$ ion is preferably coordinated by the Si- and C-bonded O atoms. Contrary to that, Na$^+$ shows stronger interaction with the C-bonded O atoms of 1,2-disila[18]crown-6. Only one of the Si/C-bonded O atoms participates in the coordination. As a result, the Me groups of the Si-based part of the ligand remain in the staggered conformation, which is also the preferred geometry of the free ligand [18]. Contrary to Compound **1**, the interaction of the Si/C-bonded O atoms with the cation does not compensate for the required change of conformation. The energy effort of 1,2-disila[18]crown-6 for adopting the geometry of the Li$^+$ and Na$^+$ complex was determined by DFT calculations. $\Delta E_{geom}$, in the case of **1**, has a value of 77.58 kJ·mol$^{-1}$, which is considerably increased. In contrast, the ligand shows with 29.24 kJ·mol$^{-1}$ smaller energy changes by coordination of Na$^+$, which can be partially attributed to the staggered arrangement of the Si-bonded methyl groups. It follows that the electrostatic attraction between the hybrid-bonded O atoms and Na$^+$ do not compensate for the required energy effort of the ecliptic arranged methyl groups. The Lewis acids therefore show a major impact on the coordinative properties of the different types of O atoms within hybrid crown ethers.

Similar coordination modes were also found in 1,2,4,5-tetrasila[18]crown-6 (**3**), which incorporates three types of O atoms: C-, Si/C-, and Si-bonded ones. Ordinary, K$^+$ perfectly fits in [18]crown-6 and 1,2-disila[18]crown-6 [18]. The presence of two disilane units leads to a further increase of the ligand such that **3** does not match with K$^+$. The completely Si-bonded O atom, which requires the highest amount of energy to adopt the complex geometry [19], does not participate in the coordination. The complexation of the heavier homologue Rb$^+$ by **3** is an issue of current investigation. In this study, no superiority in coordination ability of each of the different types of O atoms was found.

The experiment on the inverse case, e.g., small ligands with large cations in 1:1 stoichiometry, leads to the dinuclear complex (**5**), which is bridged by four triflate anions. This crystal structure represents an initial outlook on the ability of disila-crown ethers to build sandwich complexes. Therefore, reactions in 2:1 stoichiometry of ligand to salt are crucial.

**Acknowledgments:** This work was financially supported by the Deutsche Forschungsgemeinschaft (DFG).

**Author Contributions:** Kirsten Reuter performed the syntheses and analytics of Compounds **1**–**4**, DFT calculations, and interpretations and wrote the paper; Fabian Dankert conducted the synthesis and characterization of Compound **5**; Carsten Donsbach accomplished the measurement, crystal structure solution, and refinement of **5**; Carsten von Hänisch contributed to interpretation and led the over-arching research project.

## References

1. Liebau, F. Untersuchungen über die Grösse des Si–O–Si–Valenzwinkels. *Acta Cryst.* **1961**, *14*, 1103–1109. [CrossRef]
2. Almenningen, A.; Bastiansen, O.; Ewing, V.; Hedberg, K.; Trætteberg, M. The Molecular Structure of Disiloxane, (SiH$_3$)$_2$O. *Acta Chem. Scand.* **1963**, *17*, 2455–2460. [CrossRef]
3. Stone, F.G.A.; Seyferth, D. The Chemistry of Silicon Involving Probable Use of *d*-Type Orbitals. *J. Inorg. Nucl. Chem.* **1955**, *1*, 112–118. [CrossRef]
4. Craig, D.P.; Maccoll, A.; Nyholm, R.S.; Orgel, L.E. Chemical bonds involving *d*-orbitals. Part I. *J. Chem. Soc.* **1954**, 332–353. [CrossRef]
5. Eméleus, H.J.; Onyszchuk, M. The Reaction of Methyldisoxanes and 1:1 Dimethyldisilthiane with Boron and Hydrogen Halides. *J. Chem. Soc.* **1958**, 604–609. [CrossRef]
6. Pitt, C.G. Hyperconjugation and its Role in Group IV Chemistry. *J. Organomet. Chem.* **1973**, *61*, 49–70. [CrossRef]
7. Shambayati, S.; Schreiber, S.L.; Blake, J.F.; Wierschke, S.G.; Jorgenson, W.L. Structure and basicity of silyl ethers: A crystallographic and ab initio inquiry into the nature of silicon-oxygen interactions. *J. Am. Chem. Soc.* **1990**, *112*, 697–703. [CrossRef]
8. Cypryk, M.; Apeloig, Y. Ab Initio Study of Silyloxonium Ions. *Organometallics* **1997**, *16*, 5938–5949. [CrossRef]
9. Gillespie, R.J.; Johnson, S.A. Study of Bond Angles and Bond Lenghts in Disiloxane and Related Molecules in Terms of the Topology of the Electron Density and Its Laplacian. *Inorg. Chem.* **1997**, *36*, 3031–3039. [CrossRef] [PubMed]

10. Grabowski, S.J.; Hesse, M.F.; Paulmann, C.; Luger, P.; Beckmann, J. How to Make the Ionic Si–O Bond More Covalent and the Si–O–Si Linkage a Better Acceptor for Hydrogen Bonding. *Inorg. Chem.* **2009**, *48*, 4384–4393. [CrossRef] [PubMed]

11. Passmore, J.; Rautiainen, J.M. On The Lower Basicity of Siloxanes Compared to Ethers. *Eur. J. Inorg. Chem.* **2012**, 6002–6010. [CrossRef]

12. Cameron, T.S.; Decken, A.; Krossing, I.; Passmore, J.; Rautiainen, J.M.; Wang, X.; Zeng, X. Reactions of a Cyclodimethyldisiloxane $(Me_2SiO)_6$ with Silver Salts of Weakly Coordinating Anions; Crystal Structures of $[Ag(Me_2SiO)_6][Al]$ ([Al] = FAl{OC(CF_3)_3}_3], [Al{OC(CF_3)_3}_4]$) and Their Comparison with $[Ag(18\text{-Crown-}6)]_2[SbF_6]_2$. *Inorg. Chem.* **2013**, *52*, 3113–3126. [CrossRef] [PubMed]

13. Decken, A.; Passmore, J.; Wang, W. Cyclic Dimethylsiloxanes as Pseudo Crown Ethers: Syntheses and Characterization of $Li(Me_2SiO)_5[Al{OC(CF_3)_3}_4]$, $Li(Me_2SiO)_6[Al{OC(CF_3)_3}_4]$ and $Li(Me_2SiO)_6[Al{OC(CF_3)_2Ph}_4]$. *Angew. Chem. Int. Ed.* **2006**, *45*, 2773–2777. [CrossRef] [PubMed]

14. Ritch, J.S.; Chivers, T. Silicon Analogues of Crown Ethers and Cryptands: A New Chapter in Host–Guest Chemistry? *Angew. Chem. Int. Ed.* **2007**, *46*, 4610–4613. [CrossRef] [PubMed]

15. Von Hänisch, C.; Hampe, O.; Weigend, F.; Stahl, S. Stepwise Synthesis and Coordination Compound of an Inorganic Cryptand. *Angew. Chem. Int. Ed.* **2007**, *46*, 4775–4779. [CrossRef] [PubMed]

16. Inoue, Y.; Ouchi, M.; Hakushi, T. Molecular Design of Crown Ethers. 3. Extraction of Alkaline Earth and Heavy Metal Picrates with 14- to 17-Crown-5 and 17- to 22-Crown-6. *Bull. Chem. Soc. Jpn.* **1985**, *58*, 525–530. [CrossRef]

17. Ouchi, M.; Inoue, Y.; Kanzaki, T.; Hakushi, T. Ring-contracted Crown Ethers: 14-Crown-5, 17-Crown-6, and Their Sila-analogues. Drastic Decrease in Cation-binding Ability. *Bull. Chem. Soc. Jpn.* **1984**, *57*, 887–888. [CrossRef]

18. Reuter, K.; Buchner, M.R.; Thiele, G.; von Hänisch, C. Stable Alkali-Metal Complexes of Hybrid Disila-Crown Ethers. *Inorg. Chem.* **2016**, *55*, 4441–4447. [CrossRef] [PubMed]

19. Reuter, K.; Thiele, G.; Hafner, T.; Uhlig, F.; von Hänisch, C. Synthesis and coordination ability of a partially silicon based crown ether. *Chem. Commun.* **2016**, *52*, 13265–13268. [CrossRef] [PubMed]

20. Dalley, N.K.; Lamb, J.D.; Nazarenko, A.Y. Crystal Structure of the Complex of 1,4,7,10,13,16-Hexacyclooctadecane with Lithium Picrate Dihydrate. *Supramol. Chem.* **1997**, *8*, 345–350.

21. Chadwick, S.; Ruhlandt-Senge, K. The Remarkable Structural Diversity of Alkali Metal Pyridine-2-thiolates with Mismatched Crown Ethers. *Chem. Eur. J.* **1998**, *4*, 1768–1780. [CrossRef]

22. Akutagawa, T.; Hasegawa, T.; Nakamura, T.; Takeda, S.; Inabe, T.; Sugiura, K.; Sakata, Y.; Underhill, A.E. Ionic Channel Structures in $[(M^+)_x([18]crown\text{-}6)][Ni(dmit)_2]_2$ Molecular Conductors. *Chem. Eur. J.* **2001**, *7*, 4902–4912. [CrossRef]

23. Dankert, F.; Reuter, K.; Donsbach, C.; von Hänisch, C. A Structural Study of Alkaline Earth Metal Complexes with Hybrid Disila-Crown Ethers. *Dalton Trans.* **2017**. [CrossRef] [PubMed]

24. Olsher, U.; Izatt, R.M.; Bradshaw, J.S.; Dalley, N.K. Coordination chemistry of lithium ion: A crystal and molecular structure review. *Chem. Rev.* **1991**, *91*, 137–164. [CrossRef]

25. Gingl, F.; Hiller, W.; Strähle, J. [Li(12-Krone-4)]Cl: Kristallstruktur und IR-Spektrum. *Z. Anorg. Allg. Chem.* **1991**, *606*, 91–96. [CrossRef]

26. Liddle, S.T.; Clegg, W. A homologous series of crown-ether-complexed alkali metal amides as discrete ion-pair species: Synthesis and structures of $[M(12\text{-crown-}4)_2][PyNPh·PyN(H)Ph]$ ($M$ = Li, Na and K). *Polyhedron* **2003**, *22*, 3507–3513. [CrossRef]

27. Feldmann, C.; Okrut, A. Two Tricyclic Polychalcogenides in $[Li(12\text{-crown-}4)_2]_2[Sb_2Se_{12}]$ and $[Li(12\text{-crown-}4)_2]_4[Te_{12}]·(12\text{-crown-}4)_2$. *Z. Anorg. Allg. Chem.* **2009**, *635*, 1807–1811. [CrossRef]

28. *Turbomole*, version 7.0; Turbomole Is a Development of University of Karlsruhe and Forschungszentrum Karlsruhe 1989–2007, Turbomole GmbH 2016. Turbomole GmbH: Karlsruhe, Germany, 2007.

29. Weigend, F.; Ahlrichs, R. Balanced bass sets of split valence, triple zeta valence and quadruple zeta valence quality for H to Rn: Design and assessment of accuracy. *Phys. Chem. Chem. Phys.* **2005**, *7*, 3297–3305. [CrossRef] [PubMed]

30. Weigend, F. Accurate Coulomb-fitting basis sers for H to Rn. *Phys. Chem. Chem. Phys.* **2006**, *8*, 1057–1065. [CrossRef] [PubMed]

31. Dolg, M.; Stoll, H.; Savin, A.; Preuss, H. Energy-adjusted pseudopotentials for the rare earth elements. *Theoret. Chim. Acta* **1989**, *75*, 173–194. [CrossRef]

32. Stoll, H.; Metz, B.; Dolg, M. Relativistic energy-consistent pseudopotentials—Recent developments. *J. Comput. Chem.* **2002**, *23*, 767–778. [CrossRef] [PubMed]

33. Grimme, S.; Antony, J.; Ehrlich, S.; Krieg, H. A consistent and accurate ab initio parametrization of density functional dispersion correction (DFT-D) for the 94 elements H–Pu. *J. Chem. Phys.* **2010**, *132*, 154104–154119. [CrossRef] [PubMed]

34. Grimme, S.; Ehrlich, S.; Goerigk, L. Effect of the damping function in dispersion corrected density functional theory. *J. Comput. Chem.* **2011**, *32*, 1456–1465. [CrossRef] [PubMed]

35. Willcott, M.R. MestRe Nova. *J. Am. Chem. Soc.* **2009**, *131*, 13180. [CrossRef]

36. Klamt, A.; Schüürmann, G. COSMO: A new approach to dielectric screening in solvents with explicit expressions for the screening energy and its gradient. *J. Chem. Soc. Perkin. Trans.* **1993**, *2*, 799–805. [CrossRef]

37. Sheldrick, G.M. *SHELXL14*; Program for the Refinement of Crystal Structures; Universität Göttingen: Göttingen, Germany, 2014.

38. Dolomanov, O.V.; Bourhis, L.J.; Hildea, R.J.; Howard, J.A.K.; Puschmann, H. *Olex2*: A complete structure solution, refinement and analysis program. *J. Appl. Crystallogr.* **2009**, *42*, 339–341. [CrossRef]

39. Putz, H.; Brandenburg, K. *Diamond—Crystal and Molecular Structure Visualization*; Crystal Impact: Bonn, Germany, 2012.

# 13

# Na$_{1+y}$VPO$_4$F$_{1+y}$ (0 ≤ y ≤ 0.5) as Cathode Materials for Hybrid Na/Li Batteries

**Nina V. Kosova * and Daria O. Rezepova**

Institute of Solid State Chemistry and Mechanochemistry SB RAS, 18 Kutateladze, Novosibirsk 630128, Russia; rezepova_do@yahoo.com

* Correspondence: kosova@solid.nsc.ru

Academic Editor: Christian M. Julien

**Abstract:** Using Rietveld-refined X-ray diffraction (XRD), Fourier transform infrared spectroscopy (FTIR) and electrochemical cycling, it was established that among sodium vanadium fluorophosphate compositions Na$_{1+y}$VPO$_4$F$_{1+y}$ (0 ≤ y ≤ 0.75), the single-phase material Na$_{1.5}$VPO$_4$F$_{1.5}$ or Na$_3$V$_2$(PO$_4$)$_2$F$_3$ with a tetragonal structure (the $P4_2/mnm$ S.G.) is formed only for y = 0.5. The samples with y < 0.5 and y > 0.5 possessed different impurity phases. Na$_3$V$_2$(PO$_4$)$_2$F$_3$ could be considered as a multifunctional cathode material for the fabrication of lithium-ion and sodium-ion high-energy batteries. The reversible discharge capacity of 116 mAh·g$^{-1}$ was achieved upon cycling Na$_3$V$_2$(PO$_4$)$_2$F$_3$ in a hybrid Na/Li cell. Decrease in discharge capacity for the other samples was in accordance with the amount of the electrochemically active phase Na$_3$V$_2$(PO$_4$)$_2$F$_3$. Na$_3$V$_2$(PO$_4$)$_2$F$_3$ showed good cycleability and a high rate of performance, presumably due to operation in the mixed Na/Li electrolyte. The study of the structure and composition of charged and discharged samples, and the analysis of differential capacity curves showed a negligible Na/Li electrochemical exchange, and a predominant sodium-based cathode reaction. To increase the degree of the Na/Li electrochemical exchange in Na$_3$V$_2$(PO$_4$)$_2$F$_3$, it needs to be desodiated first in a Na cell, and then cycled in a lithium cell. In this case, the electrolyte would be enriched with the Li ions.

**Keywords:** Na$_3$V$_2$(PO$_4$)$_2$F$_3$; NaVPO$_4$F; XRD; hybrid Na/Li batteries

---

## 1. Introduction

In recent years, investigations in the energy saving field have focused on the development of sodium-ion batteries, as sodium is much more abundant, ecologically sound, and cheaper than lithium [1]. The main issue for Na-ion based systems is their lower energy density compared to the Li-ion systems, and the smaller choice of electrode materials available for sodium technology. There has been significant interest in polyanion-based active cathode materials as safer alternatives to traditional oxide cathodes [2]. Among them, vanadium-based polyanion compounds such as sodium vanadium phosphate and sodium vanadium fluorophosphates have attracted great attention because of their high operating potential based on the inductive effects of both PO$_4^{3-}$ and F$^-$ anions, and high expected specific capacity due to the multiple accessible oxidation states of vanadium. Sodium vanadium fluorophosphates show higher energy density than Na$_3$V$_2$(PO$_4$)$_3$, owing to the substitution of one PO$_4^{3-}$ group for F$^-$ in the structure, leading to an increase in the sodium ion insertion voltage.

With respect to sodium–vanadium fluorophosphates, three phases are described in the literature: NaVPO$_4$F, Na$_3$(VO)$_2$(PO$_4$)$_2$F and Na$_3$V$_2$(PO$_4$)$_2$F$_3$. NaVPO$_4$F was first proposed as a cathode material by Barker et al. in 2003 [3]. Its theoretical specific capacity, assuming the reversible intercalation of one Na$^+$ ion per formula unit (f.u.), amounts to 143 mAh·g$^{-1}$. It is reported that NaVPO$_4$F exists in two polymorphs: the low-temperature monoclinic phase (S.G. $C2/c$), and the high-temperature tetragonal phase (S.G. $I4/mmm$) [4]. Tetragonal NaVPO$_4$F is structurally related to the known Na-ion conductor, $\alpha$-$Na_3Al_2(PO_4)_2F_3$ [5], and possesses the lattice parameters $a = b = 6.387$ Å and $c = 10.734$ Å. However, some additional peaks can be found on the XRD patterns of $NaVPO_4F$ when prepared by the ceramic method [3]. The presence of $Na_3V_2(PO_4)_3$ or other impurities is possible because sublimation of $VF_3$ can occur at high temperatures.

Sauvage et al. [6] found the "$NaVPO_4F$" stoichiometry based on a single structural consideration rather suspicious. They re-investigated the system, tuning the synthesis conditions previously reported by Barker et al. [3], and obtained a single-phase material assigned to $Na_{1.5}VOPO_4F_{0.5}$ crystallized in the $I4/mmm$ S.G. Recently, Tsirlin et al. [7] have proposed the $P42/mnm$ S.G. for the room temperature phase among various polymorphs of this material at different temperatures. Actually, $Na_{1.5}VOPO_4F_{0.5}$ or $Na_3(VO)_2(PO_4)_2F$ belongs to a family of the mixed-valence sodium vanadium fluorophosphates $Na_3V_2O_{2y}(PO_4)_2F_{3-2y}$ ($0 \leq y \leq 1$). These materials have a similar tetragonal structure, but differ by the amount of the $F^-$ atoms replaced by $O^{2-}$ in the $8j$ positions and the oxidation state of the V ions, which varies between 3+ and 4+, with a concomitant modification of the physical and electrochemical properties of the material [8–11]. Depending on $y$, the average voltage of the cell varies from 3.89 V for $y = 0$ to 3.83 V for $y = 0.5$, which is ascribed to the formation of highly covalent vanadyl-type bonds $(V=O)^{2+}$ at the apex of the bioctahedra $V_2O_8F_{3-y}O_y$ that "continuously" replace the more ionic V–F bonds as $y$ increases [9]. The extreme members of this family are $Na_3(VO)_2(PO_4)_2F$ for $y = 1$ (often written as $Na_{1.5}V^{IV}OPO_4F_{0.5}$), and $Na_3V^{III}_2(PO_4)_2F_3$ (or $Na_{1.5}V^{III}PO_4F_{1.5}$) for $y = 0$. Since the ionic radii of $V^{3+}$ and $V^{4+}$ in the octahedral coordination are 0.64 Å and 0.58 Å, respectively, the cell volume increase would imply higher $V^{3+}$ content in the samples.

The third of the sodium vanadium fluorophosphates, $Na_3V_2(PO_4)_2F_3$, was proposed by Barker et al. [12] in 2006. Its theoretical capacity of 192.4 mAh·$g^{-1}$ corresponds to the extraction of three $Na^+$ ions from $Na_3V_2(PO_4)_2F_3$, however, a reversible extraction of only two $Na^+$ ions has been achieved in practice, resulting in a capacity of about 120 mAh·$g^{-1}$. As reported by Meins et al. [5], $Na_3V_2(PO_4)_2F_3$ has a tetragonal crystal structure, with the S.G. $P4_2/mnm$ and cell parameters $a = 9.047(2)$ Å and $c = 10.705(2)$ Å. In this structure, the $V^{3+}$ ion is placed in the center of the $VO_4F_2$ octahedra, which are bridged together by one $F^-$ atom forming the $V_2O_8F_3$ bioctahedra, alternately connected by the $PO_4$ tetrahedra. This results in a stable $3D$ framework with large tunnels in the [110] and [1-10] directions. Recently, Bianchini et al. [13], using synchrotron X-ray powder diffraction, revealed a small but significant orthorhombic distortion in $Na_3V_2(PO_4)_2F_3$ ($b/a = 1.002$) and identified $Na_3V_2(PO_4)_2F_3$ in another orthorhombic crystal structure with the $Amam$ S.G. This new structure preserves the framework but modifies the distribution of the $Na^+$ ions.

An overview of the literature data related to these three sodium fluorophosphates leads to some doubt about the real existence of these three different compounds, especially that of $NaVPO_4F$. This phase and its different polymorphs [4,14] have already been questioned by some authors [6]. Besides, the diffractograms of these phases are very similar, and their charge–discharge curves display two voltage plateau of close length at the same voltages.

The resulting composition of the material, its crystal structure and electrochemical performance, may be influenced by the synthetic method. As shown in Table 1, sodium vanadium fluorophosphates can be synthesized by various methods such as the solid-state reaction [3,6,8,9,12,13,15–17], solution-based carbothermic reduction [18,19], hydrothermal synthesis [5,10,11], sol–gel [4,20] and spray-drying synthesis [21]. The solid-state methods involve high-temperature treatment with a long period of mixing (up to 24 h) in low-energy mills. On the contrary, short-time high-energy mechanical activation significantly enhances solid-state reactions, thereby reducing the duration and temperature of the subsequent heat treatment and preparation of materials in a highly dispersed state [22].

Barker et al. first described the concept of a hybrid-ion battery, where a non-lithium-containing $Na_3V_2(PO_4)_2F_3$ cathode was used in conjunction with a conventional graphite anode material and a lithium electrolyte [23]. A hybrid-ion system composed of a traditional lithium-based non-aqueous electrolyte and a metallic lithium anode would be favored for the investigation of electrochemical properties of sodium-based materials, in order to develop theoretical basics, and to provide directions for the construction of new high-energy rechargeable batteries. However, the mechanism of alkali ion insertion/deinsertion in these systems is not yet well understood [4,20,21,24–26].

We report here the energy-efficient mechanochemically-assisted solid-state synthesis of sodium vanadium fluorophosphates $Na_{1+y}VPO_4F_{1+y}$ ($0 \leq y \leq 0.75$) and the results of the comparative study of their crystal structure and electrochemistry in hybrid-ion cells.

**Table 1.** Synthetic methods for the preparation of $Na_3V_2(PO_4)_2F_3$.

| Synthetic Method | Treatment Temperature | Treatment Time | Reference |
|---|---|---|---|
| Solid-state synthesis, CTR * | 600–800 °C | 8 h | Gover et al. [12] |
| Solid-state synthesis, CTR, MA ** | 750 °C | 1.5 h | Shakoor et al. [16] |
| Solid-state synthesis, CTR, MA | 800 °C | 1 h | Bianchini et al. [13] |
| Solution-based solid-state synthesis | 650 °C | 8 h | Song et al. [18] |
| Hydrothermal reaction | 180 °C | 64 h | Le Meins et al. [5] |
| Sol–gel preparation | 650 °C | 8 h | Jiang et al. [20] |
| Spray drying | 600 °C | 2 h | Eshraghi et al. [21] |
| Solid-state synthesis, CTR, MA | 650 °C | 1 h | This work |

* Carbothermal reduction; ** mechanical activation.

## 2. Results and Discussion

### 2.1. Crystal Structure and Morphology

Figure 1 shows the XRD patterns of the as-prepared products $Na_{1+y}VPO_4F_{1+y}$, with $y = 0$; 0.25; 0.5; 0.6 and 0.75. All samples were well-crystallized, however, only the sample with $y = 0.5$ was a single-phase material, which was assigned to $Na_3V_2(PO_4)_2F_3$ (hereinafter $Na_{1.5}VPO_4F_{1.5}$). The chemical composition of the sample with $y = 0.5$ determined by the Energy dispersive X-ray spectroscopy (EDX) analysis confirmed that the Na/V ratio was close to 1.5 (1.46). The other samples contained $Na_3V_2(PO_4)_2F_3$ as a main phase, and some impurities. For instance, $Na_3V_2(PO_4)_3$ [PDF 00-053-0018], $VPO_4$ [PDF 01-076-2023] or $NaVOPO_4$ [PDF 01-089-6316] were found as impurity phases in the samples with $y < 0.5$ (Table 2). When $y$ exceeded 0.5, new weak reflections appeared, which could be assigned to other impurity phases such as $Na_3VF_6$ [PDF 00-029-1286], $NaV_2O_5$ [PDF 04-013-4702], $Na_3PO_4$ [PDF 04-015-4963].

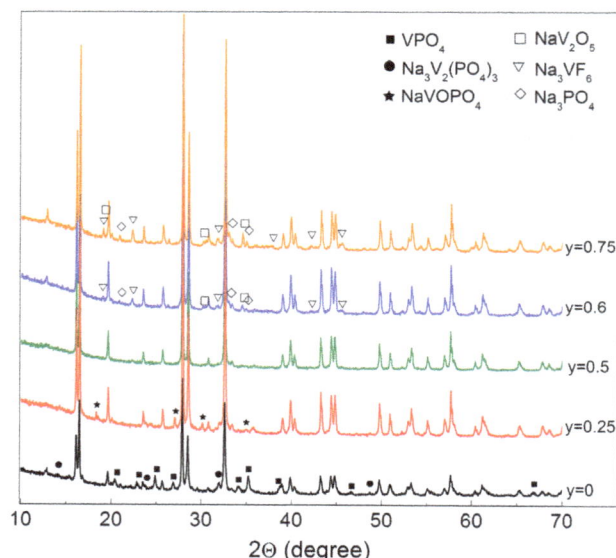

**Figure 1.** XRD patterns of the products obtained after subsequent annealing of the activated mixtures. $(1+y)NaF + VPO_4$ ($0 \leq y \leq 0.75$) at 650 °C in Ar flow.

**Table 2.** Phase composition of the as-prepared samples (%).

| Sample ($y$) | Main Phase $Na_3V_2(PO_4)_2F_3$ | Impurities | | |
|:---:|:---:|:---:|:---:|:---:|
| 0 | 80.6 | $Na_3V_2(PO_4)_3$—4.6 | $VPO_4$—14.8 | - |
| 0.25 | 92.5 | - | - | $NaVOPO_4$—7.5 |
| 0.5 | 100 | - | - | - |
| 0.6 | 86.6 | $Na_3VF_6$—2.9 | $NaV_2O_5$—4.0 | $Na_3PO_4$—6.5 |
| 0.75 | 78.9 | $Na_3VF_6$—6.6 | $NaV_2O_5$—4.8 | $Na_3PO_4$—9.7 |

As mentioned above, two structural models ($P4_2/mnm$ and $Amam$) are known for $Na_3V_2(PO_4)_2F_3$, while the $I4/mmm$ S.G. is used for $NaVPO_4F$. Different compositions of sodium vanadium fluorophosphates $Na_3V_2O_{2y}(PO_4)_2F_{3-2y}$ described in the literature are summarized in Table 3, along with their space groups and the refined lattice parameters. It is seen that there is a strong dependence of the cell volume on the y value. For the samples with $y = 0$, the calculated volume is in the range of 876–879 Å$^3$, but when y approaches 1, it decreases to 865–868 Å$^3$ due to partial substitution of the F$^-$ ions by the O$^{2-}$ ions and oxidation of the V$^{3+}$ ions to the V$^{4+}$ ions with lower ionic radius, thus indicating the formation of mixed-valence vanadium materials. In $Na_3V_2(PO_4)_2F_3$ with the $P4_2/mmm$ S.G.; the Na ions occupy the 8i sites, which have two different coordinates: two Na atoms are in the Na(1) sites and one Na atom is in the Na(2) site with the site occupancy factors (SOFs) of 1 and $\frac{1}{2}$, respectively. The $Amam$ S.G. structure preserves the tetragonal framework but strongly impacts the sodium distribution in the planes. According to this model, the Na$^+$ ions are distributed in three crystallographic sites. Na(1) is placed in the center of a pyramidal site and its refined SOF is higher than 0.95, which means that this site is essentially fully occupied. Instead, the Na(2) and Na(3) sites cannot be simultaneously occupied, because of the vicinity of the Na ions ($d = 0.93$ Å). With respect to the surrounding environment, the Na(2) and Na(3) ions are not equivalent, since the first one occupies a pyramidal site, while the second one sits in a capped prism [13]. The structure of $NaVPO_4F$ was described by the authors [3] in the $I4/mmm$ S.G.; since it showed higher symmetry than $Na_3V_2(PO_4)_2F_3$.

**Table 3.** Lattice parameters of $Na_3V_2O_{2y}(PO_4)_2F_{3-2y}$ ($0 \leq y \leq 1$) described in the literature using different space groups.

| Composition | S.G. | $a = b$, Å | $c$, Å | Volume, Å$^3$ | Ref. |
|:---:|:---:|:---:|:---:|:---:|:---:|
| NaVPO$_4$F | $I4/mmm$ | 6.387(2) | 10.734(3) | 438.1 | [3] |
| | $I4/mmm$ | 6.38 | 10.72 | 436.4 | [4] |
| $Na_3V_2(PO_4)_2F_3$ | $P4_2/mnm$ | 9.047(2) | 10.705(2) | 876.2(3) | [5] |
| | $P4_2/mnm$ | 9.0378(3) | 10.7482(4) | 877.94(6) | [12] |
| | $P4_2/mnm$ | 9.0358(2) | 10.7403(4) | 876.90(4) | [13] |
| | $P4_2/mnm$ | 9.04 | 10.74 | 877.69 | [17] |
| | $P4_2/mnm$ | 9.05 | 10.74 | 876.9 | [18] |
| | $P4_2/mnm$ | 9.04 | 10.73 | 877.0 | [19] |
| $Na_3V_2(PO_4)_2F_3$ | $Amam$ | $a = 9.0288(6)$ $b = 9.0426(6)$ | 10.7402(5) | 876.88(9) | [13] |
| $Na_{1.5}VOPO_4F_{0.5}$ | $I4/mmm$ | 6.37028(8) | 10.6365(2) | 431.63(1) | [6] |
| $Na_3V_2O_{2x}(PO_4)_2F_{3-2x}$ | $P4_2/mnm$ | 9.02548–9.04499 | 10.63184–0.62113 | 866.1–869.0 | [11] |
| $Na_{1.5}VO_{0.8}PO_4F_{0.7}$ | $P4_2/mnm$ | 9.0332(1) | 10.6297(2) | 867.37(2) | [26] |
| $Na_{1.5}VOPO_4F_{0.5}$ | $P4_2/mnm$ | 9.03051(2) | 10.62002(3) | 866.064 | [7] |

Figure 2 displays the XRD patterns of the $Na_{1+y}VPO_4F_{1+y}$ samples with $y = 0$ and $y = 0.5$ after Rietveld refinement using the $P4_2/mnm$ S.G., commonly employed to describe the structure of $Na_3V_2(PO_4)_2F_3$, and the $Amam$ S.G., which characterizes an orthorhombic distortion [13] (see Materials and Methods). Both space groups appeared to be suitable for description of the structure of the $Na_3V_2(PO_4)_2F_3$ phase, though it is not surprising that the model with the low orthorhombic symmetry fit the experimental XRD patterns of the as-prepared samples better. When using the $I4/mmm$ S.G., we were unable to index the experimental

XRD pattern because of the presence of the 12.87, 23.5, 30.90, 35.97 2Θ reflections. The refined lattice parameters of the $Na_3V_2(PO_4)_2F_3$ main phase in all as-prepared samples are presented in Table 4. As can be seen, they coincided with each other and were close to the literature data [12]. Moreover, the calculated cell volumes pointed to the formation of the compounds without considerable substitution of the $F^-$ ions by the $O^{2-}$ ions, characteristic of the mixed valence phases $Na_3V_2O_{2x}(PO_4)_2F_{3-2y}$ ($0 \leq y \leq 1$), because the cell volume of the latter was significantly lower (see Table 3).

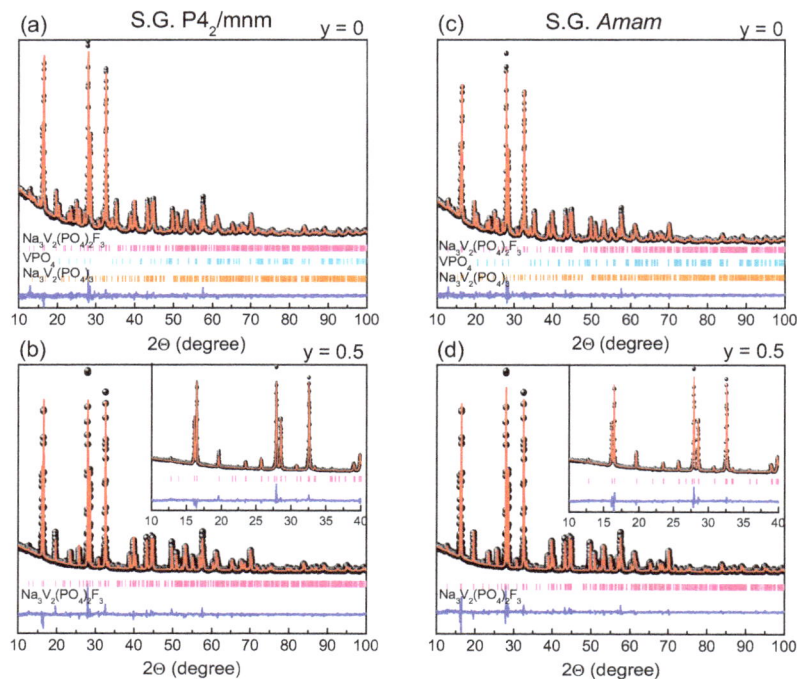

**Figure 2.** XRD patterns of the $Na_{1+y}VPO_4F_{1+y}$ samples (**a,c**) with $y = 0$ and (**b,d**) $y = 0.5$ after Rietveld refinement (**a,b**) based on the tetragonal structure with S.G. $P4_2/mnm$ and (**c,d**) on the orthorhombic structure with S.G. *Amam*.

**Table 4.** Rietveld-refined lattice parameters of the as-prepared $Na_{1+y}VPO_4F_{1+y}$ ($y = 0$; 0.25; 0.5), based on the $P4_2/mnm$ and *Amam* space groups.

| $y$ | S.G. | $a = b$, Å | $c$, Å | $V$, Å$^3$ | GOF */$R_{wp}$ |
|-----|------|-----------|--------|-----------|-----------------|
| 0.0 | $P4_2/mnm$ | 9.0376(2) | 10.7588(3) | 878.77(5) | 1.69/7.28 |
| 0.25 | $P4_2/mnm$ | 9.0372(1) | 10.7545(2) | 878.32(3) | 1.77/7.08 |
| 0.5 | $P4_2/mnm$ | 9.0393(1) | 10.7520(2) | 878.54(2) | 1.71/6.75 |
| 0.0 | *Amam* | 9.0298(5)/9.0465(5) | 10.7595(3) | 878.92(7) | 1.64/6.61 |
| 0.25 | *Amam* | 9.0296(2)/9.0450(2) | 10.7448(2) | 878.37(4) | 1.58/6.37 |
| 0.5 | *Amam* | 9.0323(2)/9.0467(2) | 10.7523(1) | 878.60(2) | 1.44/5.73 |

* Goodness of fit (GOF).

The samples were also characterized by FTIR spectroscopy. Following from Figure 3, no significant differences were observed in the FTIR spectra of the as-prepared $Na_{1+y}VPO_4F_{1+y}$ samples. In all the spectra, the characteristic vibrations of the $PO_4$ groups (the $T_d$ point group) were predominant [27]. In the spectral region of the internal modes of the $PO_4$ anion, symmetric $\nu_1$ (a singlet) and asymmetric $\nu_3$ (triply degenerated) stretching modes were located in the high-wavenumber region (900–1150 cm$^{-1}$) and were well separated from the bands due to asymmetric $\nu_4$ (triply degenerated) and symmetric $\nu_2$ (a doublet) bending vibrations that appeared in the low-wavenumber region (400–500 cm$^{-1}$). Therefore, the intense modes at 1140–1058 cm$^{-1}$ and at 558 cm$^{-1}$ on the experimental spectra corresponded to the stretching $\nu_3$ and bending $\nu_4$ vibrations of the P–O bonds in the $PO_4$ tetrahedron, respectively.

The higher intensity of the stretching mode at 944 cm$^{-1}$ for the sample with $y = 0$ indicated the presence of a significant amount of VPO$_4$ impurity.

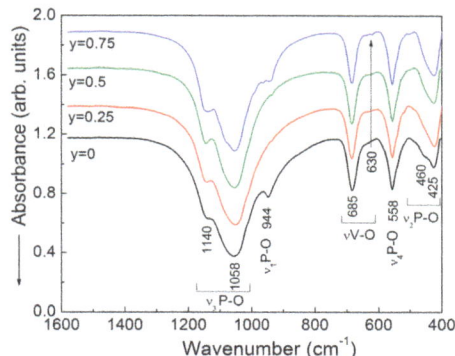

**Figure 3.** FTIR spectra of the Na$_{1+y}$VPO$_4$F$_{1+y}$ samples $(0 \leq y \leq 0.75)$.

Although all the spectra were dominated by vibrational features due to the PO$_4$ ions, transition-metal ions also registered their presence in the middle region of 600–700 cm$^{-1}$. According to [6,28], the vibrations from the V$^{3+}$–O$^{2-}$ bonds in the isolated VO$_6$ octahedra are evident at ~630 cm$^{-1}$. It is known that the signal corresponding to the characteristic short V=O vanadyl bond of the vanadium pentoxide is located at 1020 cm$^{-1}$, and this signal gradually shifts to lower wavenumbers during the reduction of V$^{5+}$ to V$^{4+}$ [29]. The occurrence of V$^{5+}$ in the VO$_6$ octahedra were not observed, indicating that the V$^{5+}$ ions were reduced to the V$^{3+}$ ions in Na$_{1+y}$VPO$_4$F$_{1+y}$.

Morphological characterization of the Na$_{1+y}$VPO$_4$F$_{1+y}$ samples carried out by scanning electron microscopy (SEM) showed that the samples consisted of a mixture of irregular-shaped fine (primary) particles with an average particle size of about 100–200 nm, and large (secondary) micron-sized particles (from 0.5 μm to several micrometers) originated as a result of agglomeration of primary particles (Figure 4). The sample with $y = 0$ had slightly smaller particle size, probably due to formation of the multi-phase composition and simultaneous suppression of the crystal growth of each phase upon annealing of the activated mixtures.

**Figure 4.** SEM images of the Na$_{1+y}$VPO$_4$F$_{1+y}$ samples with $y = 0$ and 0.5 with different magnification.

## 2.2. Electrochemistry

The electrochemical performance of the as-prepared samples was studied in hybrid Na/Li cells. Two methods were used. In the first case, the cathode material was placed in a lithium cell with a lithium anode, and the $LiPF_6$/EC+DMC (2:1) electrolyte (hereinafter Sample A), while in the second case, the cathode material was charged in a sodium cell with a sodium anode, and the $NaPF_6$/EC+DMC (2:1) electrolyte, then the cell was disassembled in a glove box, the cathode was washed with the DMC solution, and then put in a lithium cell (Sample B). Electrochemical testing was undertaken immediately following cell fabrication (typically within 10 min), thereby removing the possibility of any appreciable ion-exchange taking place between $Na_3V_2(PO_4)_2F_3$ and the electrolyte. Figure 5 shows the charge–discharge profiles and the $dQ/dV$ vs. voltage plots of the first three cycles and the 20th cycle for the samples A with $y = 0$, 025 and 0.5 within the 3.0–4.6 V voltage range at the 0.1C rate. The charge–discharge curves consisted of two plateaus with an average voltage of 3.7–3.8 V and 4.3 V. It was seen that polarization was rather small for all three samples, indicating facile reversible alkali-ion (de)insertion reactions. The potential profiles of the 20th cycle exhibited a shape very similar to that of the first cycle. This indicates that the $Li^+$ ions replaced the $Na^+$ ions at the same crystallographic sites, and this replacement did not cause significant changes to the material structure.

Three oxidation peaks and three corresponding reduction peaks were observed on the $dQ/dV$ vs. voltage plots: two close low-voltage peaks at 3.7–3.8 V, and one high-voltage peak at ~4.3 V. These values are very similar to those observed in Na cells [14,18], and the voltage difference was less than 0.3 V than expected. It was established that $Na^+$ ions at the Na2 sites are less stable than those at the Na1 sites, because they are farther shifted from the stable position [16]. Therefore, $Na^+$ ions at the Na2 sites would have a higher chemical potential than those at the Na1 sites, and thus, would be extracted at an earlier stage of charge and inserted at a later stage of discharge. According to [18], after extraction of the $Na^+$ ions from the Na2 sites, the remaining $Na^+$ ions are reorganized to the stable $Na_2V_2(PO_4)_2F_3$ configuration due to the short Na–Na distances between the ions occupying the Na1 sites.

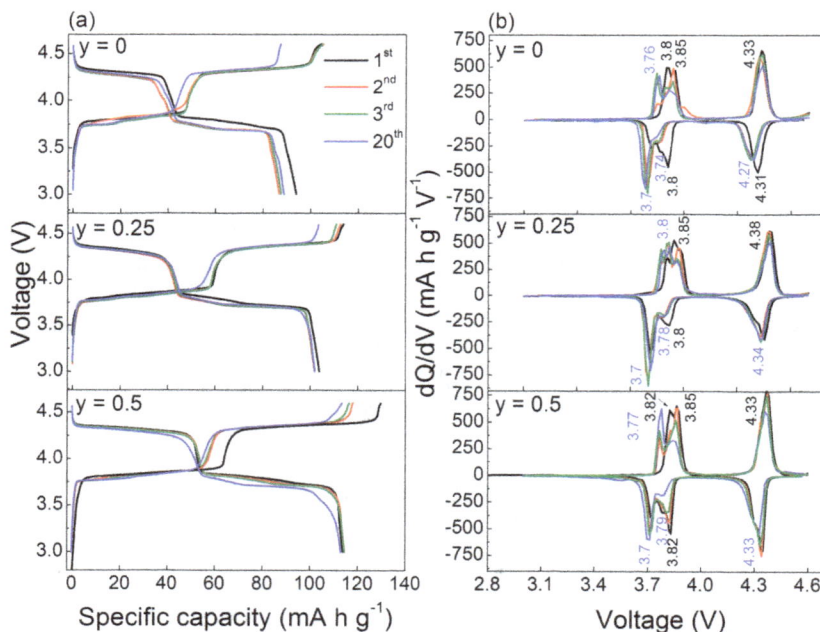

**Figure 5.** (a) Galvanostatic cycling profiles and (b) the $dQ/dV$ vs. voltage plots for the $Na_{1+y}VPO_4F_{1+y}$ ($y = 0; 0.25; 0.5$) samples in hybrid Na/Li cells. Cycling rate is C/10.

We noticed that during the few first cycles of $Na_{1+y}VPO_4F_{1+y}$ in hybrid cells, the relative intensity of the low-voltage redox peak at ~3.7 V gradually increased, presumably indicating an alternation in the overall alkali ion insertion mechanism. Such an effect was first observed by Barker et al. [25] when studying the cycling of the $Li_{4/3}Ti_{5/3}O_4 \,|\,| Na_3V_2(PO_4)_2F_3$ hybrid-ion cell. They suggested that during the first charge, the $Na^+$ ions are extracted from the fluorophosphate phase and entered into an electrolyte to create a mixed Li–Na electrolyte. The change in the intensity of the differential peak at ~3.7 V was explained as the transition from the predominant Na insertion mechanism at a cathode that is superimposed with Li in the lower voltage region, while the Na insertion dominates the higher voltage region, as can be seen from the maintenance of the intensity of the differential peak at ~4.3 V.

Ex situ XRD and EDX studies were performed to estimate the changes in structure and composition upon charge/discharge of $Na_{1+y}VPO_4F_{1+y}$ in a hybrid Na/Li cell, and the degree of the Na/Li electrochemical exchange. XRD patterns of the sample with $y = 0.5$ recorded at the end of the 12th charge and the 12th discharge in the Li cell are shown in Figure 6. On the XRD patterns of the sample at the end of the charge, some reflections of the initial phase disappeared, suggesting the occurrence of the reversible $P4_2/mnm \leftrightarrow I4/mmm$ transition, which was verified by the Rietveld refinement. The same structural transition was observed earlier during cycling of $Na_3V_2(PO_4)_2F_3$ in a Na cell [30]. The lattice parameters of the as-obtained products and the average sodium content per f.u. are shown in Table 5 in comparison with those presented in the literature [30]. It was seen that the refined lattice parameters were in a good agreement with those obtained in [30] for incompletely charged sodium vanadium fluorophosphate. According to the charge capacity and the XRD refinement, the final composition of the charged sample was ~$Na_{1.1}V_2(PO_4)_2F_3$, i.e.; it was not completely charged. Fully charged $Na_1V_2(PO_4)_2F_3$ was obtained by Bianchini et al. and indexed in the $Cmc2_1$ space group [30]. On the contrary, the XRD pattern of the sample, recorded at the end of the 12th discharge, showed the same set of reflections as the XRD pattern of the pristine sample. They were slightly shifted to larger angles, due to partial substitution of the Na ions for Li ions with lower ionic radius. The structure of the mixed Na/Li fluorophosphate was well refined with the $P4_2/mnm$ S.G., showing that the initial structure of $Na_3V_2(PO_4)_2F_3$ remained unchanged (Figure 6). The decreased occupancy of the Na sites testified that $Li^+$ ions replace the $Na^+$ ions at the same crystallographic sites. According to the EDX analysis, the Na/V ratio decreased from 1.46 for the pristine sample, to 1.26 for the mixed Na/Li fluorophosphate, thus resulting in the ~$Na_{2.52}Li_{0.48}V_2(PO_4)_2F_3$ composition, i.e.; only ~16% of Na ions were exchanged for Li ions during electrochemical cycling. This indicated that mixed Na/Li (de)intercalation occurs during cycling of $Na_3V_2(PO_4)_2F_3$ in a hybrid-ion cell, while the Na contribution is predominant. Sodium is never completely deinserted upon electrochemical cycling of $Na_3V_2(PO_4)_2F_3$. The immobile sodium therefore contributes to the structural stability of sodium vanadium fluorophosphates.

**Table 5.** Rietveld-refined lattice parameters and sodium content per formula unit for the sample with $y = 0.5$ charged and discharged in hybrid cells, compared to the literature data.

| Sample | Stage | S.G. | $a = b$, Å | $c$, Å | $V$, Å³ | $Na^+$/f.u. | GOF/$R_{wp}$ |
|---|---|---|---|---|---|---|---|
| $y = 0.5$ | charge | $I4/mmm$ | 6.2496(2) | 10.9568(7) | 427.94(4) | 0.54(2) | 1.67/10.78 |
| $y = 0.5$ (sample A) | discharge | $P4_2/mnm$ | 9.0251(3) | 10.7524(6) | 875.82(8) | 1.30(2) | 2.09/7.89 |
| $y = 0.5$ (sample B) | discharge | $P4_2/mnm$ | 9.0175(12) | 10.7000(19) | 870.07(27) | 0.82(6) | 2.43/12.65 |
| $Na_{1.3}(VPO_4)_2F_3$ [30] | charge | $I4/mmm$ | 6.2481(1) | 10.9222(2) | 426.39(1) | 0.65 | 1.31(6) |
| $Na_{1.8}(VPO_4)_2F_3$ [30] | charge | $I4/mmm$ | 6.2800(1) | 10.8493(3) | 427.88(1) | 0.9 | 1.85(7) |

The mixed Na/Li ion intercalation resulted in the high capacity and excellent long-term stability of the hybrid electrochemical cells, as can be seen in Figure 7a. The highest initial discharge capacity (116 mAh·g$^{-1}$) was observed for the sample with $y = 0.5$. It corresponded to 90% of the theoretical material utilization, based on the assumption of participation by two alkali ions per f.u. of $Na_3V_2(PO_4)_2F_3$. For the other two samples with $y = 0$ and 0.25, the reversible specific capacity of 87 and 103 mAh·g$^{-1}$, respectively, was achieved. Such a difference in the values of specific capacity correlated with the amount of the electrochemically active $Na_3V_2(PO_4)_2F_3$ phase in these samples (Table 2).

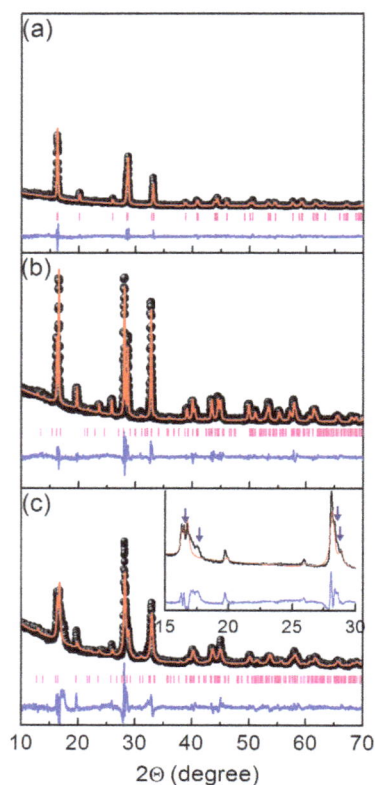

**Figure 6.** Ex situ XRD pattern of the $Na_{1.5}VPO_4F_{1.5}$ (**a,b**) samples A and (**c**) B at the difference stages of charge–discharge in the Li cell after Rietveld refinement: (**a**) 12th charge ($I4/mmm$ S.G.); (**b**) 12th discharge ($P4_2/mnm$ S.G.), (**c**) seventh discharge ($P4_2/mnm$ S.G.). Arrows indicate the reflections of an unidentified phase.

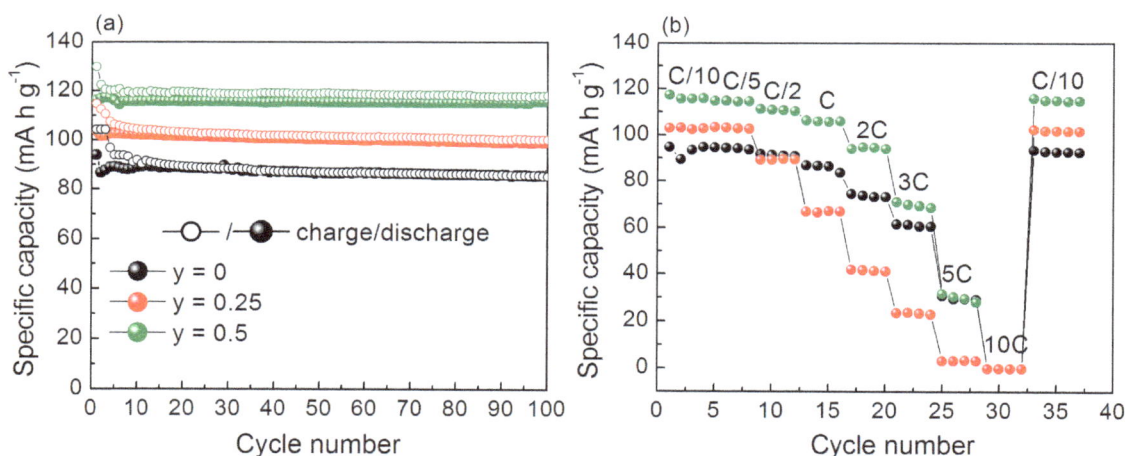

**Figure 7.** (**a**) The capacity vs. cycle number plots and (**b**) discharge capacity vs. cycling rate (C/10–10C) plots for $Na_{1+y}VPO_4F_{1+y}$ ($y = 0; 0.25; 0.5$).

According to Figure 7b, the as-prepared samples displayed a good high-rate performance. For instance, when the cycling rate increased from C/10 to 1C, the specific discharge capacity of the sample with $y = 0.5$ decreased only for 10%, from 116 to 106 mAh·g$^{-1}$. When the current density returned from 10C to C/10, the discharge capacity of the samples recovered to its initial value, proving high structural stability upon cycling.

When $Na_3V_2(PO_4)_2F_3$ was preliminarily charged to 4.6 V in the Na cell (sample B), ~1.9 Na ions per f.u. were extracted due to the oxidation of 1.9 $V^{3+}$ ions to 1.9 $V^{4+}$. During further discharging in the Li cell, the electrolyte was free from Na ions. To compare the cycleability of samples A and B, we carried out the Galvanostatic intermittent titration technique (GITT) analysis. Figure 8a represents the charge–discharge curves obtained by GITT during the second cycle for these samples, illustrating voltage dependence on the Li content under load and rest. The cells were charged and discharged at a constant current C/10 ($I_0$ = 12.8 mA·$g^{-1}$) for an interval of 20 min followed by an open-circuit stand for 40 min to allow the cell voltage relaxing to its steady-state value. It was seen that the samples had different profiles: instead of two plateaus for sample A, only one plateau at lower voltage and a sloping charge–discharge curve at higher voltage was observed for sample B, presumably evidencing a single-phase mechanism of (de)insertion. However, the relaxation spikes in the latter case became longer, evidencing slower deintercalation reaction kinetics with larger polarization and slower equilibration. Moreover, according to the differential open circuit voltage (OCV) curves (Figure 8b), only one redox peak was clearly observed at 3.7 V, instead of two peaks at 3.7 and 3.8 V for sample A, while the occurrence and the position of the second peak at 4.3 V was preserved. Besides, additional low-intensive redox peaks were observed at ~4 V, evidencing a more complicated insertion mechanism. Based on the above-mentioned hypothesis, it can be assumed that the low-voltage cathode reaction that occurred in sample B was based mostly on the Li ions, while the high-voltage reaction was Na-based. According to the EDX analysis, the resulting Na/V ratio in sample B after its cycling in the Li cell was ~0.7, corresponding to the $Na_{1.4}Li_{1.6}V_2(PO_4)_2F_3$ composition. The XRD pattern of this mixed phase after Rietveld refinement is shown in Figure 6c. As can be seen, most reflections were similar to those observed for sample A (Figure 6b). A larger degree of the $Li^+/Na^+$ ion exchange was reflected in the significant reduction of the unit cell volume. However, some additional low-intensive reflections at 16.5, 17.5, 28.3, and 28.6 2Θ degree were identified, and could not be assigned to any known phases. Thus, though both samples were desodiated to a composition close to that at the first charge, regardless of whether it was in a Li or Na cell, their electrochemical behavior and the mechanism of the alkali ion insertion appeared to be different at the first discharge. The most probable reason for this is due to the composition of the electrolyte: the mixed electrolyte for sample A due to the deinsertion of the Na ions from sodium vanadium fluorophosphate upon the charge, or the mostly single Li electrolyte for sample B.

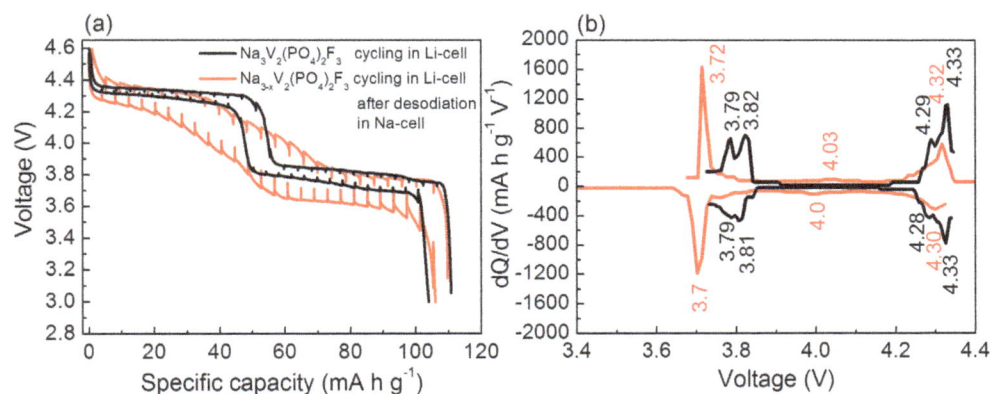

**Figure 8.** (a) GITT curves of the pristine $Na_{1.5}VPO_4F_{1.5}$ and electrochemically desodiated $Na_{1.5-x}VPO_4F_{1.5}$ in Li-cells, and (b) differential OCV curve vs. voltage plots.

## 3. Materials and Methods

A series of sodium-vanadium fluorophosphates $Na_{1+y}VPO_4F_{1+y}$ ($0 \leq y \leq 0.75$) were prepared by the two-step solid-state method, according to the following reactions using $VPO_4$ as an intermediate:

$$1/2V_2O_5 + NH_4H_2PO_4 + C \rightarrow VPO_4 + NH_3 + 3/2H_2O + CO, \qquad (1)$$

$$VPO_4 + (1+y)NaF \rightarrow Na_{1+y}VPO_4F_{1+y} \tag{2}$$

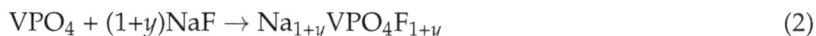

The preliminary solid-state mechanical activation (MA) of both reagent mixtures was performed by means of a high-energy AGO-2 planetary mill (~900 rpm), with stainless jars and balls in Ar atmosphere for 5 min. The activated mixtures (1) and (2) were subsequently annealed in Ar flow for 2 h at 750 °C and 650 °C, respectively, and then slowly cooled to room temperature.

The XRD patterns of the as-prepared samples were recorded by a D8 Advance Bruker diffractometer (Bruker AXS GmbH, Karlsruhe, Germany) with a high-rate detector Lynx Eye, Cu $K\alpha_{1,2}$ radiation ($\lambda_1 = 1.5406$, $\lambda_2 = 1.5445$ Å), between $2\theta = 10°$ and $100°$ with a step of $0.02°$ and an uptake time of 0.3 s. The structural refinement of the XRD data was carried out by the Rietveld method using TOPAS software [31]. First, we refined the structure of the single-phase sample ($y = 0.5$). The procedure was started with the refinement of the lattice parameters and followed by the atomic positions. The thermal displacement parameters for all atoms were refined just once, and then fixed at their final values; the thermal parameters for the atoms of the same elements were taken to be equal. Finally, the occupancy of the Na crystallographic sites was refined. The structural refinement of the main phase in the multi-phase samples was performed in a similar manner; all thermal parameters were kept fixed at the values extracted from the structural refinement of the single-phase sample. Particle size and morphology were investigated by SEM with a Hitachi S-3400 N scanning electron microscope (Hitachi High-Technologies Corporation, Tokyo, Japan). Chemical analysis of the synthesized materials was carried out by the energy-dispersive X-ray spectroscopy using an UltraDry EDX detector (Thermo Fischer Scientific Inc., Waltham, MA, USA). The FTIR spectra were recorded by means of a Tensor 27 spectrometer (Bruker Optik GmbH, Ettlingen, Germany) in the 400–4000$cm^{-1}$ range (pellets with KBr).

The electrochemical properties of the as-prepared materials were studied in hybrid Na/Li cells. The composite cathode materials were fabricated by mixing 75 wt % active material with 20 wt % Super P (Timcal, Bodio, Switzerland), and a 5 wt % PVDF/NMP binder. The mixed slurry was then pasted on the aluminum foil to obtain the working electrodes. The total amount of carbon in the cathode mass was 20 wt %; mass loading was ~2–3 mg·cm$^{-2}$ and an electrode diameter of 10 mm was used throughout. The Swagelok-type cells were assembled in an Ar filled glove box with Li metal as an anode, 1 M LiPF$_6$ solution in a mixture of ethylene carbonate (EC) and dimethyl carbonate (DMC) (2:1 by weight) as an electrolyte, and a glass fiber filter, Grade GF/C (GE Healthcare UK Ltd., Little Chalfont, UK) as a separator. The cycling was performed using a galvanostatic mode at the C/10–10C charge/discharge rates within the 3.0–4.6 V range. For ex situ experiments, the cathode mass was fabricated without a binder in a standard way: 80 wt % active material and 20 wt % Super P were ground thoroughly and then distributed uniformly at a current collector. In this case, the loading density was 20–21 mg·cm$^{-2}$. The cycling was stopped at the end of the charge or discharge, and a cell was disassembled inside an Ar glove box. The extracted electrodes were washed with DMC before being taken out of the box, dried in an argon atmosphere, and then studied by the XRD. For the GITT experiments, two methods of electrochemical measurement were used. For sample A, the initial cathode material was directly put in a lithium cell with a lithium anode and the LiPF$_6$/EC+DMC (2:1) electrolyte, while for sample B, the initial cathode material was charged in a sodium cell with a sodium anode and the NaPF$_6$/EC+DMC (2:1) electrolyte; the cell was then disassembled in a glove box, the cathode was carefully washed with the DMC solution to remove residues of the sodium electrolyte, and put in a lithium cell. GITT experiments were conducted, starting from the charge for sample A and from the discharge for sample B (after preliminary electrochemical desodation). GITT was carried out with a current pulse of 12.8 mA·g$^{-1}$ (0.1C) for 20 min followed by relaxation at an open circuit for 40 min at each step.

## 4. Conclusions

1.    It has been shown that among sodium vanadium fluorophosphate compositions $Na_{1+y}VPO_4F_{1+y}$ ($0 \leq y \leq 0.75$) prepared by the mechanochemically assisted solid-state synthesis, the single-phase material $Na_{1.5}VPO_4F_{1.5}$ or $Na_3V_2(PO_4)_2F_3$ with a tetragonal structure (the $P4_2/mnm$ S.G.) was

    formed only for $y = 0.5$. Samples with $y < 0.5$ and $y > 0.5$ possess different impurity phases. The compound with the $NaVPO_4F$ composition does not exist.

2.   Sodium vanadium fluorophosphates $Na_{1+y}VPO_4F_{1+y}$ can be considered as multifunctional cathode materials for the fabrication of lithium-ion and sodium-ion high-energy batteries. The reversible discharge capacity of 116 mAh·g$^{-1}$ for $y = 0.5$, 103 mAh·g$^{-1}$ for $y = 0.25$ and 87 mAh·g$^{-1}$ for $y = 0$ was achieved. The decrease in the discharge capacity is in accordance with the amount of the electrochemically active phase $Na_3V_2(PO_4)_2F_3$ in the samples.

3.   The ex situ XRD patterns confirm a reversible $P4_2/mnm \leftrightarrow I4/mmm$ transformation upon charging–discharging in a hybrid-ion cell, similar to the earlier observed transformation in Na-ion cells.

4.   The structural study of charged and discharged samples and the analysis of the differential capacity curves indicated a negligible Na/Li electrochemical exchange (~16%) and a predominantly sodium-based cathode reaction in the hybrid-ion cell. This is significantly lower than the ~50% exchange observed for other Na-based cathodes, such as $Na_2FePO_4F$ [32] and $Na_2FeP_2O_7$ [33] when cycled in hybrid-ion cells, showing that the properties of hybrid-ion batteries can be varied based on the alkali-ion selectivity of electrode materials. To increase the degree of the Na/Li electrochemical exchange in $Na_3V_2(PO_4)_2F_3$, it first needs to be desodiated in a Na cell, and then cycled in a Li cell with the electrolyte enriched with Li ions.

5.   After cycling in hybrid-ion cells, $Na_3V_2(PO_4)_2F_3$ showed nice cycleability and high-rate performance, presumably due to operating in the mixed Na/Li electrolyte. Thus, the hybrid-ion approach may open possibilities for many new active materials and material combinations with enhanced electrochemical performance. This approach provides an opportunity for sodium cathode materials to be used without the requirement for ion Na/Li exchange prior to cell fabrication. Since in hybrid-ion systems all anodic charge carriers originate from the electrolyte, this may limit their use in high-energy applications, where relatively thick electrodes are used in combination with thin electrolytes. However, in high power applications, this may not represent a major drawback, since thinner electrodes and thicker electrolytes are commonly used.

**Acknowledgments:** The authors are thankful to Natalia V. Bulina for registration of the XRD patterns, Alexander I. Titkov for his assistance in the EDX experiments, and the Center for Collective Use of NIOCh SB RAS.

**Author Contributions:** Nina V. Kosova conceived and designed the experiments; Daria O. Rezepova performed the experiments; both authors participated in the analysis of the experimental data and in writing the paper.

## References

1.   Yabuuchi, N.; Kubota, K.; Dahbi, M.; Komaba, S. Research development on sodium-ion batteries. *Chem. Rev.* **2014**, *114*, 11636. [CrossRef] [PubMed]

2.   Masquelier, C.; Croguennec, L. Polyanionic (phosphates, silicates, sulfates) frameworks as electrode materials for rechargeable Li (or Na) batteries. *Chem. Rev.* **2013**, *113*, 6552–6591. [CrossRef] [PubMed]

3.   Barker, J.; Saidi, M.Y.; Swoyer, J.L. A sodium-ion cell based on the fluorophosphate compound $NaVPO_4F$. *Electrochem. Solid State Lett.* **2003**, *6*, A1–A4. [CrossRef]

4.   Zhao, J.; He, J.; Ding, X.; Zhou, J.; Ma, Y.; Wu, S.; Huang, R. A novel sol–gel synthesis route to $NaVPO_4F$ as cathode material for hybrid lithium ion batteries. *J. Power Sources* **2010**, *195*, 6854–6859. [CrossRef]

5.   Le Meins, J.-M.; Crosnier-Lopez, M.-P.; Hemon-Ribaud, A.; Courbion, G. Phase transitions in the $Na_3M_2(PO_4)_2F_3$ family (M = $Al^{3+}$, $V^{3+}$, $Cr^{3+}$, $Fe^{3+}$, $Ga^{3+}$): Synthesis, thermal, structural, and magnetic studies. *J. Solid State Chem.* **1999**, *148*, 260–277. [CrossRef]

6.   Sauvage, F.; Quarez, E.; Tarascon, J.M.; Baudrin, E. Crystal structure and electrochemical properties vs. $Na^+$ of the sodium fluorophosphate $Na_{1.5}VOPO_4F_{0.5}$. *Solid State Sci.* **2006**, *8*, 1215–1221. [CrossRef]

7.   Tsirlin, A.A.; Nath, R.; Abakumov, A.M.; Furukawa, Y.; Johnston, D.C.; Hemmida, M.; Krug von Nidda, H.-A.; Loidl, A.; Geibel, C.; Rosner, H. Phase separation and frustrated square lattice magnetism of $Na_{1.5}VOPO_4F_{0.5}$. *Phys. Rev. B* **2011**, *84*, 014429. [CrossRef]

8.  Park, Y.U.; Seo, D.H.; Kim, H.; Kim, J.; Lee, S.; Kim, B.; Kang, K. A family of high-performance cathode materials for Na-ion batteries, $Na_3(VO_{1-x}PO_4)_2F_{1+2x}$ ($0 \leq x \leq 1$): Combined first-principles and experimental study. *Adv. Funct. Mater.* **2014**, *24*, 4603–4614. [CrossRef]

9.  Broux, T.; Bamine, T.; Fauth, F.; Simonelli, L.; Olszewski, W.; Marini, C.; Menetrier, M.; Carlier, D.; Masquelier, C. Strong impact of the oxygen content in $Na_3V_2(PO_4)_2F_{3-y}O_y$ ($0 \leq y \leq 0.5$) on its structural and electrochemical properties. *Chem. Mater.* **2016**, *28*, 7683–7692. [CrossRef]

10. Serras, P.; Palomares, V.; Goni, A.; Kubiak, P.; Rojo, T. Electrochemical performance of mixed valence $Na_3V_2O_{2x}(PO_4)_2F_{3-2x}$/C as cathode for sodium-ion batteries. *J. Power Sources* **2013**, *241*, 56–60. [CrossRef]

11. Serras, P.; Palomares, V.; Goni, A.; Gil de Muro, I.; Kubiak, P.; Lezama, L.; Rojo, T. High voltage cathode materials for Na-ion batteries of general formula $Na_3V_2O_{2x}(PO_4)_2F_{3-2x}$. *J. Mater. Chem.* **2012**, *22*, 22301–22308. [CrossRef]

12. Gover, R.K.B.; Bryan, A.; Burns, P.; Barker, J. The electrochemical insertion properties of sodium vanadium fluorophosphate, $Na_3V_2(PO_4)_2F_3$. *Solid State Ion.* **2006**, *177*, 1495–1500. [CrossRef]

13. Bianchini, M.; Brisset, N.; Fauth, F.; Weill, F.; Elkaim, E.; Suard, E.; Masquelier, C.; Croguennec, L. $Na_3V_2(PO_4)_2F_3$ revisited: A high-resolution diffraction study. *Chem. Mater.* **2014**, *26*, 4238–4247. [CrossRef]

14. Zhuo, H.; Wang, X.; Tang, A.; Liu, Z.; Gamboa, S.; Sebastian, P.J. The preparation of $NaV_{1-x}Cr_xPO_4F$ cathode materials for sodium-ion battery. *J. Power Sources* **2006**, *160*, 698–703. [CrossRef]

15. Lu, Y.; Zhang, S.; Li, Y.; Xue, L.; Xu, G.; Zhang, X. Preparation and characterization of carbon-coated $NaVPO_4F$ as cathode material for rechargeable sodium-ion batteries. *J. Power Sources* **2014**, *247*, 770–777. [CrossRef]

16. Shakoor, R.A.; Seo, D.-H.; Kim, H.; Park, Y.-U.; Kim, J.; Kim, S.-W.; Gwon, H.; Lee, S.; Kang, K. A combined first principles and experimental study on $Na_3V_2(PO_4)_2F_3$ for rechargeable Na batteries. *J. Mater. Chem.* **2012**, *22*, 20535–20541. [CrossRef]

17. Chihara, K.; Kitajou, A.; Gocheva, I.D.; Okada, S.; Yamaki, J. Cathode properties of $Na_3M_2(PO_4)_2F_3$ [M = Ti, Fe, V] for sodium-ion batteries. *J. Power Sources* **2013**, *227*, 80–85. [CrossRef]

18. Song, W.; Cao, X.; Wu, Z.; Chen, J.; Zhu, Y.; Hou, H.; Lan, Q.; Ji, X. Investigation of the sodium ion pathway and cathode behavior in $Na_3V_2(PO_4)_2F_3$ combined via a first principles calculation. *Langmuir* **2014**, *30*, 12438–12466. [CrossRef] [PubMed]

19. Liu, Z.; Hu, Y.-Y.; Dunstan, M.T.; Huo, H.; Hao, X.; Zou, H.; Zhong, G.; Yang, Y.; Grey, C.P. Local structure and dynamics in the Na ion battery positive electrode material $Na_3V_2(PO_4)_2F_3$. *Chem. Mater.* **2014**, *26*, 2513–2521. [CrossRef]

20. Jiang, T.; Chen, G.; Li, A.; Wang, C.; Wei, Y. Sol–gel preparation and electrochemical properties of $Na_3V_2(PO_4)_2F_3$/C composite cathode material for lithium ion batteries. *J. Alloys Compd.* **2009**, *478*, 604–607. [CrossRef]

21. Eshraghi, N.; Caes, S.; Mahmoud, A.; Cloots, R.; Vertruyen, B.; Boschini, F. Sodium vanadium (III) fluorophosphate/carbon nanotubes composite (NVPF/CNT) prepared by spray-drying: Good electrochemical performance thanks to well-dispersed CNT network within NVPF particles. *Electrochim. Acta* **2017**, *228*, 319–324. [CrossRef]

22. Kosova, N. Mechanochemical reactions and processing of nanostructured electrode materials for lithium-ion batteries. *Mater. Today Proc.* **2016**, *3*, 391–395. [CrossRef]

23. Barker, J.; Gover, R.K.B.; Burns, P.; Bryan, A.J. Hybrid-ion. A lithium-ion cell based on a sodium insertion material. *Electrochem. Solid State Lett.* **2006**, *9*, A190–A192. [CrossRef]

24. Song, W.; Liu, S. A sodium vanadium three-fluorophosphate cathode for rechargeable batteries synthesized by carbothermal reduction. *Solid State Sci.* **2013**, *15*, 1–6. [CrossRef]

25. Barker, J.; Cover, R.; Burns, P.; Bryan, A.J. $Li_{4/3}Ti_{5/3}O_4$ | | $Na_3V_2(PO_4)_2F_3$: An example of a hybrid-ion cell using a non-graphitic anode. *J. Electrochem. Soc.* **2007**, *154*, A882–A887. [CrossRef]

26. Park, Y.-U.; Seo, D.-H.; Kim, B.; Hong, K.-P.; Kim, H.; Lee, S.; Shakoor, R.A.; Miyasaka, K.; Tarascon, J.-M.; Kang, K. Tailoring a fluorophosphate as a novel 4 V cathode for lithium-ion batteries. *Sci. Rep.* **2012**, *704*, 1–7. [CrossRef] [PubMed]

27. Ait Salah, A.; Jozwiak, P.; Garbarczyk, J.; Benkhouja, K.; Zaghib, K.; Gendron, F.; Julien, C.M. Local structure and redox energies of lithium phosphates with olivine- and Nasicon-like structures. *J. Power Sources* **2005**, *140*, 370–375. [CrossRef]

28. Kosova, N.V.; Devyatkina, E.T.; Slobodyuk, A.B.; Gutakovskii, A.K. LiVPO$_4$F/Li$_3$V$_2$(PO$_4$)$_3$ nanostructured composite cathode materials prepared via mechanochemical way. *J. Solid State Electrochem.* **2014**, *18*, 1389–1399. [CrossRef]

29. Kera, Y. Infrared study of alkali tri- and hexavanadates as formed from their melts. *J. Solid State Chem.* **1984**, *51*, 205–211. [CrossRef]

30. Bianchini, M.; Fauth, F.; Brisset, N.; Weill, F.; Suard, E.; Masquelier, C.; Croguennec, L. A comprehensive investigation of the Na$_3$V$_2$(PO$_4$)$_2$F$_3$–NaV$_2$(PO$_4$)$_2$F$_3$ system by operando high resolution synchrotron X-ray diffraction. *Chem. Mater.* **2015**, *27*, 3009–3020. [CrossRef]

31. Cheary, R.W.; Coelho, A.A. A fundamental parameters approach to X-ray line-profile fitting. *J. Appl. Cryst.* **1992**, *25*, 109–121. [CrossRef]

32. Kosova, N.V.; Podugolnikov, V.R.; Devyatkina, E.T.; Slobodyuk, A.B. Structure and electrochemictry of NaFePO$_4$ and Na$_2$FePO$_4$F cathode materials prepared via mechanochemical route. *Mater. Res. Bull.* **2014**, *60*, 849–857. [CrossRef]

33. Kosova, N.V.; Rezepova, D.O.; Petrov, S.A.; Slobodyuk, A.B. Elelctrochemical and chemical Na$^+$/Li$^+$ ion exchange in Na-based cathode materials: Na$_{1.56}$Fe$_{1.22}$P$_2$O$_7$ and Na$_3$V$_2$(PO$_4$)$_2$F$_3$. *J. Electrochem. Soc.* **2017**, *164*, A6192–A6200. [CrossRef]

# 14

# Chemical Tuning and Absorption Properties of Iridium Photosensitizers for Photocatalytic Applications

**Olga S. Bokareva** *, **Tobias Möhle, Antje Neubauer** [†], **Sergey I. Bokarev, Stefan Lochbrunner** and **Oliver Kühn**

Institut für Physik, Universität Rostock, Albert-Einstein-Str. 23-24, 18059 Rostock, Germany;
tobias.moehle@uni-rostock.de (T.M.); antje-neubauer@gmx.de (A.N.); sergey.bokarev@uni-rostock.de (S.I.B.);
stefan.lochbrunner@uni-rostock.de (S.L.); oliver.kuehn@uni-rostock.de (O.K.)
* Correspondence: olga.bokareva@uni-rostock.de
† Current address: Becker & Hickl GmbH, Nahmitzer Damm 30, 12277 Berlin, Germany.

Academic Editor: Matthias Bauer

**Abstract:** Cyclometalated Ir(III) complexes are of particular interest due to the wide tunability of their electronic structure via variation of their ligands. Here, a series of heteroleptic Ir-based photosensitizers with the general formula $[Ir(C^{\wedge}N)_2(N^{\wedge}N)]^+$ has been studied theoretically by means of an optimally-tuned long-range separated density functional. Focusing on the steady-state absorption spectra, correlations between the chemical modification of both ligand types with the natures of the relevant dark and bright electronic states are revealed. Understanding such correlations builds up a basis for the rational design of efficient photocatalytic systems.

**Keywords:** photosensitizers; photochemistry; photophysics; absorption; optimally-tuned range-separated hybrid density functional

## 1. Introduction

Exploring alternative renewable energy sources represents one of the most prominent tasks for natural sciences due to the world's steadily growing energy consumption [1]. The conversion of solar energy into appropriate chemical forms and its storage, e.g., in form of hydrogen, attracts particular attention, as sun light meets all ecological and power requirements for renewable energy sources [2–4]. Although the history of sun-light driven conversion of water into hydrogen started with heterogeneous systems (see, e.g., [5,6]), homogeneous photocatalysis has been developed and extensively studied [7–10]. The key substance in the whole process is a photostable material, a photosensitizer (PS), capable to effectively absorb visible light and to live long enough in an excited state of charge-transfer nature in order to initiate separate oxidation and reduction half-reactions. The most popular PSs are noble-metal–organic complexes of ruthenium, platinum, rhenium, and iridium, although various organic dyes, porphyrins, manganese and copper complexes have been proposed as well (see reviews [9–12] and references therein). The rational design and further improvement of photocatalytic systems for hydrogen generation requires a mechanistic understanding of all underlying processes [13] and in particular, a detailed characterization of the excited electronic states of the PSs.

In the present article, we study a number of derivatives from the family of Ir(III)-based PSs with the general formula $[Ir(C^{\wedge}N)_2(N^{\wedge}N)]PF_6$ (sketched in Figure 1), where $C^{\wedge}N$ denotes (substituted) phenyl pyridine (ppy) ligands and $N^{\wedge}N$ stands for (substituted) bipyridine (bpy) ligands introduced by Bernhard et al. [14–19]. The unsubstituted $[Ir(ppy)_2(bpy)]PF_6$ is abbreviated as IrPS in the following.

Electronic transitions and charge-transfer (CT) properties of Ir(III) PSs can be flexibly tuned by chemical modification of the ligands (for a review, see [13]). The parent IrPS and its derivatives were applied in the photocatalytic scheme for water reduction half-reaction developed by Beller et al. [20]. It comprises a heteroleptic Ir(III)-based PS, triethylamine as a sacrificial reductant, and a series of iron carbonyl compounds as water reduction catalysts. A number of catalytic, EPR, spectroscopic and theoretical studies of these systems has been already published [11,21–33]. Among these works, mostly IrPS was in the focus of joint experimental/theoretical studies for investigation of the underlying mechanisms. In this communication, previously described and synthesized Ir-based PSs are studied theoretically in detail. The aim is to reveal the influence of chemical modification by ligand variation onto the characteristics of the bright and dark states of these photosensitizers.

The paper is organized as follows. First, we analyze absorption spectra and discuss peculiarities of excited states structures; Second, we provide computational and experimental details. We close the discussion with an outlook and conclusions.

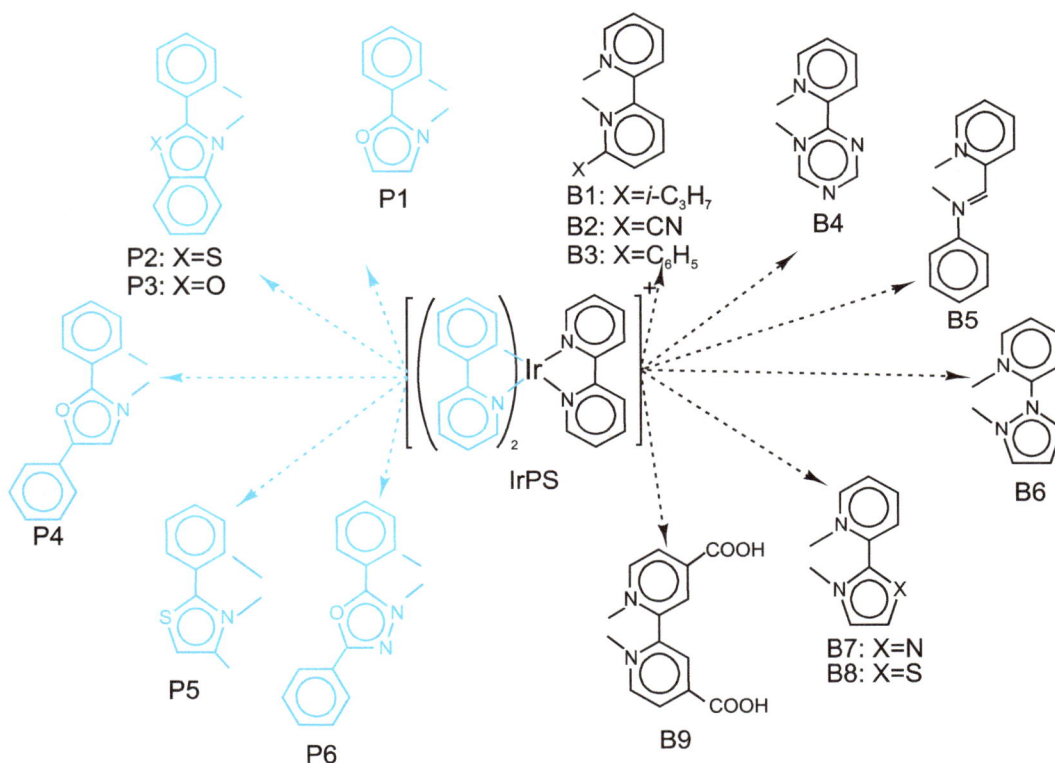

**Figure 1.** Structure formulae and notations of all photosensitizers (PSs) studied in this work.

## 2. Results and Discussion

Previously, on the basis of a multi-spectroscopic/computational study, the ppy ligand has been identified as being relevant for the properties of the first bright transition in IrPS, whereas the bpy ligand has been found to be important as an electron reservoir storing the electron after photoreduction by a sacrificial reductant [27]. Thus, chemical modification of these ligands separately could shed light on how these properties can be manipulated, thus paving the way for a rational design of future PSs. The chemical structures of the derivatives of both ppy and bpy ligands is sketched in Figure 1. The parent complex is denoted as IrPS, whereas the PSs containing modified ppy and bpy ligands are denoted as P$n$ and B$n$, respectively. For some cases, B1–B3 and B9, chemical derivation represents addition of some donor or acceptor groups to the original ligand. For most of the cases, however, the entire aromatic system of the ligand is modified by introducing heteroatoms and changing the size of the conjugated aromatic system. These derivatives were indicated as potentially perspective for photocatalytic applications, see Refs. [21,22,33].

As a first step, we have optimized the range-separation parameter. For all PSs studied in this work $\omega$ was found to be 0.17–0.19 bohr$^{-1}$. For the parent IrPS, its complexes with silver clusters, triethylamine, and iron carbonyls, a similar value of about 0.18 bohr$^{-1}$ has been obtained previously [11,28,31]. LC-BLYP with optimal $\omega$ was shown to provide the best agreement with experimental spectra of the oxidized and reduced IrPS and, simultaneously, to increase the triplet stability of the ground state solution, which is essential for the accuracy of the excited states energies [31]. The obtained value is notably smaller than standard $\omega$ parameters (0.33 bohr$^{-1}$ [34] and 0.47 bohr$^{-1}$[35]) obtained for training sets of atoms or small molecules and implemented as default values in quantum-chemical codes. The decrease of the optimal $\omega$ value with increasing system size has been previously found for numerous cases (see e.g., Refs. [36–39]), and rationalized by the fact that $\omega^{-1}$ defines a characteristic distance for switching between short- and long-range parts of the Coulomb interaction. Thus, the $\omega$ values obtained here for different species were found to be very close to the parent IrPS case, since the characteristic extent of the electron density, conjugation length, and inter-electronic distances are similar. As a note in caution we add that one should not count on such similarity to hold true in general.

In Figures 2 and 3, the calculated and experimental absorption spectra are presented for P$n$ and B$n$ derivatives. The respective stick spectra are also shown, with the sticks' colors denoting transitions of different character. The assignment was done in a fully automatized way on the basis of an analysis of configuration interaction coefficients and orbital contributions stemming from different moieties. Here, we distinguish between ligand-centered, LC, (both intraligand and ligand-to-ligand charge-transfer excitations are included), iridium ligand-field, LF, and three types of metal-to-ligand charge-transfer, MLCT, transitions. The latter includes Ir $\rightarrow \pi^*$(ppy), Ir $\rightarrow \pi^*$(bpy), and mixed MLCT transitions denoted collectively as MLCT(mixed). In addition, there is a number of transitions (denoted as mixed) having a very complicated character which cannot be classified into the above-mentioned groups. Such an assignment was done to simplify the analysis in terms of ligands acquiring an electron in course of photoexcitation to MLCT states. Since the singlet-triplet intensities were not calculated, the energetic positions of the triplet states are presented as bars below each spectrum with the same color code as for singlet states. It should be noted that the nature of the lowest-lying triplet states is of primary importance for light-emitting devices and photocatalytic reactions. However, the main focus in this communication is put on the absorption properties, where the properties of singlet states are of primary interest [40–42].

Overall, we note that there is a fairly good agreement of the calculated spectra with experiment. This holds true despite the fact that singlet-triplet contributions are omitted, vibronic structure of the transitions is not included, and a single width for broadening was applied without attempting to fit to the experiment. Thus, one may conclude that the optimally-tuned range-separation functional provides accurate information on the excited states, building a basis for further detailed analysis. This is remarkable since in the $\Delta$SCF procedure energies of only the HOMO and LUMO are self-consistently adjusted. Nevertheless, such tuning leads to a reliable prediction also of higher-lying transitions involving deeper occupied and higher unoccupied orbitals.

The photochemical fate of the photosensitizer is determined first of all by the character of the lowest bright transition as well as by lower-lying dark states. Concerning the performance, the first bright transition is of particular importance because the sun's spectrum has minute overlap with the high-energy part of the absorption spectrum. Thus, one of the conditions to improve the efficiency of PSs is to ensure a notable spectral overlap, what requires shifting of the bright transitions below 3.7 eV. However, this energy should be larger than the theoretical limit of 1.23 eV required to reduce $H_3O^+$ to hydrogen in course of the subsequent steps of the catalytic reaction. Difference densities for the mentioned transitions are shown in Figures 2 and 3. More detailed information on the natures of first dark and first bright singlet-singlet transitions is presented in Table 1.

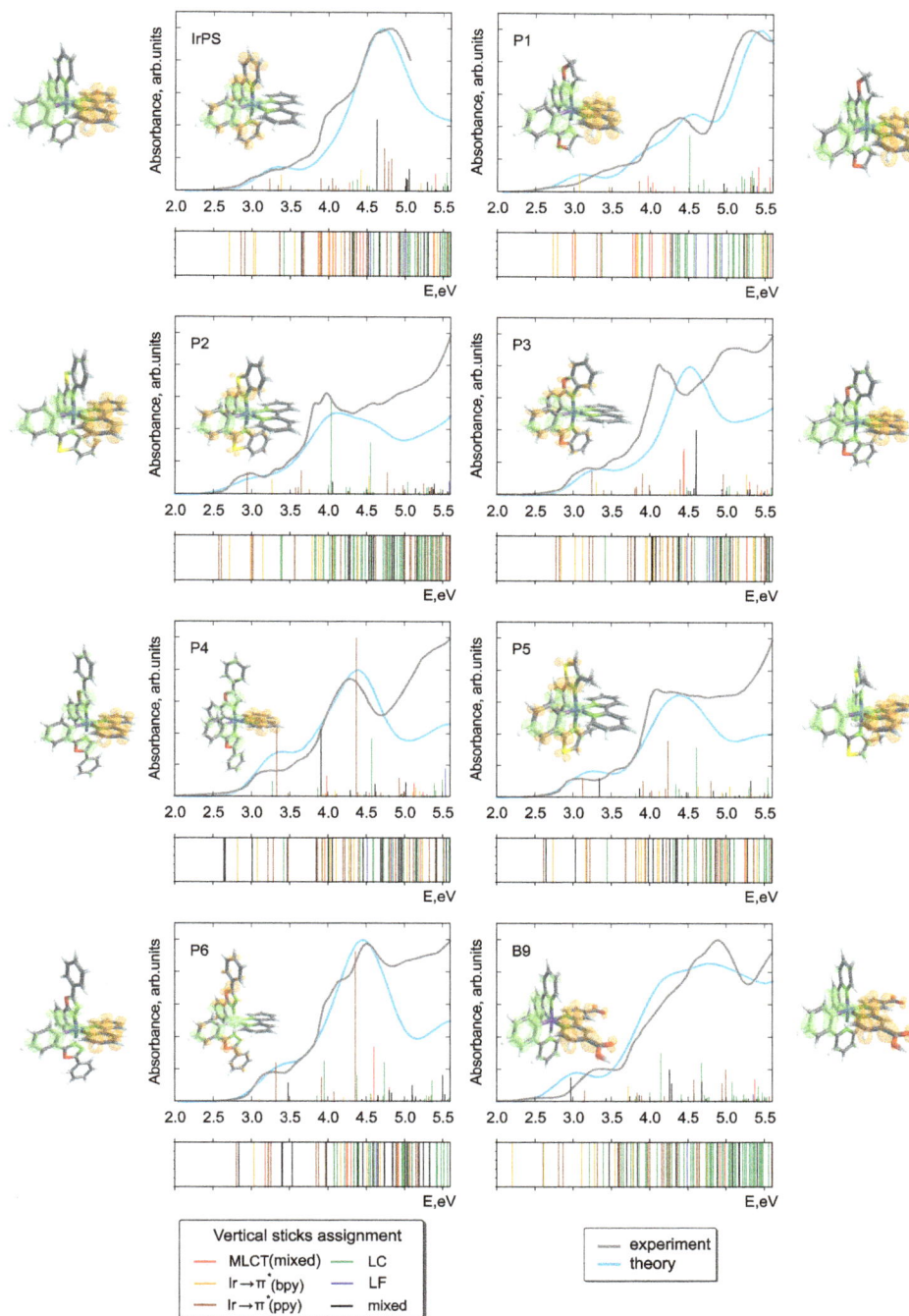

**Figure 2.** Theoretical singlet-singlet and experimental absorption spectra of IrPS and all its P*n*-derivatives (one bpy-derivative is shown as well). The calculated singlet-triplet transitions are presented as bars below the respective graphs. The color of the stick spectra denote different types of electronic transitions, see text. Also shown are transition density differences for the first dark (aside) and first bright (inset) singlet-singlet transitions. The green and orange colors denote the areas of electron lack and gain upon the respective excitation.

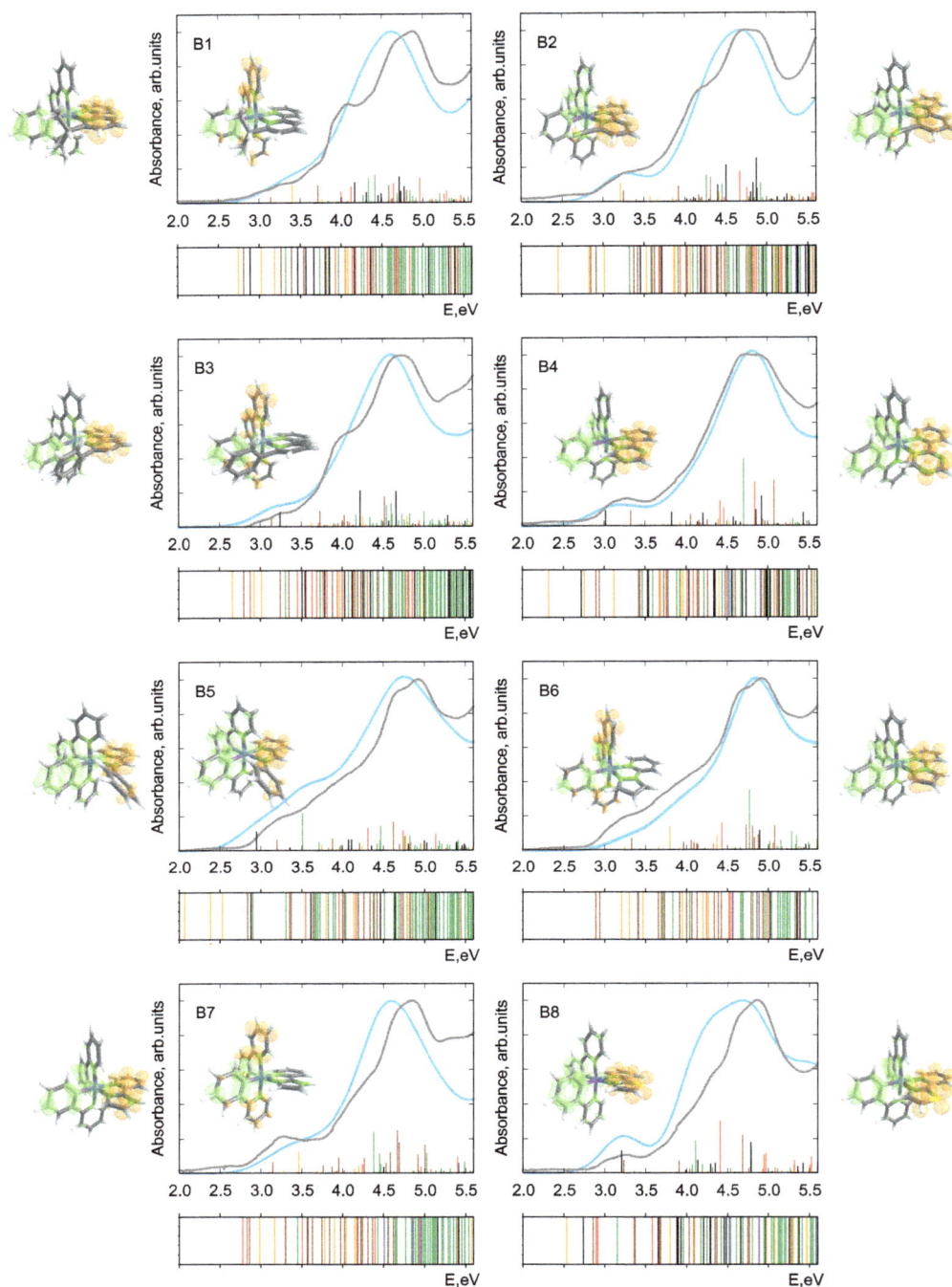

**Figure 3.** Theoretical and experimental absorption spectra of all B*n*-derivatives of the parent IrPS. For notation see caption of Figure 2.

Regardless of the chemical derivation, the HOMO and LUMO orbitals always correspond to $d_{x^2-y^2}$ and $\pi^*$(bpy) MOs, respectively. The first dark transition can be unambiguously assigned to a HOMO-LUMO transition in all cases (see Table 1 and compare density difference plots given as the right-most and left-most columns in Figures 2 and 3), whereas the bright state shows more variability (see density differences in the insets in Figures 2 and 3). Although for the parent IrPS the $d_{x^2-y^2} \rightarrow \pi^*$(ppy) transition is the lowest bright one, the modification of both ligands can change its nature. For about half of the derivatives, the excitation ends up at a $\pi^*$(bpy) orbital and the donor orbitals change to $d_{xz}$ or $d_{yz}$ of Ir, $\pi$(ppy) or even to mixture of both cases. Albeit the fact that the molecules are in general asymmetric, i.e., no symmetry-based selection rules can be applied, the dark states correspond to transitions between orbitals lying in the same plane, as, e.g., $d_{x^2-y^2} \rightarrow \pi^*$(bpy). In turn, bright states correspond to transitions where donor and acceptor orbitals are not coplanar. For instance, this holds true for the transitions $d_{x^2-y^2} \rightarrow \pi^*$(ppy), $d_{xz/yz} \rightarrow \pi^*$(bpy), and $\pi$(ppy) $\rightarrow \pi^*$(bpy).

**Table 1.** Energies and assignment of the first dark and first bright singlet-singlet transitions for all compounds. The character is given in terms of leading orbitals with respect to highest occupied molecular orbitals (HOMO) (H) and lowest unoccupied molecular orbitals (LUMO) (L).

| Compound | E, eV (Osc. str.) | Assignment | Character |
|---|---|---|---|
| IrPS | 2.75 (0.0009) | $d_{x^2-y^2} \rightarrow \pi^*$(bpy) | H–L |
|  | 3.23 (0.0719) | $d_{x^2-y^2} \rightarrow \pi^*$(ppy) | H–(L+1) |
| B1 | 2.81 (0.0052) | $d_{x^2-y^2} \rightarrow \pi^*$(bpy) | H–L |
|  | 3.14 (0.0312) | $d_{x^2-y^2} \rightarrow \pi^*$(ppy) | H–(L+1) |
| B2 | 2.49 (0.0005) | $d_{x^2-y^2} \rightarrow \pi^*$(bpy) | H–L |
|  | 3.22 (0.1051) | $d_{xz}/d_{yz} \rightarrow \pi^*$(bpy) | (H-2)/(H-3)–L |
| B3 | 2.72 (0.0030) | $d_{x^2-y^2} \rightarrow \pi^*$(bpy) | H–L |
|  | 3.14 (0.0478) | $d_{x^2-y^2} \rightarrow \pi^*$(ppy) | H–(L+1) |
| B4 | 2.37 (0.0003) | $d_{x^2-y^2} \rightarrow \pi^*$(bpy) | H–L |
|  | 3.02 (0.0864) | $d_{xz}/\pi$(ppy) $\rightarrow \pi^*$(bpy) | (H-3)/(H-1)–L |
| B5 | 2.15 (0.0005) | $d_{x^2-y^2} \rightarrow \pi^*$(bpy) | H–L |
|  | 2.95 (0.1088) | $d_{xz}/\pi$(ppy) $\rightarrow \pi^*$(bpy) | (H-3)/(H-1)–L |
| B6 | 3.25 (0.0162) | $d_{x^2-y^2} \rightarrow \pi^*$(bpy) | H–L |
|  | 3.33 (0.0658) | $d_{x^2-y^2} \rightarrow \pi^*$(ppy) | H–(L+1) |
| B7 | 2.87 (0.0018) | $d_{x^2-y^2} \rightarrow \pi^*$(bpy) | H–L |
|  | 3.14 (0.0632) | $d_{x^2-y^2} \rightarrow \pi^*$(ppy) | H–(L+1) |
| B8 | 2.59 (0.0005) | $d_{x^2-y^2} \rightarrow \pi^*$(bpy) | H–L |
|  | 3.21 (0.1278) | $d_{xz}/\pi$(ppy) $\rightarrow \pi^*$(bpy) | (H-3)/(H-1)–L |
| B9 | 2.25 (0.0010) | $d_{x^2-y^2} \rightarrow \pi^*$(bpy) | H–L |
|  | 2.97 (0.1461) | $d_{xz}/\pi$(ppy) $\rightarrow \pi^*$(bpy) | (H-3)/(H-1)–L |
| P1 | 2.82 (0.0008) | $d_{x^2-y^2} \rightarrow \pi^*$(bpy) | H–L |
|  | 3.07 (0.1123) | $d_{xz} \rightarrow \pi^*$(bpy) | (H-1)–L |
| P2 | 2.78 (0.0009) | $d_{x^2-y^2} \rightarrow \pi^*$(bpy) | H–L |
|  | 2.94 (0.1319) | $d_{x^2-y^2} \rightarrow \pi^*$(ppy) | H–(L+1) |
| P3 | 2.89 (0.0017) | $d_{x^2-y^2} \rightarrow \pi^*$(bpy) | H–L |
|  | 3.24 (0.1433) | $d_{x^2-y^2} \rightarrow \pi^*$(ppy) | H–(L+1) |
| P4 | 2.87 (0.0027) | $d_{x^2-y^2} \rightarrow \pi^*$(bpy) | H–L |
|  | 3.27 (0.0942) | $\pi$(ppy) $\rightarrow \pi^*$(bpy) | (H-1)–L |
| P5 | 2.80 (0.0012) | $d_{x^2-y^2} \rightarrow \pi^*$(bpy) | H–L |
|  | 3.13 (0.1200) | $d_{x^2-y^2} \rightarrow \pi^*$(ppy) | H–(L+1) |
| P6 | 3.10 (0.0091) | $d_{x^2-y^2} \rightarrow \pi^*$(bpy) | H–L |
|  | 3.32 (0.2397) | $d_{x^2-y^2} \rightarrow \pi^*$(ppy) | H–(L+1) |

The nature of the relevant transitions can be roughly correlated with the type of the chemical modification of the original ligands. Apart from adding side donor or acceptor groups ($-i$-$C_3H_7$, $-$Ph, $-$COOH, $-$CN) to the initial bpy or ppy ligands, the aromatic system is modified by introducing heteroatoms. The $sp^2$ pyridinic nitrogen atoms slightly increase the aromatic stabilization energy, whereas the sp$^3$ pyrolic nitrogen or sp$^2$ oxygen and sulfur heteroatoms decrease it. For the P$n$ series, this implies that apart from P2 and P3 the energy of the $d_{x^2-y^2} \to \pi^*$(ppy) transition shifts to the blue and may become higher than the bright $d_{xz/yz} \to \pi^*$(bpy) ones. In such situations, the nature of the lowest bright transitions changes. For the B$n$ series, the introduction of acceptor groups or the increase of aromaticity of the "bpy" ligand leads to a change of character of the first bright transition to $d_{xz/yz}/\pi$(ppy) $\to \pi^*$(bpy), see Table 1. The only exception is B8.

The energies of the bright states slightly vary in the series of PSs. For most species, a slight red shift up to 0.3 eV with respect to the parent IrPS can be observed. Exceptions being B6, P4, and P6 where a small blue shift up to 0.1 eV occurs. This implies that the spectral overlap with the sun light should be enhanced upon certain chemical modifications, but a dramatic increase of the absorption efficiency cannot be achieved without a principal change of the class of PSs. The transition energies of the lowest dark state mostly vary with changing the bpy ligand, dropping by up to 0.06 eV in case of B5, although the derivation of the ppy ligand may also influence its energy as in case of P6. The relative intensities of the two considered transitions for each compound vary by a factor from 6 to 270.

Interestingly, the modification of one ligand may notably influence the transitions involving other parts of the complex. This can be caused by general changes in the orbital shapes due to the changing donor or acceptor character of the ligands. As an illustration one can look at the character of doubly-occupied 5d-orbitals of IrPS. In Table 2, the percentage of localization on the central Ir-atom is presented. The extent of d-character lies between 35% and 67% pointing to the partial charge delocalization even without electronic excitation. For some cases, even 4 or 5 orbitals in the occupied subspace near the HOMO can be ascribed as d-ones mixed with some $\pi$ orbitals of the ligands.

**Table 2.** Percentage of localization of electron density on d orbitals of Ir for all compounds. In parenthesis, the character is given in terms of leading orbitals with respect to HOMO (H).

| Compound | $d_{x^2-y^2}$ | $d_{yz}$ | $d_{xz}$ |
|---|---|---|---|
| IrPS | 0.48 (H) | 0.67 (H-2) | 0.48 (H-3) |
| B1 | 0.51 (H) | 0.58 (H-2) | 0.61 (H-3) |
| B2 | 0.46 (H) | 0.55 (H-2) | 0.43 (H-3) |
| B3 | 0.46 (H) | 0.61 (H-2) | 0.60 (H-3) |
| B4 | 0.46 (H) | 0.56 (H-2) | 0.50 (H-3) |
| B5 | 0.47 (H) | 0.62 (H-2) | 0.50 (H-3) |
| B6 | 0.47 (H) | 0.57 (H-2) | 0.47 (H-3) |
|  |  |  | 0.40 (H-1) |
| B7 | 0.48 (H) | 0.55 (H-2) | 0.41 (H-3) |
|  |  |  | 0.53 (H-1) |
| B8 | 0.47 (H) | 0.66 (H-2) | 0.49 (H-3) |
| B9 | 0.47 (H) | 0.63 (H-2) | 0.51 (H-3) |
| P1 | 0.53 (H) | 0.56 (H-2) | 0.51 (H-1) |
| P2 | 0.46 (H) | 0.30 (H-1) | 0.48 (H-3) |
| P3 | 0.46 (H) | 0.48 (H-2) | 0.30 (H-1) |
|  |  | 0.35 (H-4) | 0.35 (H-3) |
| P4 | 0.36 (H) | 0.65 (H-2) | 0.53 (H-3) |
|  | 0.39 (H-1) |  |  |
| P5 | 0.45 (H) | 0.35 (H-2) | 0.56 (H-3) |
|  |  | 0.55 (H-4) |  |
| P6 | 0.45 (H) | 0.42 (H-2) | 0.37 (H-5) |
|  |  | 0.46 (H-4) |  |

The variation of CT character can be most easily seen on the example of triplet states given as bars under the respective spectra in Figures 2 and 3. For most derivatives, a high density of the LC states is observed for energies above 4 eV. However, for P2, B5, and B9 they systematically appear at notably lower energies. LC states in general can be expected to produce charge separation in the molecule insufficient to initiate further photocatalytic redox reactions. Such changes can also explain the strong variation of transition intensities mentioned above.

Note also that the environment, due to, e.g., solvatochromic and rigidochromic effects, could also influence the relative positions of LC and MLCT states, see Ref. [43]. Although for emission from a single or a small group of states this can have a pronounced effect, we found that the general shape of the absorption spectrum does not change much when e.g., comparing acetonitrile with tetrahydrofuran for the case of IrPS (data not shown). The positions of individual lines can vary somewhat but due to their large number and significant broadening this influence is mitigated.

## 3. Computational Details

Density functional theory (DFT) together with the long-range corrected functional LC-BLYP [34,44,45] was chosen as the main theoretical approach. Although DFT is a well-known approach for investigation of ground and excited-state properties of relatively large systems [46] the CT properties of metal–organic complexes were shown to be strongly dependent on the exchange-kernel [25,47]. In case of widely used local and hybrid functionals, DFT suffers from incorrect description of long-range CT states [48,49], which are of primary interest for photocatalytic applications of metal–organic complexes, since the catalytic cycle necessarily includes intra- or inter-molecular charge separation. The problem might be solved by using long-range corrected functionals, where exact Hartree-Fock exchange is used at long distances providing the correct asymptotic behavior [50–55]. This scheme introduces an additional parameter, $\omega$, determining the range of switching between short- and long-range exchange contributions. In general, this parameter depends on the electronic density and, thus, is system-dependent. A fully ab initio procedure for tuning the range-separation parameter is provided by the $\Delta$SCF method. It ensures that Koopmans' theorem is fulfilled for the highest occupied and lowest unoccupied molecular orbitals (HOMO and LUMO) [53,56,57]. The details of the particular implementation of the protocol applied here can be found elsewhere [31].

All PSs have been studied using the following computational strategy. First, the ground state geometry in vacuum has been optimized with the standard value of the range-separation parameter ($0.47$ bohr$^{-1}$). For this geometry, the optimal tuning procedure has been applied. Second, the geometry has been re-optimized with the corresponding optimal value of $\omega$ and including the solvent (THF) via the polarizable continuum model (PCM). All minimal configurations have been checked for the absence of imaginary frequencies by harmonic frequency analysis. For the calculations of the lowest 200 singlet and 200 triplet states, the standard linear-response time-dependent DFT (TDDFT) formalism was utilized. However, the intensities of the spin-forbidden singlet-triplet transitions were not calculated, although Ir has large spin-orbit coupling constant. Recently, it was shown that singlet/triplet mixing in similar Ir photosensitizers has only a minor impact on the absorption spectra and results in small red shifts of the bands in the visible range and a decrease of intensities as triplet states possess small oscillator strengths [40–42]. As we have not aimed at fitting spectral widths for various PSs and transitions of different nature, a Gaussian broadening with a universal parameter (0.25 eV) has been applied to all stick-spectra. For DFT/TDDFT calculations no symmetry restrictions were applied. All calculations have been performed using the LANL2DZ ECP basis set for Ir and the 6-31G(d) basis set for all other atoms. Solvent effects on transitions have been included within the PCM approach [58]. All calculations were done with the Gaussian09 suite of programs [59]. Pre- and post-processing of data has been done with homemade programs.

## 4. Experimental Details

Steady-state UV/Vis absorption spectra of P$n$ compounds were recorded with a Specord 50 spectrophotometer (Analytik Jena, Jena, Germany) at room temperature. The photosensitizers were dissolved with a concentration of about $10^{-5}$ M in tetrahydroforan of Uvasol-quality (Merck,

Darmstadt, Germany). The solutions were filled in a fused silica cuvette with an optical path length of 1 cm. Experimental spectra of B1–B8 and B9 in acetonitrile were taken from Refs. [21,33], respectively.

## 5. Conclusions

In the present work, we have explored ground and excited state properties of a series of Ir-based photosensitizers, which are derivatives of the parent IrPS extensively investigated before in catalytic, spectroscopic, and theoretical studies [11,21–33]. To facilitate an accurate description of the excited states, the optimally-tuned range-separated density functional approach has been chosen. This method has demonstrated a very good agreement of calculated absorption spectra with experiments and thus provides a good basis for theoretical analysis. This holds true especially for CT excitations, where conventional density functionals are known to fail. Chemical modification of the parent IrPS demonstrates a number of trends, with some of them being non-trivial. For instance, modification of one ligand can notably influence the transitions involving other parts of the photosensitizer. In general, the most important bright singlet-singlet transition is of $Ir \rightarrow \pi^*(ppy)$ type and chemical derivation may shift it both to the red (what is a favorable scenario to enhance absorption efficiency due to the better overlap with the sun spectrum) or to the blue. In the latter case the nature of the lowest bright transition may change.

To summarize, we have demonstrated that the chosen method can provide an in-depth understanding of the correlations between chemical structure and photochemical properties of photosensitizers. This is an important prerequisite for the rational design of efficient photocatalytic systems.

**Acknowledgments:** We are grateful to Nils Rockstroh for experimental absorption spectra of B1–B8 and Stefanie Tschierlei for providing experimental spectrum for B9. We would also like to thank Felix Gärtner, Daniela Cozzula, Stefania Denurra, Sebastian Losse, Gopinatan Anilkumar, Henrik Junge and Matthias Beller from Leibniz Institute for Catalysis for synthesis of the studied PSs.

**Author Contributions:** Olga S. Bokareva and Tobias Möhle performed theoretical computations, Antje Neubauer measured absorption spectra, Sergey I. Bokarev, Oliver Kühn and Stefan Lochbrunner designed and coordinated the project. All authors were involved in writing the manuscript.

## References

1.  Schiermeier, Q.; Tollefson, J.; Scully, T.; Witze, A.; Morton, O. Energy alternatives: Electricity without carbon. *Nature* **2008**, *454*, 816–823.

2.  Esswein, A.J.; Nocera, D.G. Hydrogen production by molecular photocatalysis. *Chem. Rev.* **2007**, *107*, 4022–4047.

3.  Blankenship, R.E.; Tiede, D.M.; Barber, J.; Brudvig, G.W.; Fleming, G.; Ghirardi, M.; Gunner, M.R.; Junge, W.; Kramer, D.M.; Melis, A.; et al. Comparing photosynthetic and photovoltaic efficiencies and recognizing the potential for improvement. *Science* **2011**, *332*, 805–809.

4.  Hambourger, M.; Moore, G.F.; Kramer, D.M.; Gust, D.; Moore, A.L.; Moore, T.A. Biology and technology for photochemical fuel production. *Chem. Soc. Rev.* **2009**, *38*, 25–35.

5.  Burschka, J.; Pellet, N.; Moon, S.J.; Humphry-Baker, R.; Gao, P.; Nazeeruddin, M.K.; Grätzel, M. Sequential deposition as a route to high-performance perovskite-sensitized solar cells. *Nature* **2013**, *499*, 316–319.

6.  Ismail, A.A.; Bahnemann, D.W. Photochemical splitting of water for hydrogen production by photocatalysis: A review. *Sol. Ener. Mat. Sol. Cells* **2014**, *128*, 85–101.

7.  Wang, M.; Na, Y.; Gorlov, M.; Sun, L. Light-driven hydrogen production catalysed by transition metal complexes in homogeneous systems. *Dalton Trans.* **2009**, *33*, 6458–6467.

8.  Teets, T.S.; Nocera, D.G. Photocatalytic hydrogen production. *Chem. Comm.* **2011**, *47*, 9268–9274.

9.  Hagfeldt, A.; Boschloo, G.; Sun, L.; Kloo, L.; Pettersson, H. Dye-sensitized solar cells. *Chem. Rev.* **2010**, *110*, 6595–6663.

10. Eckenhoff, W.T.; Eisenberg, R. Molecular systems for light driven hydrogen production. *Dalton Trans.* **2012**, *41*, 13004–13021.

11.  Bokarev, S.I.; Bokareva, O.S.; Kühn, O. A theoretical perspective on charge transfer in photocatalysis. The example of Ir-based systems. *Coord. Chem. Rev.* **2015**, *304–305*, 133–145.

12.  Junge, H.; Rockstroh, N.; Fischer, S.; Brückner, A.; Ludwig, R.; Lochbrunner, S.; Kühn, O.; Beller, M. Light to Hydrogen: Photocatalytic Hydrogen Generation from Water with Molecularly-Defined Iron Complexes. *Inorganics* **2017**, *5*, 14.

13.  You, Y.; Nam, W. Photofunctional triplet excited states of cyclometalated Ir(III) complexes: Beyond electroluminescence. *Chem. Soc. Rev.* **2012**, *41*, 7061–7084.

14.  Goldsmith, J.I.; Hudson, W.R.; Lowry, M.S.; Anderson, T.H.; Bernhard, S. Discovery and high-throughput screening of heteroleptic iridium complexes for photoinduced hydrogen production. *J. Am. Chem. Soc.* **2005**, *127*, 7502–7510.

15.  Lowry, M.S.; Goldsmith, J.I.; Slinker, J.D.; Rohl, R.; Pascal, R.A.; Malliaras, G.G.; Bernhard, S. Single-Layer Electroluminescent Devices and Photoinduced Hydrogen Production from an Ionic Iridium(III) Complex. *Chem. Mater.* **2005**, *17*, 5712–5719.

16.  Tinker, L.L.; McDaniel, N.D.; Curtin, P.N.; Smith, C.K.; Ireland, M.J.; Bernhard, S. Visible light induced catalytic water reduction without an electron relay. *Chem. Eur. J.* **2007**, *13*, 8726–8732.

17.  Cline, E.D.; Adamson, S.E.; Bernhard, S. Homogeneous catalytic system for photoinduced hydrogen production utilizing iridium and rhodium complexes. *Inorg. Chem.* **2008**, *47*, 10378–10388.

18.  Tinker, L.L.; Bernhard, S. Photon-driven catalytic proton reduction with a robust homoleptic iridium(III) 6-phenyl-2,2′-bipyridine complex ([Ir(C–N–N)$_2$)]$^+$). *Inorg. Chem.* **2009**, *48*, 10507–10511.

19.  Curtin, P.N.; Tinker, L.L.; Burgess, C.M.; Cline, E.D.; Bernhard, S. Structure-activity correlations among iridium(III) photosensitizers in a robust water-reducing system. *Inorg. Chem.* **2009**, *48*, 10498–10506.

20.  Gärtner, F.; Sundararaju, B.; Surkus, A.E.; Boddien, A.; Loges, B.; Junge, H.; Dixneuf, P.H.; Beller, M. Light-driven hydrogen generation: Efficient iron-based water reduction catalysts. *Angew. Chem. Int. Ed.* **2009**, *48*, 9962–9965.

21.  Gärtner, F.; Cozzula, D.; Losse, S.; Boddien, A.; Anilkumar, G.; Junge, H.; Schulz, T.; Marquet, N.; Spannenberg, A.; Gladiali, S.; et al. Synthesis, characterisation and application of iridium(III) photosensitisers for catalytic water reduction. *Chem. Eur. J.* **2011**, *17*, 6998–7006.

22.  Gärtner, F.; Boddien, A.; Barsch, E.; Fumino, K.; Losse, S.; Junge, H.; Hollmann, D.; Brückner, A.; Ludwig, R.; Beller, M. Photocatalytic hydrogen generation from water with iron carbonyl phosphine complexes: Improved water reduction catalysts and mechanistic insights. *Chemistry* **2011**, *17*, 6425–6436.

23.  Hollmann, D.; Gärtner, F.; Ludwig, R.; Barsch, E.; Junge, H.; Blug, M.; Hoch, S.; Beller, M.; Brückner, A. Insights into the mechanism of photocatalytic water reduction by DFT-supported in situ EPR/Raman spectroscopy. *Angew. Chem. Int. Ed.* **2011**, *50*, 10246–10250.

24.  Gärtner, F.; Denurra, S.; Losse, S.; Neubauer, A.; Boddien, A.; Gopinathan, A.; Spannenberg, A.; Junge, H.; Lochbrunner, S.; Blug, M.; et al. Synthesis and characterization of new iridium photosensitizers for catalytic hydrogen generation from water. *Chem. Eur. J.* **2012**, *18*, 3220–3225.

25.  Bokarev, S.I.; Bokareva, O.S.; Kühn, O. Electronic excitation spectrum of the photosensitizer [Ir(ppy)$_2$(bpy)]$^+$. *J. Chem. Phys.* **2012**, *136*, 214305.

26.  Bokareva, O.S.; Bokarev, S.I.; Kühn, O. Electronic excitation spectra of the [Ir(ppy)$_2$(bpy)]$^+$ photosensitizer bound to small silver clusters Ag$_n$ (n = 1–6). *Phys. Chem. Chem. Phys.* **2012**, *14*, 4977–4984.

27.  Bokarev, S.I.; Hollmann, D.; Pazidis, A.; Neubauer, A.; Radnik, J.; Kühn, O.; Lochbrunner, S.; Junge, H.; Beller, M.; Brückner, A. Spin density distribution after electron transfer from triethylamine to an [Ir(ppy)$_2$(bpy)]$^+$ photosensitizer during photocatalytic water reduction. *Phys. Chem. Chem. Phys.* **2014**, *16*, 4789–4796.

28.  Bokareva, O.S.; Kühn, O. DFT-D investigation of the interaction between Ir (III) based photosensitizers and small silver clusters Ag$_n$ (n = 2–20, 92). *Chem. Phys.* **2014**, *435*, 40–48.

29.  Neubauer, A.; Grell, G.; Friedrich, A.; Bokarev, S.I.; Schwarzbach, P.; Gärtner, F.; Surkus, A.E.; Junge, H.; Beller, M.; Kühn, O.; et al. Electron- and Energy-Transfer Processes in a Photocatalytic System Based on an Ir(III)-Photosensitizer and an Iron Catalyst. *J. Phys. Chem. Lett.* **2014**, *5*, 1355–1360.

30.  Fischer, S.; Hollmann, D.; Tschierlei, S.; Karnahl, M.; Rockstroh, N.; Barsch, E.; Schwarzbach, P.; Luo, S.P.; Junge, H.; Beller, M.; et al. Death and Rebirth: Photocatalytic Hydrogen Production by a Self-Organizing Copper–Iron System. *ACS Catal.* **2014**, *4*, 1845–1849.

31.  Bokareva, O.S.; Grell, G.; Bokarev, S.I.; Kühn, O. Tuning Range-Separated Density Functional Theory for Photocatalytic Water Splitting Systems. *J. Chem. Theor. Comp.* **2015**, *11*, 1700–1709.

32.  Bokareva, O.S.; Kühn, O. Quantum chemical study of the electronic properties of an Iridium-based photosensitizer bound to medium-sized silver clusters. *Chem. Phys.* **2015**, *457*, 1–6.

33.  Tschierlei, S.; Neubauer, A.; Rockstroh, N.; Karnahl, M.; Schwarzbach, P.; Junge, H.; Beller, M.; Lochbrunner, S. Ultrafast excited state dynamics of iridiumIII complexes and their changes upon immobilisation onto titanium dioxide layers. *Phys. Chem. Chem. Phys.* **2016**, *18*, 10682–10687.

34.  Tawada, Y.; Tsuneda, T.; Yanagisawa, S.; Yanai, T.; Hirao, K. A long-range-corrected time-dependent density functional theory. *J. Chem. Phys.* **2004**, *120*, 8425–8433.

35.  Song, J.W.; Hirosawa, T.; Tsuneda, T.; Hirao, K. Long-range corrected density functional calculations of chemical reactions: Redetermination of parameter. *J. Chem. Phys.* **2007**, *126*, 154105.

36.  Stein, T.; Eisenberg, H.; Kronik, L.; Baer, R. Fundamental Gaps in Finite Systems from Eigenvalues of a Generalized Kohn-Sham Method. *Phys. Rev. Lett.* **2010**, *105*, 266802.

37.  Refaely-Abramson, S.; Baer, R.; Kronik, L. Fundamental and excitation gaps in molecules of relevance for organic photovoltaics from an optimally tuned range-separated hybrid functional. *Phys. Rev. B* **2011**, *84*, 075144.

38.  Körzdörfer, T.; Sears, J.S.; Sutton, C.; Brédas, J.L. Long-range corrected hybrid functionals for $\pi$-conjugated systems: Dependence of the range-separation parameter on conjugation length. *J. Chem. Phys.* **2011**, *135*, 204107.

39.  Salzner, U.; Baer, R. Koopmans' springs to life. *J. Chem. Phys.* **2009**, *131*, 231101.

40.  Li, X.; Minaev, B.; Ågren, H.; Tian, H. Theoretical Study of Phosphorescence of Iridium Complexes with Fluorine-Substituted Phenylpyridine Ligands. *Eur. J. Inorg. Chem.* **2011**, *2011*, 2517–2524.

41.  Minaev, B.; Baryshnikov, G.; Agren, H. Principles of phosphorescent organic light emitting devices. *Phys. Chem. Chem. Phys.* **2014**, *16*, 1719–1758.

42.  Brahim, H.; Daniel, C. Structural and spectroscopic properties of Ir(III) complexes with phenylpyridine ligands: Absorption spectra without and with spin–orbit-coupling. *Comp. Theor. Chem.* **2014**, *1040–1041*, 219–229.

43.  Colombo, M.; Hauser, A.; Güdel, H. Competition between ligand centered and charge transfer lowest excited states in bis cyclometalated $Rh^{3+}$ and $Ir^{3+}$ complexes. In *Electronic and Vibronic Spectra of Transition Metal Complexes I Topics in Current Chemistry*; Springer-Verlag Berlin Heidelberg: Heidelberg, Germany, 1994; Volume 171, pp. 143–171.

44.  Iikura, H.; Tsuneda, T.; Yanai, T.; Hirao, K. A long-range correction scheme for generalized-gradient-approximation exchange functionals. *J. Chem. Phys.* **2001**, *115*, 3540–3544.

45.  Chiba, M.; Tsuneda, T.; Hirao, K. Excited state geometry optimizations by analytical energy gradient of long-range corrected time-dependent density functional theory. *J. Chem. Phys.* **2006**, *124*, 144106.

46.  Sousa, S.F.; Fernandes, P.A.; Ramos, M.J.A. General performance of density functionals. *J. Phys. Chem. A* **2007**, *111*, 10439–10452.

47.  Laurent, A.D.; Jacquemin, D. TD-DFT benchmarks: A review. *Int. J. Quant. Chem.* **2013**, *113*, 2019–2039.

48.  Dreuw, A.; Head-Gordon, M. Single-reference ab initio methods for the calculation of excited states of large molecules. *Chem. Rev.* **2005**, *105*, 4009–4037.

49.  Peach, M.J.G.; Helgaker, T.; Salek, P.; Keal, T.W.; Lutnaes, O.B.; Tozer, D.J.; Handy, N.C. Assessment of a Coulomb-attenuated exchange-correlation energy functional. *Phys. Chem. Chem. Phys.* **2006**, *8*, 558–562.

50.  Gerber, I.C.; Ángyán, J.G. Hybrid functional with separated range. *Chem. Phys. Lett.* **2005**, *415*, 100–105.

51.  Zhao, Y.; Truhlar, D.G. Density Functional for Spectroscopy: No Long-Range Self-Interaction Error, Good Performance for Rydberg and Charge-Transfer States, and Better Performance on Average than B3LYP for Ground States. *J. Phys. Chem. A* **2006**, *110*, 13126–13130.

52.  Peach, M.J.G.; Benfield, P.; Helgaker, T.; Tozer, D.J. Excitation energies in density functional theory: An evaluation and a diagnostic test. *J. Chem. Phys.* **2008**, *128*, 44118.

53.  Livshits, E.; Baer, R. A well-tempered density functional theory of electrons in molecules. *Phys. Chem. Chem. Phys.* **2007**, *9*, 2932–2941.

54.  Chai, J.D.; Head-Gordon, M. Systematic optimization of long-range corrected hybrid density functionals. *J. Phys. Chem.* **2008**, *128*, 084106.

55.  Mori-Sánchez, P.; Cohen, A.J.; Yang, W. Self-interaction-free exchange-correlation functional for thermochemistry and kinetics. *J. Chem. Phys.* **2006**, *124*, 91102.

56.  Stein, T.; Kronik, L.; Baer, R. Reliable Prediction of Charge Transfer Excitations in Molecular Complexes Using Time-Dependent Density Functional Theory. *J. Am. Chem. Soc.* **2009**, *131*, 2818–2820.

57.  Stein, T.; Kronik, L.; Baer, R. Prediction of charge-transfer excitations in coumarin-based dyes using a range-separated functional tuned from first principles. *J. Chem. Phys.* **2009**, *131*, 244119.

58.  Tomasi, J.; Mennucci, B.; Cammi, R. Quantum mechanical continuum solvation models. *Chem. Rev.* **2005**, *105*, 2999–3093.

59.  Frisch, M.J.; Trucks, G.W.; Schlegel, H.B.; Scuseria, G.E.; Robb, M.A.; Cheeseman, J.R.; Scalmani, G.; Barone, V.; Mennucci, B.; Petersson, G.A.; et al. *Gaussian 09 Revision C.1*; Gaussian Inc.: Wallingford, CT, USA, 2009.

# 15

# Visible Light-Activated PhotoCORMs

**Emmanuel Kottelat and Fabio Zobi ***

Department of Chemistry, University of Fribourg, Chemin du Musée 9, CH-1700 Fribourg, Switzerland;
emmanuel.kottelat@unifr.ch
* Correspondence: fabio.zobi@unifr.ch

Academic Editor: Lígia M. Saraiva

**Abstract:** Despite its well-known toxicity, carbon monoxide (CO) is now recognized as a potential therapeutic agent. Its inherent toxicity, however, has limited clinical applications because uncontrolled inhalation of the gas leads to severe systemic derangements in higher organisms. In order to obviate life-threatening effects and administer the gas by bypassing the respiratory system, CO releasing molecules (CORMs) have emerged in the last decades as a plausible alternative to deliver controlled quantities of CO in cellular systems and tissues. As stable, solid-storage forms of CO, CORMs can be used to deliver the gas following activation by a stimulus. Light-activated CORMs, known as photoCORMs, are one such example. This class of molecules is particularly attractive because, for possible applications of CORMs, temporal and spatial control of CO delivery is highly desirable. However, systems triggered by visible light are rare. Most currently known photoCORMs are activated with UV light, but red light or even infrared photo-activation is required to ensure that structures deeper inside the body can be reached while minimizing photo-damage to healthy tissue. Thus, one of the most challenging chemical goals in the preparation of new photoCORMs is the reduction of radiation energy required for their activation, together with strategies to modulate the solubility, stability and nontoxicity of the organic or organometallic scaffolds. In this contribution, we review the latest advances in visible light-activated photoCORMs, and the first promising studies on near-infrared light activation of the same.

**Keywords:** photoCORMs; visible light; manganese

---

## 1. Introduction

Carbon monoxide (CO) has been known for a long time as a dangerous gas to mammals [1]. Indeed, due to its toxicity, colorless, odorless and tasteless nature, the gas is commonly referred to as "the silent killer". Carbon monoxide toxicity partially originates from the high affinity of the molecule for the iron of hemoglobin, with which it strongly interacts, thereby reducing the protein's ability to shuttle and transfer oxygen into tissues. However, it has also been demonstrated that hemoglobin acts as detoxifying protein and that it is rather the small amount of unbound CO that escapes it that is likely responsible for the main toxicity of the gas [2].

Despite its toxicity CO, is produced endogenously in mammals by the *Heme Oxygenase* family of enzymes during heme catabolism. Studies on its role in mammals have revealed that CO plays a fundamental part as a gaseous signaling neurotransmitter, together with nitric oxide (NO) and dihydrogen sulfide ($H_2S$) [3]. Lack of endogenous CO production leads to systemic function disorders, including diabetes [4], inflammation [5–8], or tissue cellular apoptosis [4]. The discovery of the unexpected beneficial role of CO has led, on one hand, to its evaluation as a therapeutic agent in clinics and hospitals and, on the other hand, to an increasing interest in the chemistry of CO. The former has most certainly fueled the latter as mounting biological and medicinal evidence indicates that CO can prevent or alleviate a variety of diseases, including, e.g., cardiovascular inflammation [4,9,10], hepatic ischemia [11], cell proliferation [4] or have antiatherogenic [12] or cytoprotective effects [4,13], *inter alia*.

In order to exert its therapeutic effects, a proper amount of CO must be delivered in a controlled manner to avoid tissue hypoxia and severe toxic side effects. In its gaseous form, however, the molecule is difficult to handle and to administer in a precise concentration at a specific location within the organism. As a way to deliver the gas by bypassing the respiratory system, CO releasing molecules (CORMs) have emerged as an important research area, bridging disciplines ranging from organometallic and organic chemistry to pharmacology and medicine. The rising interests in these molecules are due to the fact that they, in principle, allow for the controlled and targeted delivery of the gas into wounded or injured tissues [14–18], depending on the nature and the specificity of the CORM used. To date, most CORMs are transition metal complexes bearing at least one carbonyl ligand. It is this ligand that once released from the molecule with a defined stimulus, acts as the endogenously generated CO.

The majority of CORMs release CO thermodynamically, i.e., spontaneously by dissolution in the aqueous media [19,20]. The rate of CO delivery can be calculated kinetically by spectrophotometry or infrared analysis and can be modulated by modification of the ancillary ligands of the releasing complex [21,22]. Nevertheless, in the case of spontaneous CO liberation, the starting time of action can be difficult to control. Thus, a stimulus that can activate CORMs at a precise moment would be preferred. Enzyme-triggered release [23–26] or electromagnetic heating [27] are some of the ways used to initiate CO release. However, the most commonly employed stimulus for CORM activation is light. Several transition metal carbonyl complexes are known to be photosensitive, and they have been naturally studied to this end. Indeed, the number of publications on photoCORMs has increased exponentially in the last decade [28,29]. Early studies in this area resulted in promising compounds that were activated by UV light irradiation [30–34]. Unfortunately, the shortest wavelengths have poor penetration and are potentially harmful to healthy tissues. Thus, research in the area has also moved towards the discovery and preparation of photoCORMs that can be activated with radiations of longer wavelengths. These efforts have resulted in systems showing controlled CO release with red light activation and, in some instances, with radiations approaching the boundaries of near infrared excitation [21,35].

These systems are the subject of this brief review. Specifically, we have limited our contribution to papers that have appeared since 2010 and in particular, to photoCORMs activated with light of a wavelength >450 nm. However, when necessary for our discussion, we allowed for certain flexibility to these stringent selection criteria. In any case, this review is not comprehensive of all contributions in the CORM field in the mentioned timeframe and for a more wide-ranging view of CORMs and photoCORMs, the reader may refer to other reviews [2,4,14–18,36–40]. Before introducing the latest advances in visible and near-infrared activated photoCORMs, and in order to guide readers unfamiliar with the subject matter, we will begin with a brief historical overview of photoCORMs.

## 2. PhotoCORMs: Identification and Design

The term photoCORM was introduced by Ford and coworkers, who examined the delivery of small gasotransmitters (nitric oxide and carbon monoxide) from transition metal complexes, notably a tungsten complex (**1**, Figure 1) showing CO release and water solubility [41]. Motterlini and co-workers reported the first biological use of light-induced CO release from chemical compounds, namely the pure metal carbonyl complexes $Fe(CO)_5$ (**2**, Figure 1) and $Mn_2(CO)_{10}$ (**3**, Figure 1) [19]. Due to poor aqueous solubility and toxicity, most notably for the Fe compound, the development of these CORMs was only possible by modifying the ancillary ligands around the metal center. Over the years the *fac*-$\{Mn(CO)_3\}^+$ core became the prominent moiety in the advancement of photoCORMs. The first ligand used in such *fac*-$\{Mn(CO)_3\}^+$-based photoactivatable CO delivery systems was the tridentate tris(pyrazolyl)methane published by Schatzschneider (**4**, Figure 1). The resulting photoCORM released 2 equivalents of CO per Mn unit [31,32].

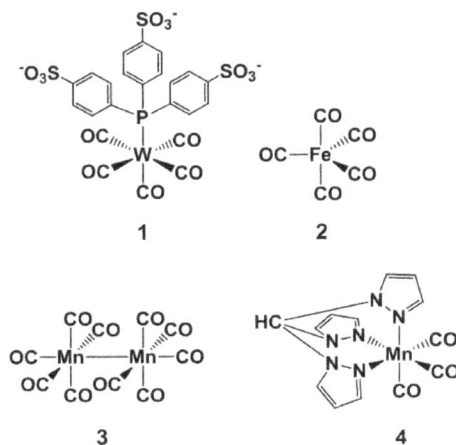

**Figure 1.** Early photoCORM structures.

Generally speaking, the mechanism of CO labilization and loss from transition metal carbonyl complexes is nowadays well understood. Due to the high π-accepting capability of the carbonyl ligand, CO induces a strong ligand field enhanced by synergistic CO σ-donation to an empty metal $d$-orbital and π-backbonding from an occupied $d$-orbtital to the π* antibonding orbital of CO. Irradiation of the photoCORM with sufficient energy can promote excitation of $d$-orbital electrons increasing, on one hand, the formal oxidation state of the central metal ion and, consequently, reducing the synergistic M–CO σ/σ interaction. The net result of irradiation is that of weakening the metal–carbonyl bond, thereby promoting CO release. This simplified picture is also useful for understanding photo-induced CO release ensuing from metal-to-ligand charge transfer (MLCT) processes where electrons are excited (i.e., transferred) from a metal $d$-orbital to the π* anti-bonding orbital of an ancillary ligand. This last process is one of the most prominent pathways leading to the photo-labialization of CO and examples of it will be discussed in more details in the ensuing sections.

The role of the ancillary ligands is decisive for tuning MLCT, the irradiation wavelength, the rate of CO loss and the toxicity of the photoproducts. In order to achieve visible light photoactivation, the energy gap of the orbital involved in electron transfer (e.g., HOMO–LUMO energy levels) needs to be as narrow as possible. Ancillary ligands with hyperconjugation and π-acceptors groups on the ligand frame are used for this purpose, along with π-donor atoms directly bound to the metal center [21]. Bromide and chloride anions are often the first choice of the latter type of ligands [22]. Tuning the wavelength of activation of photoCORMs is only one aspect to render the species possibly valuable for further medical applications. In principle, any photoCORM has to be biocompatible with animal systems to be further investigated. Thus, solubility in water, in biological media (phosphate buffer solution, PBS) or water (≤0.5%) DMSO mixture, the non-toxicity of the CORM and the photoproducts, as well as their elimination from the organism, are all key requirements for a pharmaceutically useful photoCORM. Without light irradiation, stability in water and under aerobic conditions is also necessary for the tested compounds. For adequate use as therapeutic agents, only light irradiation must provide CO release and the decomposition into photoproducts.

An ideal photoCORM needs to release CO in a controlled manner and in correct amounts. At least one CO equivalent should be delivered per metal center. Thus, confirmation of the photoactivity of the CO releasing molecule is necessary. The most commonly used experiment to assess photo-CO release is the myoglobin assay, a spectrophotometric assay that allows monitoring the formation of carboxyMyoglobin (MbCO) [42]. Typically, a myoglobin solution of known concentration is prepared, and sodium dithionite is used to reduce ferric to ferrous iron in the heme of the metalloprotein before the addition of 1 (or more) equivalent(s) of the tested compound. If the latter is not soluble in pure water, a final percentage of 0.5% of DMSO is generally safe. Binding of CO to the ferrous heme (in this case referred to a deoxyMyoglobin, dMb) is then monitored by UV analysis. If CO is indeed released from the photoCORM, the visible absorption band of dMb at 560 nm is progressively replaced by two bands at 540 and 579 nm (typical of MbCO); the Soret band shows also a shift, from ca. 435 to 424 nm [18]. While this assay has proven extremely valuable in the identification and screening of CORMs, it should be kept in mind that the same assay is not perfect. McLean, Mann and Poole have shown that in some instances, and depending on the type of CORM being evaluated, sodium dithionite can influence CO release and give rise to artefacts [43]. Alternative experiments for CO release assessment have been developed; these include gas chromatography [41], portable CO sensors [44], IR [45] and fluorescence [46] spectroscopy.

## 3. UV Light Photoactivated CORMs

The conception of new photoactivatable CO releasing molecules, which are nontoxic, stable in physiological media and safe and efficient CO deliverers upon visible light illumination, remains a challenge. To date, the vast majority of organic and organometallic photoCORMs provide CO only upon UV light irradiation [47–52]. The tris(pyrazolyl)methane photoCORM characterized by Schatzschneider (**4**, Figure 1) and co-workers is a typical example. This *fac*-{Mn(CO)$_3$}$^+$-based complex is cationic in nature; it shows good water-solubility and anticancer activity [32] and could also be used as a tag in medical imaging applications due to the characteristic and intense IR bands of the carbonyl stretching frequencies [34]. Furthermore, the tripodal ligand could be used to tune the outer coordination sphere of the photoCORM without altering the photoactivity of the compound. Peptide functionality or binding to SiO$_2$ surface were thus reported for similar compounds showing similar CO releasing properties [30].

Other examples of *fac*-{Mn(CO)$_3$}$^+$-based photoCORMs with tripodal ligands have also been reported. Berends and Kurz replaced a pyrazolyl ring by an acetate unit, thus conferring a neutral charge to the complex (**5** and **6**, Figure 2) [53]. They demonstrated a stepwise CO loss process and hypothesized the formation of intermediate photoproducts before the oxidation of the metal center and the formation of a Mn–O–Mn moiety. Moreover, this complex had better CO release capabilities than the tris-pyrazolyl CORM of Schatzschneider. Recently, the group of Kelebekli reported novel tricarbonyl Mn complexes bearing bipyridiyl and imidazole derivatives (**7** and **8**, Figure 2) [54,55]. This set of complexes showed effective cancer treatment and cytotoxic activity against breast cancer cells and the MCF-7 human cell line. Antioxidant capability and redox properties were also revealed, by scavenging 2,2-diphenyl-1-picrylhydrazyl, superoxide and nitroxyl radicals. Taken together, the results showed promising steps for the use of photoCORMs as drug templates for the treatment of breast cancer. Tinajero and coworkers reported an Mn-based photoactivated CORM which reduces growth and viability of *Escherichia coli*, only by light induction [56]. They demonstrated that CO binds to specific intracellular sites, namely, respiratory oxidases and a globin protein expressed in a particular strain, while the Mn metal center is not accumulated in the cells. This study is notably the first investigation of the antimicrobial activity of a photoCORM against a resistant pathogen. The understanding of the mechanisms of the CORM toxicity is in fact the key challenge for the increase in the use of such compounds as therapeutic agents.

As solubility and nontoxicity are amongst the required elements for a valuable therapeutic CORM, the use of various biocompatible scaffolds is displaying a marked increase in publications (*vide infra*). For example, Yang and coworkers synthesized and characterized sawhorse-shaped ruthenium complexes with diverse amino and carboxylic acid derivatives on the outer coordination sphere of the CORM [57]. Cell viability investigation of **10** (Figure 2) indicated the absence of toxicity with or without irradiation. While the compound was stable in the dark for hours, CO release was proven by the common myoglobin assay, making it a potential lead structure for further investigations.

Mascharak and coworkers reported recently an interesting use of photoCORMs as luminescent trackers [58]. The orange color assigned to the intact CORM (**11**, Figure 2) was used to successfully track the entry of the prodrug into the cancer cells, while the shift to deep blue fluorescence after CO loss was used for the assessment of the delivery in specific cellular sites. The utility of the theranostic photoCORM was proven in breast cancer cells with the help of flow cytometry and fluorescence microscopy, attesting the therapeutic potential of the compound.

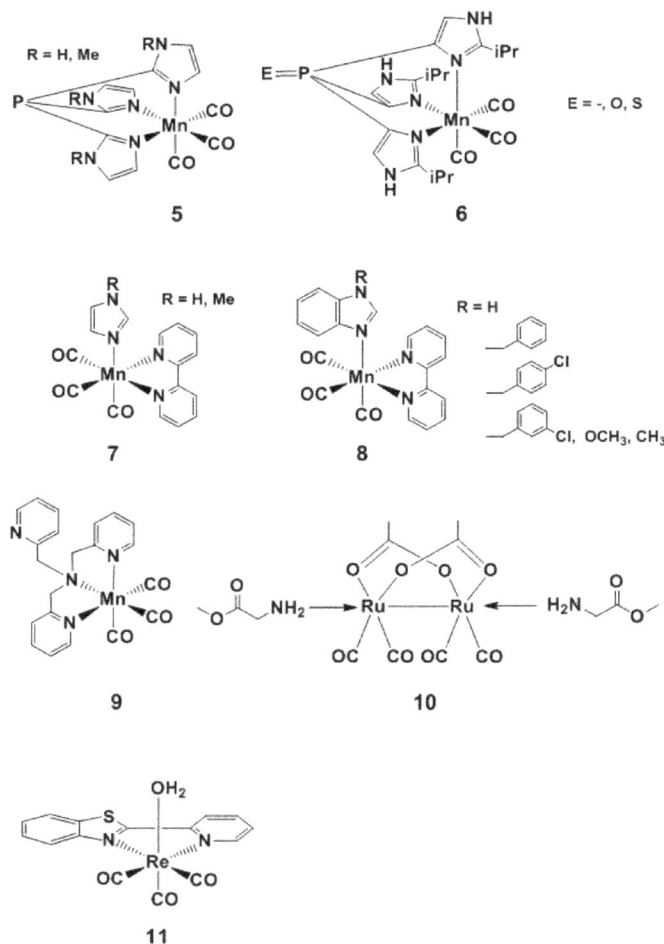

**Figure 2.** Selected examples of UV-activated photoCORMs.

## 4. Visible Light and Near Infrared-Light (NIR) Photoactivated CORMs

### 4.1. Inorganic and Organometallic PhotoCORMs

Westerhausen and co-workers reported in 2011 a dicarbonylbis(cysteamine)iron(II) complex (**12**, Figure 3), named CORM-S1 [59]. To our knowledge, this is the first photoCORM activated by visible light (i.e., >450 nm). Indeed, 470 nm light irradiation of **12** promotes constant CO release over several minutes, whereas exposition to broadband white light leads to rapid and complete CO liberation in a physiological medium. This property together with the high water solubility of the complex argues for its biocompatibility. The cysteamine ligand does not seem provide easy possibilities for further modifications of the inner- and outer-coordination sphere of the molecule, in order to finely tune the excitation wavelength needed for the CO liberation. Nonetheless, Motterlini and coworkers [60] did derivatize the corresponding ligand into cysteinate. More recently, Westerhausen [61] proposed the thiolato-bridged $[\{(OC)_3Mn\}_2(\mu\text{-}SCH_2CH_2NH_3)_3]$ dimer, obtained from the reaction of pentacarbonyl manganese with cysteamine (**13**, Figure 3). The complex was shown to be highly soluble in water, stable under anaerobic conditions and released CO when irradiated at 365 nm and 470 nm. Although CO liberation was faster with UV-light, both irradiation wavelengths promoted delivery of the all six COs. Another interesting example of biocompatible photoCORM is the [FeFe]-hydrogenase derivative proposed by Fan and co-workers (**14**, Figure 3) [62]. The carboxylate groups on the side chain provide a polar water soluble complex, and COs are completely released by white light irradiation. Detailed investigations with a specific irradiation at 390 nm were also performed. Cell proliferation assay on MCF-10A epithelial cells indicates that the salt $Na_2[(\mu\text{-}SCH_2CH_2COO)\text{-}Fe(CO)_3]_2$ (i.e., $Na^+$ salt of **14**) is not cytotoxic.

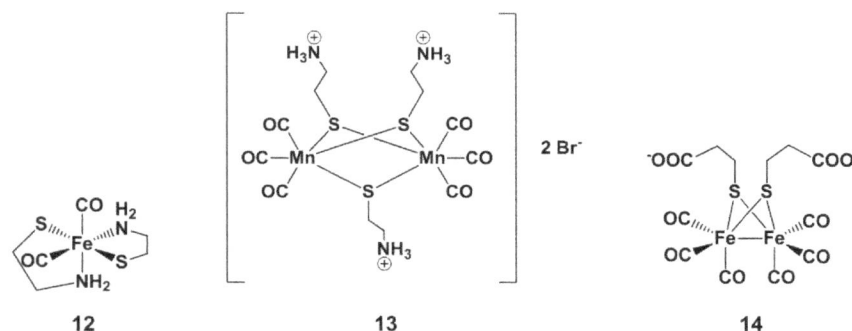

**Figure 3.** CORM-S1 (**12**) thiolato-bridged $Mn(CO)_3$ based (**13**) and [FeFe]-hydrogenase based photoCORM (**14**).

Early in 2012, Mascharak reported the use of (2-pyridylmethyl)(2-quinolylmethyl)amine as tripodal based ligand (**15**, Figure 4) for $fac\text{-}\{Mn(CO)_3\}^+$ complexes [63]. This photoCORM was activated by visible light. Later that year, the same group reported the most prominent step towards visible light activation of photoCORMs, using highly conjugated Schiff bases as ligands for similar cores (**16–18**, Figure 4) [64]. Noticeable in the series were the [MnBr(CO)$_3$qmtpm] (**17**, Figure 4) and [MnBr(CO)$_3$pmtpm] (**18**, Figure 5) complexes, which exhibit MLCT maximal absorbance at 535 and 500 nm, respectively. As described above, the small energy transition could be explained by the combined effect of hyperconjugation of the bidentate ligand, the $\pi$-donor character of the bromo ancillary ligand and the electron donating nature of the –SMe group. These considerations led the same group to the design and synthesis of azobipyridine type complexes (**19**, Figure 4) reported in 2014 [65]. The complex with Mn as the metal center and bromide as the sixth ancillary ligand exhibits MLCT transition at 585 nm, whereas the same complex with an Re center shows a MLCT band at 530 nm. As this later compound also displays MLCT under visible light, one might have expected it to be the first CO releasing molecule with a $Re^I$ center. The myoglobin assay, however, does not indicate any CO delivery. DFT calculations performed to understand this discrepancy indicate that the spin-orbit coupling (prominent in heavy metals [66]) promotes intersystem $^1$MLCT–$^3$MLCT crossing in the rhenium complex, thereby dissipating the excitation energy without inducing Re–CO bond rupture. As expected, both complexes with PPh$_3$ as the ancillary ligand show low MLCT transition energy due to the high $\pi$-acceptor character of PPh$_3$.

More recently, the group of Mascharak published visible light-induced CO release by a biocompatible tris-carbonyl Mn-based photoCORM showing good water-solubility [67]. Three $\alpha$-diimines ligand-based complexes were prepared (**20–22**, Figure 4) and one of them (**21**, Figure 4) demonstrated carbon monoxide delivery to myoglobin in PBS solution. In these complexes, one of the imine functions is not part of a conjugated rigid ring, leading to higher molar absorbance and faster CO release, compared to imines in a rigid $\alpha$-diimine ring system of bipyridine or phenanthroline type ligands [68]. In order to enhance the accumulation on the targeted specific site, cellular uptake, non-toxicity of byproducts and water-solubility of these CORMs, Mascharak and his group proposed ligand-appended adamantylamine derivatives [33,69]. Aminoadamantane-containing compounds are commercially available in the pharmaceutical market and are used for the treatment of ovarian carcinoma or human prostate and colon cancer [70,71]. Side effects of released adamantine-based ligands following photo-activation of CORMs and their cellular uptake were also studied in detail and indicated promising results [71]. Of the series of molecules, only complex **21** dissolved in pure aqueous media and physiological PBS conditions and remained stable for at least 24 h if kept in the dark. The three CORMs showed CO delivery upon illumination with light $\geq$ 450 nm. The CO release rates increased according to the order **20** < **21** < **22** in $CH_2Cl_2$, and UV analysis indicated that **21** released 2 eq. of CO upon light irradiation under aerobic conditions in water. Furthermore, the photo-activation of **21** appears to be insensitive to dithionite or glutathione reductants.

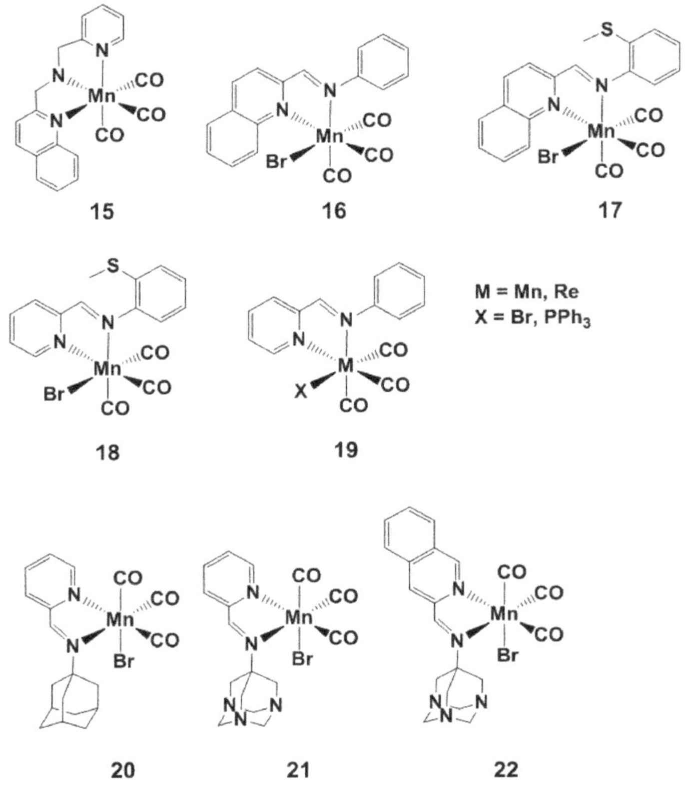

**Figure 4.** Selected structures of photoCORMs developped by Mascharak et al. [63–65,67].

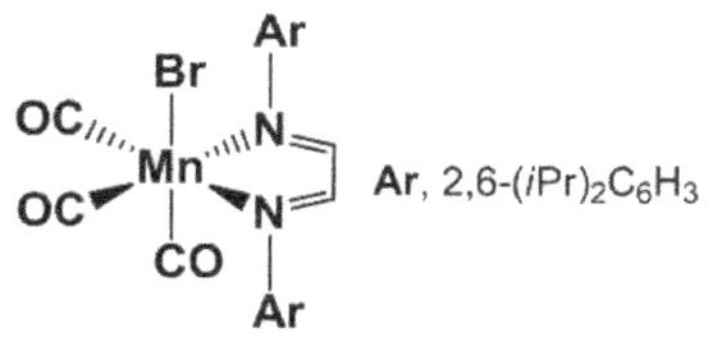

**Figure 5.** Structure of Bengali's photoCORM.

Since hyperconjugated diimines ligands coordinated to *fac*-{Mn(CO)$_3$}$^+$ complexes permit photoCORM activation and liberation of carbon monoxide with visible light, the group of Bengali proposed compound **23** (Figure 5) [72]. The MLCT transition of this species was observed at 582 nm, and 560 nm light irradiation provided rapid CO release by substitution of the carbonyl with a solvent molecule (CH$_3$CN or THF). DFT calculations on **23** and the corresponding diimine tetracarbonyl manganese cation indicate that the bulky steric environment around the carbonyl ligands is the predominant cause of the weakening of the metal–carbonyl bond interaction.

An approach to visible-light activated CORMs, which also took into consideration the toxicity of the ensuing photoproducts, was described by Mansour and Shehab, who reported the tazarotene (TZ) and metamizole (MZ) [MnBr(CO)$_3$TZ] (**24**, Figure 6) and [MnBr(CO)$_3$MZ] (**25**, Figure 6) species [73].

Both TZ and MZ are approved drugs in medical treatments [74,75], and following photo-CO release and dissociation from the *fac*-{Mn(CO)$_3$}$^+$ core ensured no toxicity of the photoproducts. TZ interacts with the Mn center as a bidentate (N-pyridine– and C≡C–) ligand while MZ behaves as tridentate ligand. Although these photoCORMs (i.e., **24** and **25**) demonstrated poor stability in DMSO, they are able to release CO upon exposure to blue light (468 nm). Under these conditions, complex **24** released 1 eq. of CO after 42 min of irradiation while **25** only 0.5 eq. after 70 min. When **24** and **25** were irradiated at longer wavelengths (535 nm) and tested via the myoglobin (Mb) assays, no evidence of CO loss or of Mb–CO binding could be observed. The antimicrobial activity of TZ, MZ and their corresponding CO releasing molecules **24** and **25** was also tested against two bacterial strains. Both TZ and MZ drugs are ineffective against *Staphylococcus aureus* and *Escherichia coli*, but the authors mentioned that **24**, and to a lower extent **25**, demonstrates higher antimicrobial effects. The authors suggested, therefore, that the coordination of the two drugs to the *fac*-{Mn(CO)$_3$}$^+$ moiety results in an increased antibacterial effect.

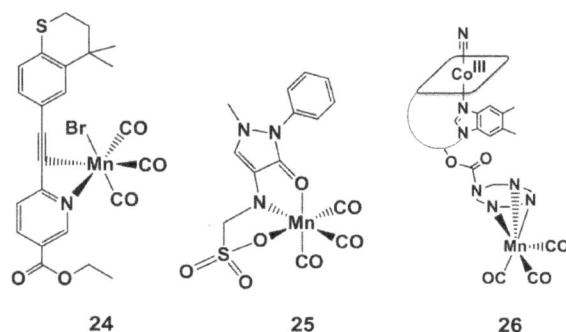

**Figure 6.** PhotoCORM structures of Mansour (**24**, **25**) and Zobi's B12-MnCORM (**26**).

A bio-compatible vitamin B$_{12}$-conjugate bearing the same *fac*-[Mn(CO)$_3$]$^+$ core (**26**, Figure 6) was the subject of a study published by our group in 2013. IR spectromicroscopy was used to characterize the intracellular uptake and reactivity of the photosensitive compound (in 3T3 fibroblasts), whose CO releasing properties could be triggered by illumination with visible light (470 nm). The intracellular reactivity of the compound, evaluated as a function of the release of the CO ligands following light exposure, was elucidated. The distribution of the complex could be described by IR mapping and, taking into consideration the role of cellular topography in the quantitative interpretation of the maps, a perinuclear distribution, in the nucleus and/or in its proximity, appeared to be the most likely interpretation of the images obtained. Intracellular photoinduced CO release of **26** prevents fibroblasts from dying under conditions of hypoxia and metabolic depletion, conditions that may occur in vivo during insufficient blood supply to oxygen-sensitive tissues such as the heart or brain.

### 4.2. Organic PhotoCORMs

An interesting novel feature of a class of photoCORMs bearing flavonato derivatives was demonstrated by Berreau and coworkers [76]. They reported two types of Zn(II) flavonolato complexes that exhibit red-shift energy absorbance and enhanced CO release compared to the free corresponding flavonols. These ligands (**27–30**, Figure 7) were known to deliver CO in organic or organic/water media when induced by visible light [77]. In this latter publication, the tuning on the phenyl moiety of the flavonol structures was demonstrated to be relevant for the red-shift of the absorption maximum of the free ligand. Whereas **27** exhibits an absorption centered at 409 nm, the dialkylamine subunit on **28** shifted the same absorption band to 442 nm. With both dialkylamine and thione present (**30**), the maximum was observed at 544 nm. The characterization of the flavonolato complexes (**31–34**, Figure 7) indicated a spectral absorption band in the therapeutic window (>650 nm) and a solid-state CO release reactivity comparable to the activity seen in solution when irradiated with 546 nm light. The quantum yield was also significantly higher for the Zn compounds with respect to the neutral flavonols and reported organic photoCORMs [78,79]. A 1 eq. of CO per Zn unit is delivered from **31–34** when dissolved in pyridine. Similar to the free flavonols, the MLCT maximum of the corresponding complexes shifted gradually from 480 nm (for **31**) to 600 nm (for **34**), showing that the tunability of the ligand framework impacts the MLCT without influencing the amount of CO release. Furthermore, the bis-flavonato based complexes (**35–38**, Figure 7) demonstrated release of 2 eq. of CO per Zn(II). Surprisingly, **35** and **36** do exhibit also 2 eq. of CO release in the solid state, while **37** and **38** do not. The MLCT maximum of **35–38** was comparable to the corresponding **31–34** complexes. It should also be mentioned that CO release of both free ligands **27–30** and complexes **31–38** exhibit dioxygenase-type ($O_2$-dependent) CO release reactivity and the flavonato species are degraded by dioxygenase-type enzymes when CO is released, either by thermal reactivity [80] or enzyme catalysis [81]. Moreover, compound **35**, when coated as a film on a flask, showed the delivery of CO in oxidative palladium-catalyzed carbonylation processes. A total of 1 eq. of CO was delivered after 24 h of irradiation of the evaluated solid film. Overall, these results suggest that flavonolato derivatives are adjustable molecules for the design of and development of both inorganic and organic photoCORMs, and these solid bis-flavonolato derivatives can be used in oxidative catalyzed alkoxycarbonylations as CO releasing agents. However, further studies on efficiency and applications of this novel feature need to be assessed.

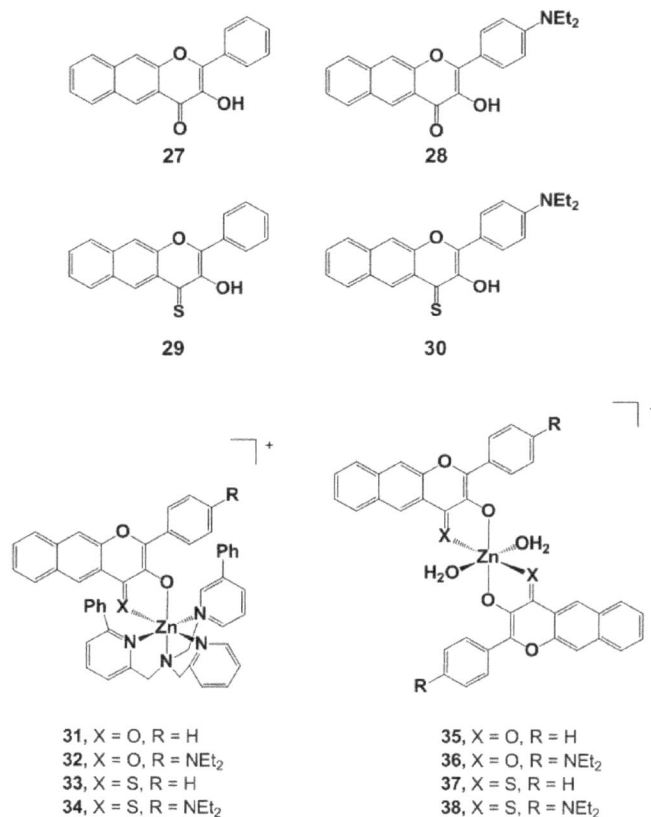

**31**, X = O, R = H
**32**, X = O, R = NEt$_2$
**33**, X = S, R = H
**34**, X = S, R = NEt$_2$

**35**, X = O, R = H
**36**, X = O, R = NEt$_2$
**37**, X = S, R = H
**38**, X = S, R = NEt$_2$

**Figure 7.** Structures of flavonols **27–30** (top) and their Zn complexes **31–38**.

The main advance towards visible light induction of organic photoCORMs was accomplished by Klán and coworkers [79]. Two boron dipyrromethene (BODIPY) derivatives were used as non-metallic CORMs. BODIPY compounds are widely known as strong chromophores. The CO releasing properties of molecules **39** and **40** (Figure 8) their biocompatilibity and toxicity were evaluated by Klán at several irradiation wavelengths (350–730 nm). Molecule **39** exhibits photochemical decomposition with 500 nm light irradiation. Compound **40** was designed with an extended $\pi$-system in order to decrease the LUMO level energy by $\pi$ delocalization. Under aerobic condition in PBS, **40** showed

CO release upon irradiation at 358 and 652 nm (both absorption maxima of the molecule) and also at 732 nm, at the absorption tail. However, under aerobic conditions the quantum yield of **39–40** drops to ca. 45% of the maximum. Thus, in opposition to the flavonol compounds, the BODIPY derivatives operate best in anaerobic environments. In in vivo experiments, hairless mice were separated in three groups, (i) without any treatment; (ii) with intraperitoneal application of **40** left in the dark; (iii) with intraperitoneal application of **40** followed by irradiation for 4 h with white light. This last group exhibited accumulation of carboxyhaemoglobin in blood, and high CO levels in kidney and hepatic tissues. Additionally, an in vitro blood sample containing **40** was irradiated and there also, CO could be easily detected. Furthermore, irradiation of a long-lasting sample was evaluated and revealed that CO release stopped when the illumination was suspended, granting control over the CO delivery. Finally, the mechanism of CO loss was assessed as being provided by a photoinduced electron transfer (PET) from the carboxylate to the BODIPY moiety, leading to an intersystem crossing from an excited singlet state to the triplet state. Finally, these compounds showed no in vivo toxicity in mice making this class of near-infrared light activated photoCORMs a major advance towards longer wavelength induction. However, whereas the organic flavonol derivatives yield, for the most part, well characterized (and non-toxic) products, the products of the BODIPY CO release reaction are less defined.

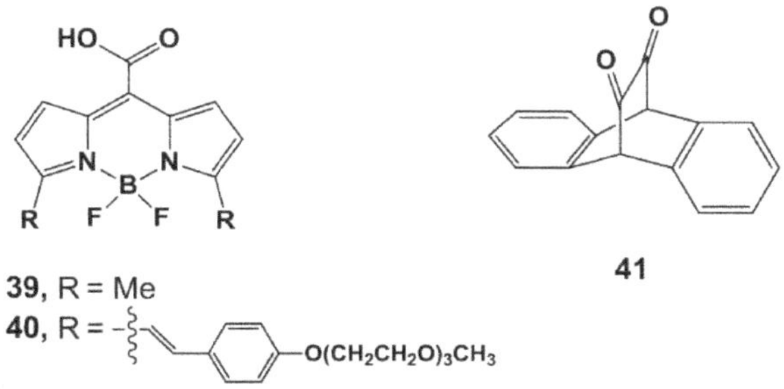

**Figure 8.** Structures of the nonmetallic CORMs **39**, **40** (BODIPY derivatives) and **41** (9,10-dihydro-9,10-ethanoanthracene-11,12-dione).

Another class of nonmetallic photoCORMs reported by Liao and coworkers allowed a non-invasive and simple approach to track the degree of photo-induced release of CO via fluorescence measurements [82]. These cyclic diketone compounds (**41**, Figure 8) require a hydrophobic carrier to be activated, and they were thus incorporated into micelles. In a study showing the first use of this CO releasing composite material for engineered tissue applications, Bashur and coworkers published photoCORM materials incorporated into electrospun scaffolds for engineered vascular tissues [83]. Tissue engineering is a promising approach as bypass graft overcoming coronary heart disease, but has demonstrated poor durability due to aneurysm and thrombosis development. The use of CO for the regulation of the tissue functionalities is thus being investigated [84]. The CO releasing material, 9,10-dihydro-9,10-ethanoanthracene-11,12-dione (**41**, Figure 8), improves vascular cellular functions in small-diameter grafts by CO liberation. As a photoactivatable and nontoxic CORM, **41** showed CO delivery when incorporated (ca. 2% $w/w$) in an electrospun poly-caprolactone scaffold under 470 nm light irradiation. Photo-activation of **41** is reproducible under both dry and cell culture conditions. Under dry conditions, **41**-derived CO nearly saturates myoglobin in the assay reaching up to 92% of the theoretical capacity of the CORM-loaded scaffold. Under cell culture conditions, a photo-induction period of 30 min activation before cellular incubation and fluorescence analysis indicated maximal CO release. CORM **41** does not appear to affect cell viability up to a 100 µM concentration, but the cell phenotypic response of rat smooth muscle cells (SMC) does not indicate a significant impact of the CO delivery. As this could be due to the slow release of CO during the activation time, a higher CORM-loading could be needed (although an increased dose of CO may also be toxic). This obstacle may be surmounted by increasing the hydrophobicity of the environment encapsulating **41**, thereby accelerating the activation rate of CO loss during incubation.

### 4.3. PhotoCORM Materials

An alternative strategy of photoCORM incorporation into material matrices was reported by Schiller and coworkers, via the use of optical device remote-controlled activation of a tetranuclear Mn-based complex (**42**, Figure 9) embedded on poly-lactide (**42a**) and polymethacrylate (**42b**) non-woven fabrics [44]. The same group reported previously an efficient poly-lactide matrix for light-induced CO delivery from the water-insoluble photoactivatable $Mn_2(CO)_{10}$ CORM [85]. This matrix-incorporated compound demonstrated antimicrobial activity against *Staphylococcus aureus* upon light irradiation at 405 nm. Experiments performed on dry compounds (405 nm irradiation) indicated a total amount of CO being released in the range of 8.1–8.3 $\mu$mol/mg of material for **42a** and 11.0–11.7 $\mu$mol/mg for **42b** after 30 min illumination. When the same experiment was performed with an incident radiation of 365 nm, 10.7 $\mu$mol/mg and 11.5 $\mu$mol/mg were measured respectively for **42a** and **42b**. Surprisingly, the rate of CO delivery of **42a** was faster upon irradiation at 405 nm as compared to 365 nm, which is in contrast with common expectations. The investigation of both structures with the fiber optical device was also reported. Technically, a glass sphere was connected to the extremity of a fiber optic in order to extend the irradiated surface of sample areas, in combination with a laser. CO release was determined to reach 1.2 $\mu$mol/mg after 30 min at 405 nm light exposition. This lower value in comparison to dry tests could be explained as a result of a narrower spatial irradiation. In this device, the intensity of the irradiating light plays also a role in CO release. The concentration of delivered CO increases with a higher irradiation intensity, at a similar wavelength. **42a** was also determined to be nontoxic against 3T3 mouse fibroblast cells, similar to the control polymer matrix.

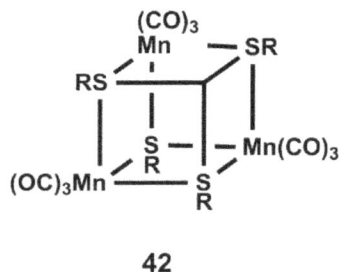

**42**

**Figure 9.** Schematic structure of Schiller's tetranuclear photoCORM.

Ueno and coworkers recently published the characterization of engineered protein crystals containing the $Mn(CO)_5Br$ photoCORM and as a CO releasing material with the aim of modulating nuclear factor activation [86]. The choice of this particular carrier matrix was driven by the opportunity of using the inner pores of the protein crystals, acting as solvent channels [87,88], and by the natural production of those crystals in insects after infection by a cypovirus. In order to increase the amount of photoCORMs loaded in the carrier, and thus the relative CO concentration, a mutant of the protein containing a hexa-histidine tag was prepared. This composite protein-CORM matrix showed CO liberation by illumination at 456 nm, while the release was considerably lower without irradiation. Various analyses suggested moreover that Mn ions are retained into the protein crystal after CO release, thereby reducing the possible toxicity of Mn metabolites. Since it is widely known that the intensity and the rate of CO production into cells are necessary for the activation of the nuclear factor NF-κB [89], the activity of this NF was checked against the presence of the CORM-embedded protein crystal [90]. The activity of the κB-Fluc transfected HEK293 cells cultured with the photoCORM was evaluated by investigation of the bioluminescence intensity of the cells after 12 h of incubation. The data indicate that the mutant protein enhanced photoactivation of NF-κB compared to the wild type protein after 10 min of light irradiation. There were no notable differences at 0, 5 and 20 min of illumination. The detailed mechanism of biological mechanism of this type of composite is currently being investigated by this group. Another strategy to overcome the potential toxicity of the remaining metal-ligand fragment was reported by the group of Smith [91]. This photoCORM system is composed of four and eight manganese cores attached via a dendritic structure (**43**, Figure 10). The formed dendrimers exhibit stability in the dark and release CO when irradiated at 410 nm. By contrast, the monomer model system (**44**, Figure 10) demonstrated a delivery twice as fast in otherwise similar conditions. The total amount of CO delivered with the eight core dendrimer reached 15 CO eq. The quantum yield and the half-life time were similar for both the four and the eight metallic centers, demonstrating no dependent behavior from one end to another.

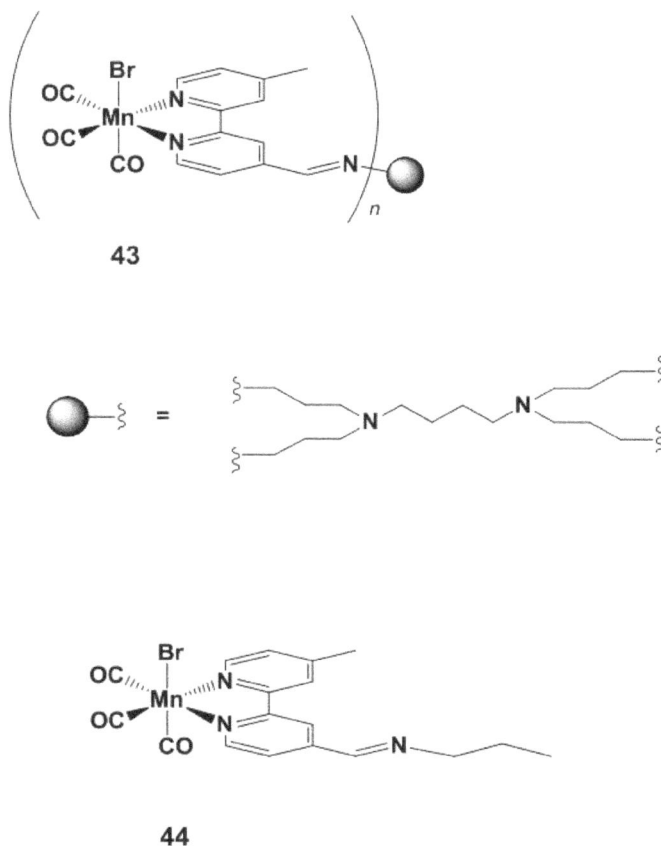

**43**

**44**

**Figure 10.** Dendritic photoCORM (**43**) and its monomer model system (**44**) reported by Smith et al. [91].

## 4.4. NIR PhotoCORMs

Similar to manganese, the *fac*-{Re(CO)$_3$}$^+$ core was used by Ford and coworkers to synthesize a water-soluble and luminescent photoactivated CORM [92] (**45**, Figure 11). Specifically, the –CH$_2$OH substituents of the phosphinetriyltrimethanol ligand allowed conferring water solubility of the entire complex without modification on the overall charge of the compound. One CO was delivered when irradiated at 405 nm, and the luminescent properties of the photoproduct (following H$_2$O–CO ligand exchange) allowed monitoring cellular uptake of **45** via confocal fluorescence microscopy. Based on the same ligand, Ford reported in 2015 a water-soluble nanocarrier with upconversion nanoparticles embedded in an amphiphilic polymer containing a water-insoluble Mn-based photoCORM (**46**, Figure 11) [35]. In such an ensemble, the up-converted nanoparticles absorb NIR light at 980 nm. CO is thus released by '**45** by reabsorption of the emitted light from the up-converted particles. Previously, Ford and others used NIR wavelengths to uncage lanthanide ion doped with up-converted nanoparticles [93–96]. While the nanoparticles consist of

NaGDF$_4$ moieties doped with ytterbium and thulium, the amphiliphic matrix is prepared from phospholipid-functionalized poly(ethylene glycol). When **46** is dissolved in CH$_2$Cl$_2$, it exhibits an MLCT band around 470 nm, thus the 490 nm light emitted from the particles can excite MLCT transition and thus promote CO release. In myoglobin experiments, the CO delivering properties of the composite was assessed. With 365 nm illumination, CO is delivered to myoglobin by photo-labilization and migration through the matrix. As expected, a 980 nm irradiation also induces CO release via up-converted nanoparticles. Furthermore, the tuning on the polymer capsule could be used to modify the circulation, specificity and elimination of the composite.

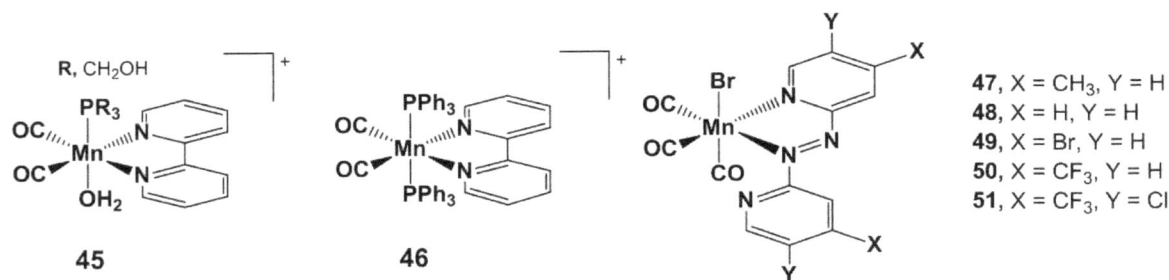

**Figure 11.** Structure of Ford's bipyridine based photoCORMs (**45** and **46**) and of Zobi's azobipyridine complexes (**47–51**).

With the same purpose, Zobi and coworkers described substituted azobipyridine-based photoactivatable CORMs that release CO directly upon NIR irradiation(**47–51**, Figure 11) [21]. By investigating a series of azobipyridine derivatives, they reported rational fine tuning of the maximal absorbance of the complexes by modification of the $\pi$-frame of the ligand. A similar ligand was previously reported by Mascharak, with a strategy based on two complementary approaches: (a) the stabilization of the LUMO involved in the MLCT transition by increasing the conjugation of aromatic bidentate ligands; (b) the presence of $\sigma$-donor ancillary ligands in order to elevate the energy HOMO-2 orbital (also involved in MLCT) by increasing the electronic density on the metal center [22]. The series of ligands synthesized and characterized by the group of Zobi consists of a symmetric azobipyridine. The introduction of a second pyridine was rationalized so as to lower the energy level of the $\pi^*$ MO, given the higher electronegativity of nitrogen compared to carbon. Electron-donating and electron-withdrawing substituents on the $\pi$-frame are expected to finely modulate the MLCT wavelength of absorbance. Electron poor $\pi$-ligands would facilitate the transfer of the electron density from the metal center to the $\pi^*$ orbital of the ligand, and thus promote the photo-labilization of the corresponding CO bond. To confirm these assumptions, the stability of complexes **47–51** in CH$_2$Cl$_2$ was assessed in absence of light, with irradiation at the maximal absorbance wavelength in the visible region (MLCT) of each complex ($\lambda_{max}$), and with NIR light irradiation (810 nm). The results demonstrated that **47** and **48** are stable in the absence of light for several hours. Conversely, **49–51** exhibit decomposition at room temperature as evidenced by hypochromic shift of the $\lambda_{max}$. With illumination at $\lambda_{max}$, all CORMs shows faster decomposition, and as the MLCT band of the complexes tails beyond the visible spectral region, photodecomposition of **51** is triggered at 810 nm. The myoglobin assay confirmed that the photodegradation is followed by CO release. Similarly, CO release is faster with $\lambda_{max}$ activation than in the dark. Moreover, DFT calculations confirmed the red-shift of the MLCT absorption band with electron withdrawing substituents (from 625 to 695 nm, for **47** to **51**, respectively) observed by the experimental procedure. Although the toxicity of those compounds needs to be investigated, this study represents a major step towards direct NIR photoactivation of metallic CORMs.

## 5. Discussion

In the past ten years, developments in photoCORM chemistry followed two main approaches: (i) the tuning of ligands in the primary coordination sphere of metal-based CORMs to obtain a precise wavelength sensitivity in terms of photoactivation, (ii) the use of materials such as matrices, scaffolds, crystals or upconversion particles to either enhance photoCORMs water-solubility, or provide encapsulation of possible toxic metal-based photoproducts. While important steps forwards have been archived in the last decade, the common use of CO releasing molecule in medical environments is yet to be realized and the key chemical requirements to attain this goal are currently still being mapped. In general, more comprehensive studies regarding the performance of photoCORMs in in vitro and in vivo experiments need to be addressed. Cytotoxicity evaluations of new photoCORMs are not commonly realized while anti-inflammatory assays are rarely reported. In general the biocompatibility of photoCORMs remains a crucial point. This is particularly so with metal-based CORMs, while organic photoCORMs allow for the opportunity of a more straightforward functionalization for the specificity and tunability of their biological properties. In this respect, future investigations on organic photoCORMs are certainly poised to contribute to the next key advances in the field. Furthermore, the appreciation of the new reported photoCORM materials in a complex biological system remains to be evaluated. In addition, one major question is emerging from the NIR activatable photoCORMs: how far can one, or needs one, go to lengthen irradiation wavelengths? As the photoactivation limits are pushed, issues such as thermostability of the photoCORMs become important. It would appear to us that an excitation "limit" will be reached soon. The same perhaps has already been reached. Nevertheless, the large existing panel of tunable photoCORMs will help the scientific community to determine the adequate light sensitivity and will help the researchers to succeed in the transition between the laboratories and the common therapeutic administration.

**Acknowledgments:** Financial support from the Swiss National Science Foundation (Grant# PP00P2_144700) is gratefully acknowledged.

**Author Contributions:** Emmanuel Kottelat searched and selected the literature, and wrote the first draft of the review, whereas Fabio Zobi edited and reviewed the manuscript.

# References

1.  Widdop, B. Analysis of Carbon Monoxide. *Ann. Clin. Biochem.* **2002**, *39*, 378–391. [CrossRef] [PubMed]
2.  Romao, C.C.; Blättler, W.A.; Seixas, J.D.; Bernardes, G.J. Developing Drug Molecules for Therapy with Carbon Monoxide. *Chem. Soc. Rev.* **2012**, *41*, 3571–3583. [CrossRef] [PubMed]
3.  Fukuto, J.M.; Carrington, S.J.; Tantillo, D.J.; Harrison, J.G.; Ignarro, L.J.; Freeman, B.A.; Chen, A.; Wink, D.A. Small Molecule Signaling Agents: The Integrated Chemistry Biochemistry of Nitrogen Oxides, Oxides of Carbon, Dioxygen, Hydrogen Sulfide, and Their Derived Species. *Chem. Res. Toxicol.* **2012**, *25*, 769–793. [CrossRef] [PubMed]
4.  Motterlini, R.; Otterbein, L.E. The Therapeutic Potential of Carbon Monoxide. *Nat. Rev. Drug. Discov.* **2010**, *9*, 728–743. [CrossRef] [PubMed]
5.  Halilovic, A.; Patil, K.A.; Bellner, L.; Marrazzo, G.; Castellano, K.; Cullaro, G.; Dunn, M.W.; Schwartzman, M.L. Knockdown of Heme Oxygenase-2 Impairs Corneal Epithelial Cell Wound Healing. *J. Cell. Physiol.* **2011**, *226*, 1732–1740. [CrossRef] [PubMed]
6.  Motterlini, R.; Gonzales, A.; Foresti, R.; Clark, J.E.; Green, C.J.; Winslow, R.M. Heme Oxygenase-1-Derived Carbon Monoxide Contributes to the Suppression of Acute Hypertensive Responses in Vivo. *Circ. Res.* **1998**, *83*, 568–577. [CrossRef] [PubMed]
7.  Otterbein, L.E. Carbon Monoxide: Innovative Anti-Inflammatory Properties of an Age-Old Gas Molecule. *Antioxid. Redox Signal.* **2002**, *4*, 309–319. [CrossRef] [PubMed]
8.  Nakao, A.; Kaczorowski, D.J.; Sugimoto, R.; Billiar, T.R.; McCurry, K.R. Serial Review Application of Heme Oxygenase-1, Carbon Monoxide and Biliverdin for the Prevention of Intestinal Ischemia/Reperfusion Injury. *J. Clin. Biochem. Nutr.* **2008**, *42*, 78–88. [CrossRef] [PubMed]
9.  Otterbein, L.E.; Bach, F.H.; Alam, J.; Soares, M.; Lu, H.T.; Wysk, M.; Davis, R.J.; Flavell, R.A.; Choi, A.M. Carbon Monoxide Has Anti-Inflammatory Effects Involving the Mitogen-Activated Protein Kinase Pathway. *Nat. Med.* **2000**, *6*, 422–428. [PubMed]

10.  Wu, M.L.; Ho, Y.C.; Yet, S.F. A Central Role of Heme Oxygenase-1 in Cardiovascular Protection. *Antioxid. Redox Signal.* **2011**, *15*, 1835–1846. [CrossRef] [PubMed]

11.  Neto, J.S.; Nakao, A.; Kimizuka, K.; Romanosky, A.J.; Stolz, D.B.; Uchiyama, T.; Nalesnik, M.A.; Otterbein, L.E.; Murase, N. Protection of Transplant-Induced Renal Ischemia-Reperfusion Injury with Carbon Monoxide. *Am. J. Physiol. Renal* **2004**, *287*, F979–F989. [CrossRef] [PubMed]

12.  Otterbein, L.E.; Zuckerbraun, B.S.; Haga, M.; Liu, F.; Song, R.; Usheva, A.; Stachulak, C.; Bodyak, N.; Smith, R.N.; Csizmadia, E. Carbon Monoxide Suppresses Arteriosclerotic Lesions Associated with Chronic Graft Rejection and with Balloon Injury. *Nat. Med.* **2003**, *9*, 183–190. [CrossRef] [PubMed]

13.  Katori, M.; Busuttil, R.W.; Kupiec-Weglinski, J.W. Heme Oxygenase-1 System in Organ Transplantation1. *Transplantation* **2002**, *74*, 905–912. [CrossRef] [PubMed]

14.  Heinemann, S.H.; Hoshi, T.; Westerhausen, M.; Schiller, A. Carbon Monoxide–Physiology, Detection and Controlled Release. *Chem. Commun.* **2014**, *50*, 3644–3660. [CrossRef] [PubMed]

15.  Rimmer, R.D.; Pierri, A.E.; Ford, P.C. Photochemically Activated Carbon Monoxide Release for Biological Targets. Toward Developing Air-Stable Photocorms Labilized by Visible Light. *Coord. Chem. Rev.* **2012**, *256*, 1509–1519. [CrossRef]

16.  Schatzschneider, U. Photocorms: Light-Triggered Release of Carbon Monoxide from the Coordination Sphere of Transition Metal Complexes for Biological Applications. *Inorg. Chim. Acta* **2011**, *374*, 19–23. [CrossRef]

17.  Schatzschneider, U. Novel Lead Structures Activation Mechanisms for CO-releasing molecules (CORMs). *Br. J. Pharmacol.* **2015**, *172*, 1638–1650. [CrossRef] [PubMed]

18.  Zobi, F. CO and CO-Releasing Molecules in Medicinal Chemistry. *Future Med. Chem.* **2013**, *5*, 175–188. [CrossRef] [PubMed]

19.  Motterlini, R.; Clark, J.E.; Foresti, R.; Sarathchandra, P.; Mann, B.E.; Green, C.J. Carbon Monoxide-Releasing Molecules Characterization of Biochemical and Vascular Activities. *Circ. Res.* **2002**, *90*, e17–e24. [CrossRef] [PubMed]

20.  Hewison, L.; Crook, S.H.; Mann, B.E.; Meijer, A.J.; Adams, H.; Sawle, P.; Motterlini, R.A. New Types of CO-Releasing Molecules (CO-RMs), Based on Iron Dithiocarbamate Complexes and [Fe(CO)$_3$I(S$_2$COEt)]. *Organometallics* **2012**, *31*, 5823–5834. [CrossRef]

21.  Kottelat, E.; Ruggi, A.; Zobi, F. Red-Light Activated Photocorms of Mn(I) Species Bearing Electron Deficient 2,2′-Light Activated. *Dalton Trans.* **2016**, *45*, 6920–6927. [CrossRef] [PubMed]

22.  Chakraborty, I.; Carrington, S.J.; Mascharak, P.K. Design Strategies to Improve the Sensitivity of Photoactive Metal Carbonyl Complexes (PhotoCORMs) to Visible Light and Their Potential as CO-Donors to Biological Targets. *Accounts Chem. Res.* **2014**, *47*, 2603–2611. [CrossRef] [PubMed]

23.  Romanski, S.; Kraus, B.; Schatzschneider, U.; Neudörfl, J.M.; Amslinger, S.; Schmalz, H.G. Acyloxybutadiene Iron Tricarbonyl Complexes as Enzyme-Triggered CO-Releasing Molecules (ET-CORMs). *Angew. Chem. Int. Ed.* **2011**, *50*, 2392–2396. [CrossRef] [PubMed]

24.  Romanski, S.; Rücker, H.; Stamellou, E.; Guttentag, M.; Neudörfl, J.M.; Alberto, R.; Amslinger, S.; Yard, B.; Schmalz, H.G. Iron Dienylphosphate Tricarbonyl Complexes as Water-Soluble Enzyme-Triggered CO-Releasing Molecules (ET-CORMs). *Organometallics* **2012**, *31*, 5800–5809. [CrossRef]

25.  Romanski, S.; Kraus, B.; Guttentag, M.; Schlundt, W.; Rücker, H.; Adler, A.; Neudörfl, J.M.; Alberto, R.; Amslinger, S.; Schmalz, H.G. Acyloxybutadiene Tricarbonyl Iron Complexes as Enzyme-Triggered CO-Releasing Molecules (ET-CORMs): A Structure–Activity Relationship Study. *Dalton Trans.* **2012**, *41*, 13862–13875. [CrossRef] [PubMed]

26.  Botov, S.; Stamellou, E.; Romanski, S.; Guttentag, M.; Alberto, R.; Neudörfl, J M.; Yard, B.; Schmalz, H.G. Synthesis and Performance of Acyloxy-Diene-Fe (CO)$_3$ Complexes with Variable Chain Lengths as Enzyme-Triggered Carbon Monoxide-Releasing Molecules. *Organometallics* **2013**, *32*, 3587–3594. [CrossRef]

27.  Kunz, P.C.; Meyer, H.; Barthel, J.; Sollazzo, S.; Schmidt, A.M.; Janiak, C. Metal Carbonyls Supported on Iron Oxide Nanoparticles to Trigger the CO-Gasotransmitter Release by Magnetic Heating. *Chem. Commun.* **2013**, *49*, 4896–4898. [CrossRef] [PubMed]

28.  Gonzales, M.A.; Mascharak, P.K. Photoactive Metal Carbonyl Complexes as Potential Agents for Targeted CO Delivery. *J. Inorg. Biochem.* **2014**, *133*, 127–135. [CrossRef] [PubMed]

29.  Wright, M.A.; Wright, J.A. PhotoCORMs: CO Release Moves into the Visible. *Dalton Trans.* **2016**, *45*, 6801–6811. [CrossRef] [PubMed]

30. Dördelmann, G.; Pfeiffer, H.; Birkner, A.; Schatzschneider, U. Silicium Dioxide Nanoparticles as Carriers for Photoactivatable CO-Releasing Molecules (PhotoCORMS). *Inorg. Chem.* **2011**, *50*, 4362–4367. [CrossRef] [PubMed]

31. Niesel, J.; Pinto, A.; NDongo, H.W.P.; Merz, K.; Ott, I.; Gust, R.; Schatzschneider, U. Photoinduced CO Release, Cellular Uptake and Cytotoxicity of a Tris(Pyrazolyl)Methane (tpm) Manganese Tricarbonyl Complex. *Chem. Commun.* **2008**, 1798–1800. [CrossRef] [PubMed]

32. Rudolf, P.; Kanal, F.; Knorr, J.; Nagel, C.; Niesel, J.; Brixner, T.; Schatzschneider, U.; Nuernberger, P. Ultrafast Photochemistry of a Manganese-Tricarbonyl CO-Releasing Molecule (CORM) in Aqueous Solution. *J. Phys. Chem. Lett.* **2013**, *4*, 596–602. [CrossRef] [PubMed]

33. Pfeiffer, H.; Rojas, A.; Niesel, J.; Schatzschneider, U. Sonogashira "Click" Reactions for the N-Terminal and Side-Chain Functionalization of Peptides with [Mn(CO)$_3$(tpm)]$^+$-Based CO Releasing Molecules (tpm = Tris(Pyrazolyl)Methane). *Dalton Trans.* **2009**, 4292–4298. [CrossRef] [PubMed]

34. Meister, K.; Niesel, J.; Schatzschneider, U.; Metzler-Nolte, N.; Schmidt, D.A.; Havenith, M. Label-Free Imaging of Metal–Carbonyl Complexes in Live Cells by Raman Microspectroscopy. *Angew. Chem. Int. Ed.* **2010**, *49*, 3310–3312. [CrossRef] [PubMed]

35. Pierri, A.E.; Huang, P.J.; Garcia, J.V.; Stanfill, J.G.; Chui, M.; Wu, G.; Zheng, N.; Ford, P.C. A PhotoCORM Nanocarrier for CO Release Using NIR Light. *Chem. Commun.* **2015**, *51*, 2072–2075. [CrossRef] [PubMed]

36. Crespy, D.; Landfester, K.; Schubert, U.S.; Schiller, A. Potential Photoactivated Metallopharmaceuticals: From Active Molecules to Supported Drugs. *Chem. Commun.* **2010**, *46*, 6651–6662. [CrossRef] [PubMed]

37. García-Gallego, S.; Bernardes, G. Carbon-Monoxide-Releasing Molecules for the Delivery of Therapeutic CO In Vivo. *Angew. Chem. Int. Ed.* **2014**, *53*, 9712–9721. [CrossRef] [PubMed]

38. Mann, B.E. Carbon monoxide: An essential signalling molecule. In *Medicinal Organometallic Chemistry*; Gerard, J., Metzler-Note, N., Eds.; Springer: New York, NY, USA, 2010; Volume 32, pp. 247–285.

39. Marhenke, J.; Trevino, K.; Works, C. The Chemistry, Biology and Design of Photochemical CO Releasing Molecules and the Efforts to Detect CO for Biological Applications. *Coord. Chem. Rev.* **2016**, *306*, 533–543. [CrossRef]

40. Ji, X.; Damera, K.; Zheng, Y.; Yu, B.; Otterbein, L.E.; Wang, B. Toward Carbon Monoxide-Based Therapeutics: Critical Drug Delivery and Developability Issues. *J. Pharm. Sci.* **2016**, *105*, 406–416. [CrossRef] [PubMed]

41. Rimmer, R.D.; Richter, H.; Ford, P.C. A Photochemical Precursor for Carbon Monoxide Release in Aerated Aqueous Media. *Inorg. Chem.* **2009**, *49*, 1180–1185. [CrossRef] [PubMed]

42. Zobi, F.; Degonda, A.; Schaub, M.C.; Bogdanova, A.Y. CO Releasing Properties and Cytoprotective Effect of *cis-trans*-[ReII(CO)$_2$Br$_2$L$_2$]$_n$ Complexes. *Inorg. Chem.* **2010**, *49*, 7313–7322. [CrossRef] [PubMed]

43. McLean, S.; Mann, B.E.; Poole, R.K. Sulfite Species Enhance Carbon Monoxide Release from CO-Releasing Molecules: Implications for the Deoxymyoglobin Assay of Activity. *Anal. Biochem.* **2012**, *427*, 36–40. [CrossRef] [PubMed]

44. Gläser, S.; Mede, R.; Görls, H.; Seupel, S.; Bohlender, C.; Wyrwa, R.; Schirmer, S.; Dochow, S.; Reddy, G.U.; Popp, J. Remote-Controlled Delivery of CO Via Photoactive CO-Releasing Materials on a Fiber Optical Device. *Dalton Trans.* **2016**, *45*, 13222–13233. [CrossRef] [PubMed]

45. Klein, M.; Neugebauer, U.; Gheisari, A.; Malassa, A.; Jazzazi, T.M.; Froehlich, F.; Westerhausen, M.; Schmitt, M.; Popp, J.R. IR Spectroscopic Methods for the Investigation of the CO Release from CORMs. *J. Phys. Chem. A* **2014**, *118*, 5381–5390. [CrossRef] [PubMed]

46. Michel, B.W.; Lippert, A.R.; Chang, C.J. A Reaction-Based Fluorescent Probe for Selective Imaging of Carbon Monoxide in Living Cells Using a Palladium-Mediated Carbonylation. *J. Am. Chem. Soc.* **2012**, *134*, 15668–15671. [CrossRef] [PubMed]

47. Chapman, O.; Wojtkowski, P.; Adam, W.; Rodriguez, O.; Rucktäschel, R. Photochemical transformations. XLIV. Cyclic peroxides. Synthesis and chemistry of .alpha.-lactones. *J. Am. Chem. Soc.* **1972**, *94*, 1365–1367. [CrossRef]

48. Kuzmanich, G.; Garcia, P.; Adam, W.; Rodriguez, O.; Rucktäschel, R. Ring strain release as a strategy to enable the singlet state photodecarbonylation of crystalline 1,4-cyclobutanediones. *J. Phys. Org. Chem.* **2011**, *24*, 883–888. [CrossRef]

49. Poloukhtine, A.; Popik, V.V. Mechanism of the Cyclopropenone Decarbonylation Reaction. A Density Functional Theory and Transient Spectroscopy Study. *J. Phys. Chem. A* **2006**, *110*, 1749–1757. [CrossRef] [PubMed]

50. Poloukhtine, A.; Popik, V.V. Highly Efficient Photochemical Generation of a Triple Bond: Synthesis, Properties, and Photodecarbonylation of Cyclopropenones. *J. Org. Chem.* **2003**, *68*, 7833–7840. [CrossRef] [PubMed]

51. Poloukhtine, A.A.; Mbua, N.E.; Wolfert, M.A.; Boons, G.J.; Popik, V.V. Selective Labeling of Living Cells by a Photo-Triggered Click Reaction. *J. Am. Chem. Soc.* **2009**, *131*, 15769–15776. [CrossRef] [PubMed]

52. Kuzmanich, G.; Gard, M.N.; Garcia-Garibay, M.A. Photonic Amplification by a Singlet-State Quantum Chain Reaction in the Photodecarbonylation of Crystalline Diarylcyclopropenones. *J. Am. Chem. Soc.* **2009**, *131*, 11606–11614. [CrossRef] [PubMed]

53. Berends, H.M.; Kurz, P. Investigation of Light-Triggered Carbon Monoxide Release from Two Manganese PhotoCORMs by IR, UV-vis EPR Spectroscopy. *Inorg. Chim. Acta* **2012**, *380*, 141–147. [CrossRef]

54. Üstün, E.; Ayvaz, M.Ç.; Çelebi, M.S.; Aşcı, G.; Demir, S.; Özdemir, İ. Structure, CO-Releasing Property, Electrochemistry, DFT Calculation, and Antioxidant Activity of Benzimidazole Derivative Substituted [Mn(CO)$_3$(bpy)L]PF$_6$ Type Novel Manganese Complexes. *Inorg. Chim. Acta* **2016**, *450*, 182–189. [CrossRef]

55. Üstün, E.; Özgür, A.; Coşkun, K.A.; Demir, S.; Özdemir, İ; Tutar, Y. CO-Releasing Properties and Anticancer Activities of Manganese Complexes with Imidazole/Benzimidazole Ligands. *J. Coord. Chem.* **2016**, *69*, 3384–3394.

56. Tinajero-Trejo, M.; Rana, N.; Nagel, C.; Jesse, H.E.; Smith, T.W.; Wareham, L.K.; Hippler, M.; Schatzschneider, U.; Poole, R.K. Antimicrobial Activity of the Manganese Photoactivated Carbon Monoxide-Releasing Molecule [Mn(CO)$_3$(tpa-$\kappa^3$N)]$^+$ against a Pathogenic *Escherichia coli* That Causes Urinary Infections. *Antioxid. Redox Signal.* **2016**, *24*, 765–780. [CrossRef] [PubMed]

57. Yang, S.; Chen, M.; Zhou, L.; Zhang, G.; Gao, Z.; Zhang, W. Photo-Activated Co-Releasing Molecules (Photocorms) of Robust Sawhorse Scaffolds [$\mu^2$-OOCR$^1$, $\eta^1$-NH$_2$CHR$^2$(C=O]OCH$_3$, Ru(I)$_2$CO$_4$. *Dalton Trans.* **2016**, *45*, 3727–3733. [CrossRef] [PubMed]

58. Carrington, S.J.; Chakraborty, I.; Bernard, J.M.; Mascharak, P.K. A Theranostic Two-Tone Luminescent Photocorm Derived from Re (I) and (2-Pyridyl)-Benzothiazole: Trackable CO Delivery to Malignant Cells. *Inorg. Chem.* **2016**, *55*, 7852–7858. [CrossRef] [PubMed]

59. Kretschmer, R.; Gessner, G.; Görls, H.; Heinemann, S.H.; Westerhausen, M. Dicarbonyl-Bis(Cysteamine) Iron (II): A Light Induced Carbon Monoxide Releasing Molecule Based on Iron (CORM-S1). *J. Inorg. Biochem.* **2011**, *105*, 6–9. [CrossRef] [PubMed]

60. Hewison, L.; Johnson, T.R.; Mann, B.E.; Meijer, A.J.; Sawle, P.; Motterlini, R. A Re-Investigation of [Fe(L-Cysteinate)$_2$(CO)$_2$]$^{2-}$: An Example of Non-Heme CO Coordination of Possible Relevance to CO Binding to Ion Channel Receptors. *Dalton Trans.* **2011**, *40*, 8328–8334. [CrossRef] [PubMed]

61. Mede, R.; Klein, M.; Claus, R.A.; Krieck, S.; Quickert, S.; Görls, H.; Neugebauer, U.; Schmitt, M.; Gessner, G.; Heinemann, S.H. CORM-EDE1: A Highly Water-Soluble and Nontoxic Manganese-Based PhotoCORM with a Biogenic Ligand Sphere. *Inorg. Chem.* **2015**, *55*, 104–113. [CrossRef] [PubMed]

62. Poh, H.T.; Sim, B.T.; Chwee, T.S.; Leong, W.K.; and Fan, W.Y. The Dithiolate-Bridged Diiron Hexacarbonyl Complex Na$_2$[(M-SCH$_2$CH$_2$COO)Fe(CO)$_3$]$_2$ as a Water-Soluble PhotoCORM. *Organometallics* **2014**, *33*, 959–963. [CrossRef]

63. Gonzalez, M.A.; Yim, M.A.; Cheng, S.; Moyes, A.; Hobbs, A.J.; Mascharak, P.K. Manganese Carbonyls Bearing Tripodal Polypyridine Ligands as Photoactive Carbon Monoxide-Releasing Molecules. *Inorg. Chem.* **2011**, *51*, 601–608. [CrossRef] [PubMed]

64. Gonzalez, M.A.; Carrington, S.J.; Fry, N.L.; Martinez, J.L.; Mascharak, P.K. Syntheses, Structures, and Properties of New Manganese Carbonyls as Photoactive CO-Releasing Molecules: Design Strategies That Lead to CO Photolability in the Visible Region. *Inorg. Chem.* **2012**, *51*, 11930–11940. [CrossRef] [PubMed]

65. Chakraborty, I.; Carrington, S.J.; Mascharak, P.K. Photodelivery of CO by Designed Photocorms: Correlation between Absorption in the Visible Region and Metal–CO Bond Labilization in Carbonyl Complexes. *ChemMedChem* **2014**, *9*, 1266–1274. [CrossRef] [PubMed]

66. Vlček, A., Jr.; Farrell, I.R.; Liard, D.J.; Matousek, P.; Towrie, M.; Parker, A.W.; Grills, D.C.; George, M.W. Early Photochemical Dynamics of Organometallic Compounds Studied by Ultrafast Time-Resolved Spectroscopic Techniques. *Dalton Trans.* **2002**, 701–712.

67. Jimenez, J.; Chakraborty, I.; Carrington, S.J.; Mascharak, P.K. Light-Triggered CO Delivery by a Water-Soluble and Biocompatible Manganese PhotoCORM. *Dalton Trans.* **2016**, *45*, 13204–13213. [CrossRef] [PubMed]

68. Jimenez, J.; Chakraborty, I.; Mascharak, P.K. Synthesis and Assessment of Coered Cls, D.C.; George, M.W.; Early Photochemical Dynamics of α-imenez, J.; Chakof Varied Complexity. *Eur. J. Inorg. Chem.* **2015**, *2015*, 5021–5026. [CrossRef] [PubMed]

69. Santoro, G.; Beltrami, R.; Kottelat, E.; Blacque, O.; Bogdanova, A.Y.; Zobi, F. N-Nitrosamine-{cis-Re[CO]$_2$}$^{2+}$ Cobalamin Conjugates as Mixed Co/No-Releasing Molecules. *Dalton Trans.* **2016**, *45*, 1504–1513. [CrossRef] [PubMed]

70. Blanářová, O.V.; Jelínková, I.; Szőőr, Á.; Skender, B.; Souček, K.; Horváth, V.; Vaculová, A.; Sova, P.; Szöllősi, J. Cisplatin and a Potent Platinum (IV) Complex-Mediated Enhancement of Trail-Induced Cancer Cells Killing Is Associated with Modulation of Upstream Events in the Extrinsic Apoptotic Pathway. *Carcinogenesis* **2011**, *32*, 42–51. [CrossRef] [PubMed]

71. Horváth, V.; Blanářová, O.; Švihálková-Šindlerová, L.; Souček, K.; Hofmanová, J.; Sova, P.; Kroutil, A.; Fedoročko, P.; Kozubík, A. Platinum (IV) Complex with Adamantylamine Overcomes Intrinsic Resistance to Cisplatin in Ovarian Cancer Cells. *Gynecol. Oncol.* **2006**, *102*, 32–40. [CrossRef] [PubMed]

72. Yempally, V.; Kyran, S.J.; Raju, R.K.; Fan, W.Y.; Brothers, E.N.; Darensbourg, D.J.; Bengali, A.A. Thermal and Photochemical Reactivity of Manganese Tricarbonyl and Tetracarbonyl Complexes with a Bulky Diazabutadiene Ligand. *Inorg. Chem.* **2014**, *53*, 4081–4088. [CrossRef] [PubMed]

73. Mansour, A.M.; Shehab, O.R. Experimental and Quantum Chemical Calculations of Novel Photoactivatable Manganese (I) Tricarbonyl Complexes. *J. Organomet. Chem.* **2016**, *822*, 91–99. [CrossRef]

74. Mansour, A.M. Crystal Structure, DFT, Spectroscopic and Biological Activity Evaluation of Analgin Complexes with Co (II), Ni (II) and Cu (II). *Dalton Trans.* **2014**, *43*, 15950–15957. [CrossRef] [PubMed]

75. Mansour, A.M. Tazarotene Copper Complexes: Synthesis, Crystal Structure, DFT and Biological Activity Evaluation. *Polyhedron* **2016**, *109*, 99–106. [CrossRef]

76. Anderson, S.N.; Larson, M.T.; Berreau, L.M. Solution or Solid—It Doesn't Matter: Visible Light-Induced CO Release Reactivity of Zinc Flavonolato Complexes. *Dalton Trans.* **2016**, *45*, 14570–14580. [CrossRef] [PubMed]

77. Anderson, S.N.; Richards, J.M.; Esquer, H.J.; Benninghoff, A.D.; Arif, A.M.; Berreau, L.M. A Structurally-Tunable 3-Hydroxyflavone Motif for Visible Light-Induced Carbon Monoxide-Releasing Molecules (CORMs). *ChemistryOpen* **2015**, *4*, 590–594. [CrossRef] [PubMed]

78. Wang, D.; Viennois, E.; Ji, K.; Damera, K.; Draganov, A.; Zheng, Y.; Dai, C.; Merlin, D.; Wang, B. A Click-and-Release Approach to CO Prodrugs. *Chem. Commun.* **2014**, *50*, 15890–15893. [CrossRef] [PubMed]

79. Palao Utiel, E.; Slanina, T.; Muchova, L.; Šolomek, T.; Vitek, L.; Klán, P. Transition-Metal-Free CO-Releasing Bodipy Derivatives Activatable by Visible to NIR Light as Promising Bioactive Molecules. *J. Am. Chem. Soc.* **2016**, 126–133. [CrossRef] [PubMed]

80. Pap, J.S.; Kaizer, J.; Speier, G. Model Systems for the CO-Releasing Flavonol 2,4-Dioxygenase Enzyme. *Coord. Chem. Rev.* **2010**, *254*, 781–793. [CrossRef]

81. Fetzner, S. Ring-Cleaving Dioxygenases with a Cupin Fold. *Appl. Environ. Microbiol.* **2012**, *78*, 2505–2514. [CrossRef] [PubMed]

82. Peng, P.; Wang, C.; Shi, Z.; Johns, V.K.; Ma, L.; Oyer, J.; Copik, A.; Igarashi, R.; Liao, Y. Visible-Light Activatable Organic CO-Releasing Molecules (PhotoCORMs) That Simultaneously Generate Fluorophores. *Org. Biomol. Chem.* **2013**, *11*, 6671–6674. [CrossRef] [PubMed]

83. Michael, E.; Abeyrathna, N.; Patel, A.V.; Liao, Y.; Bashur, C.A. Incorporation of Photo-Carbon Monoxide Releasing Materials into Electrospun Scaffolds for Vascular Tissue Engineering. *Biomed. Mater.* **2016**, *11*, 025009. [CrossRef] [PubMed]

84. Bauer, I.; Pannen, B.H. Bench-to-Bedside Review: Carbon Monoxide—from Mitochondrial Poisoning to Therapeutic Use. *Crit. Care* **2009**, *13*, 1. [CrossRef] [PubMed]

85. Bohlender, C.; Gläser, S.; Klein, M.; Weisser, J.; Thein, S.; Neugebauer, U.; Popp, J.; Wyrwa, R.; Schiller, A. Light-Triggered CO Release from Nanoporous Non-Wovens. *J. Mater. Chem. B* **2014**, *2*, 1454–1463. [CrossRef]

86. Tabe, H.; Shimoi, T.; Boudes, M.; Abe, S.; Coulibaly, F.; Kitagawa, S.; Mori, H.; Ueno, T. Photoactivatable CO Release from Engineered Protein Crystals to Modulate NF-κB Activation. *Chem. Commun.* **2016**, *52*, 4545–4548. [CrossRef] [PubMed]

87. Tabe, H.; Fujita, K.; Abe, S.; Tsujimoto, M.; Kuchimaru, T.; Kizaka-Kondoh, S.; Takano, M.; Kitagawa, S.; Ueno, T. Preparation of a Cross-Linked Porous Protein Crystal Containing Ru Carbonyl Complexes as a CO-Releasing Extracellular Scaffold. *Inorg. Chem.* **2014**, *54*, 215–220. [CrossRef] [PubMed]

88. Tabe, H.; Shimoi, T.; Fujita, K.; Abe, S.; Ijiri, H.; Tsujimoto, M.; Kuchimaru, T.; Kizaka-Kondo, S.; Mori, H.; Kitagawa, S. Design of a CO-Releasing Extracellular Scaffold Using in Vivo Protein Crystals. *Chem. Lett.* **2015**, *44*, 342–344. [CrossRef]

89. Chlopicki, S.; Lomnicka, M.; Fedorowicz, A.; Grochal, E.; Kramkowski, K.; Mogielnicki, A.; Buczko, W.; Motterlini, R. Inhibition of Platelet Aggregation by Carbon Monoxide-Releasing Molecules (CO-RMs): Comparison with NO Donors. *Naunyn Schmiedebergs Arch. Pharmacol.* **2012**, *385*, 641–650. [CrossRef] [PubMed]

90. Brasier, A.; Tate, J.; Habener, J. Optimized Use of the Firefly Luciferase Assay as a Reporter Gene in Mammalian Cell Lines. *Biotechniques* **1988**, *7*, 1116–1122.

91. Govender, P.; Pai, S.; Schatzschneider, U.; Smith, G.S. Next Generation PhotoCORMs: Polynuclear Tricarbonylmanganese (I)-Functionalized Polypyridyl Metallodendrimers. *Inorg. Chem.* **2013**, *52*, 5470–5478. [CrossRef] [PubMed]

92. Pierri, A.E.; Pallaoro, A.; Wu, G.; Ford, P.C. A Luminescent Biocompatible PhotoCORM. *J. Am. Chem. Soc.* **2012**, *134*, 18197–18200. [CrossRef] [PubMed]

93. Carling, C.J.; Nourmohammadian, F.; Boyer, J.C.; Branda, N.R. Remote-Control Photorelease of Caged Compounds Using Near-Infrared Light and Upconverting Nanoparticles. *Angew. Chem. Int. Ed.* **2010**, *122*, 3870–3873. [CrossRef]

94. Zheng, Q.; Bonoiu, A.; Ohulchanskyy, T.Y.; He, G.S.; Prasad, P.N. Water-Soluble Two-Photon Absorbing Nitrosyl Complex for Light-Activated Therapy through Nitric Oxide Release. *Mol. Pharm.* **2008**, *5*, 389–398. [CrossRef] [PubMed]

95. Garcia, J.V.; Yang, J.; Shen, D.; Yao, C.; Li, X.; Wang, R.; Stucky, G.D.; Zhao, D.; Ford, P.C.; Zhang, F. Nanostructured Materials. *Small* **2012**, *8*, 3800–3805. [CrossRef] [PubMed]

96. Wecksler, S.; Mikhailovsky, A.; Ford, P.C. Photochemical Production of Nitric Oxide Via Two-Photon Excitation with NIR Light. *J. Am. Chem. Soc.* **2004**, *126*, 13566–13567. [CrossRef] [PubMed]

# 16

# Potassium C–F Interactions and the Structural Consequences in N,N'-Bis(2,6- difluorophenyl) formamidinate Complexes

**Daniel Werner [1], Glen B. Deacon [2,\*] and Peter C. Junk [3,\*]**

[1]  Institut für Anorganische Chemie, University of Tübingen (EKUT) Auf der Morgenstelle 18, 72076 Tübingen, Germany; daniel.werner@uni-tuebingen.de
[2]  School of Chemistry, Monash University, Clayton, Victoria 3800, Australia
[3]  College of Science & Engineering, James Cook University, Townsville, Queensland 4811, Australia
\*  Correspondence: glen.deacon@monash.edu (G.B.D.); peter.junk@jcu.edu.au (P.C.J.)

Academic Editor: Matthias Westerhausen

**Abstract:** Treatment of $K[N(SiMe_3)_2]$ with N,N'-bis(2,6-difluorophenyl)formamidine (DFFormH) in toluene, resulted in the formation of $[K(DFForm)]_\infty$ (**1**) as a poorly soluble material. Upon dissolution in thf and layering with n-hexane, **1** was crystallised and identified as a two-dimensional polymer, in which all fluorine and nitrogen atoms, and also part of one aryl group, bridge between four symmetry equivalent potassium ions, giving rise to a completely unique $\mu_4$-(N,N',F,F'):(N,N'):$\eta^4$(Ar-C(2,3,4,5,6)):(F'',F''') DFForm coordination. The two-dimensional nature of the polymer could be deconstructed to one dimension by crystallisation from neat thf at $-35\,^\circ C$, giving $[K_2(DFForm)_2(thf)_2]_\infty$ (**2**), where the thf molecules bridge the monomeric units. Complete polymer dissociation was observed when **1** was crystallised from toluene/n-hexane mixtures in the presence of 18-crown-6, giving [K(DFForm)(18-crown-6)] (**3**), which showed unprecedented $\kappa(N,C_{ispo},F)$ DFForm coordination, rather than the expected $\kappa(N,N')$ coordination.

**Keywords:** potassium; formamidinate; C–F bond; coordination chemistry

---

## 1. Introduction

With the ability to adopt numerous coordination modes, flexible N,N'-bis(aryl)formamidinates (and by extension aryl-functionalised amidinates) have earned a special place in coordination chemistry [1–5]. Not only does the anionic NCHN bite provide a variety of different nitrogen-based coordination modes (e.g., monodentate $\kappa(N)$, bidentate $\kappa(N,N')$, or various bridging modes e.g., $\mu$-1$\kappa(N)$:2$\kappa(N')$, $\mu$-1$\kappa(N,N')$:2$\kappa(N,N')$ to list a few) [6,7], the nitrogen-bound aromatic substituents can also provide additional coordination modes. The potential to form metal–arene interactions, such as $\eta^6$ coordination, has been largely observed in group one chemistry [8–13], though some examples are known in f-block chemistry [14]. In almost all examples of this aromatic coordination, the phenyl rings contained alkyl-substituents in either the 2,6 positions (e.g., iPr, Et, Me), or in the 2,4,6 positions (e.g., Me). This is likely due to a combination of steric pressure, which starves the metal centre from coordination of additional donors, and increased electron donation from the aromatic ring caused by the alkyl substituents. Another means to engage the aromatic component in coordination is through the addition of donor functionalities (e.g., OMe, F), especially in the ortho-positions, thereby transforming the formamidinate ligand into a tri- [15–17], or tetra-dentate (e.g., N,N',X or N,N',X,X') [17], chelate, with examples across a variety of different metal classes [18]. For s-block chemistry however, the use of such ligands has been restricted to very few examples, namely the use of N,N'-bis(2-fluorophenyl)formamidine (FForm) [19].

Nearly 15 years ago, FForm was complexed to the group one metals Li, Na, and K [19]. Akin to the transition metal complexes of Cotton and co-workers [20], the presence of the fluorine atom on the *ortho*-position of the aromatic rings permitted an additional coordinating site. This further led to partial, or complete, exclusion of bound donor molecules (e.g., $Et_2O$, thf), by the formation of either binuclear, or for potassium, polymeric constructs (e.g., $[Na(FForm)(Et_2O)]_2$ or $[K(FForm)]_\infty$) [19]. This contrasts the group one complexes of the non-fluorinated $N,N'$-di(aryl)formamidinate ligands [21–23], which readily retain coordinating solvent. Since then, we have expanded the use of fluorinated formamidinate ligands to *f*-block chemistry [9,16,17,24–28], in a variety of different contexts [5]. One of the fluorinated formamidinate ligands used was $N,N'$-bis(2,6-difluorophenyl)formamidinate (DFForm) in both trivalent [16,17], and divalent [16] rare-earth complexes. Despite the presence of the additional fluorine atoms, the observation of any M–F interaction was rare, typically only occurred in unsolvated species, and interactions were displaced on coordination of donor solvents [16,17]. It is likely that the smaller ionic radii of the trivalent rare-earths, compared with the larger potassium ion [29], create a significant strain in the NCHN bite of the DFForm ligand when it coordinates the fluorine atoms, and therefore donor solvent coordination is preferred. Although DFForm has been used in some transition metal complexes, it has no precedent in *s*-block chemistry. We hypothesised that the additional two fluorine atoms over FForm could engage in further coordination chemistry, generating different coordination modes from FForm, and quite spectacular results have been obtained by way of new formamidinate binding modes.

## 2. Results and Discussion

Treatment of $K[N(SiMe_3)_2]$ with DFFormH in toluene resulted in the formation of a colourless, poorly soluble white powder. Upon dissolution in thf, concentration, and layering with *n*-hexane, white crystals of targeted $[K(DFForm)]_\infty$ (**1**, Scheme 1i) were obtained. The structure of **1** was determined by X-ray crystallography, revealing that **1** is a two-dimensional polymer. The binding of the DFForm ligand in **1** is complex, and is discussed starting from the asymmetric unit (ASU), and then extending in both dimensions of the polymeric network.

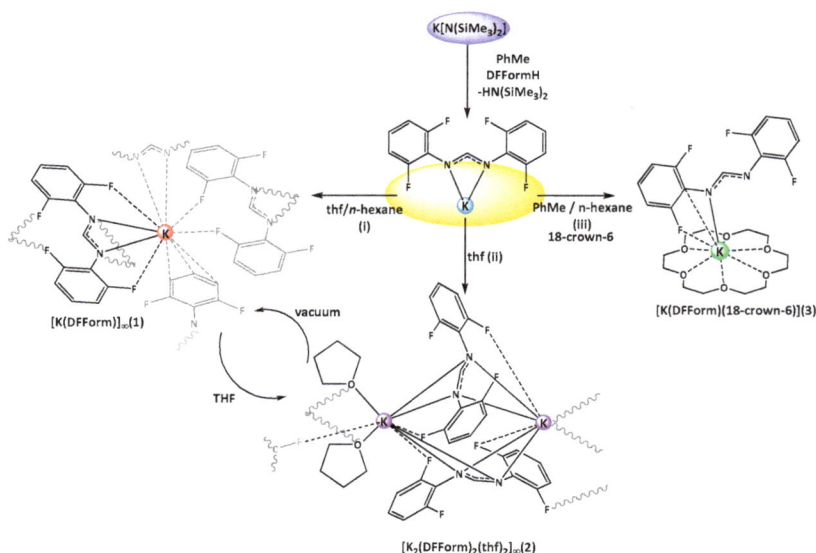

**Scheme 1.** Synthesis of K(DFForm) complexes (**1–3**) by protonolysis and crystallisation from different solvent mixtures. (**i**) thf, *n*-hexane, at room temperature; (**ii**) neat thf, crystallisation at −35 °C; (**iii**) toluene, *n*-hexane, crystallisation at room temperature. The diagram further indicates the different bonding modes of the DFForm ligand in complexes, such as the ($F,N,N',F'$) or arene–K interactions in **1**, the twisted $1\kappa(F,N,N',):2\kappa(N,N',F')$ DFForm coordination in **2**, or the unusual ($N,C_{ipso},F$) coordination in **3**.

Complex **1** crystallised in the triclinic space group *P*-1, with only one potassium ion and one DFForm ligand in the ASU (Figure 1A). The DFForm ligand of the ASU is bound (*F,N,N',F'*) to the ASU potassium ion. This tetradentate binding of the DFForm ligand contrasts that of the FForm ligand in [K(FForm)]$_\infty$ [19], where the ASU contains one FForm ligand bound $\eta^4(N,(Ar\text{-}C6,5),F)$ to potassium. As the K ion is bound by the DFForm NCHN bite in an almost symmetrical manner, and does not favour one nitrogen donor (as observed in [K(FForm)]), the K···F–C bonding is weak, and thus the C–F bonds in **1** (of either C1/F1 or C9/F3, Figure 1) are almost unchanged from those of DFFormH (C–F: 1.3596(17)–1.3625(18)) [30]. By contrast, the asymmetrical NCHN binding of FForm to K in [K(FForm)]$_\infty$ (along with the coordination across the aromatic component), brings the fluorine atom into a closer proximity to the potassium atom (K–F: 3.029(4)), and weakens the C–F bond (C–F: 1.377(6) Å) [19]. Another example of such tetradentate (*F,N,N',F'*) DFForm coordination was observed in the homoleptic cerium DFForm complex, [Ce(DFForm)$_3$] [17], where one of three DFForm ligands is tetradentate, with the other two being tri-dentate (*F,N,N'*). In this example, all Ce···F–C interactions were identical at 2.92 Å (range: 2.9187(13)–2.9213(13)), and consequentially each C–F bond was also strained to a similar degree (range: 1.374(1)–1.376(1) Å). However, considering that a ten-coordinate cerium(III) is smaller than a nine-coordinate potassium (difference in ionic radii: −0.3 Å) [29], the tetradentate DFForm ligand for the cerium complex had to bend the aryl-rings towards the cerium ion to bring the fluorine atoms into proximity, causing a strain on the C$_{ipso}$–N–CH angle (range: 126.4(2)°–128.3(2)°). However, due to the larger ionic radii of potassium, this phenomenon is not observed in **1** (range: 120.39(9)°–120.75(9)°).

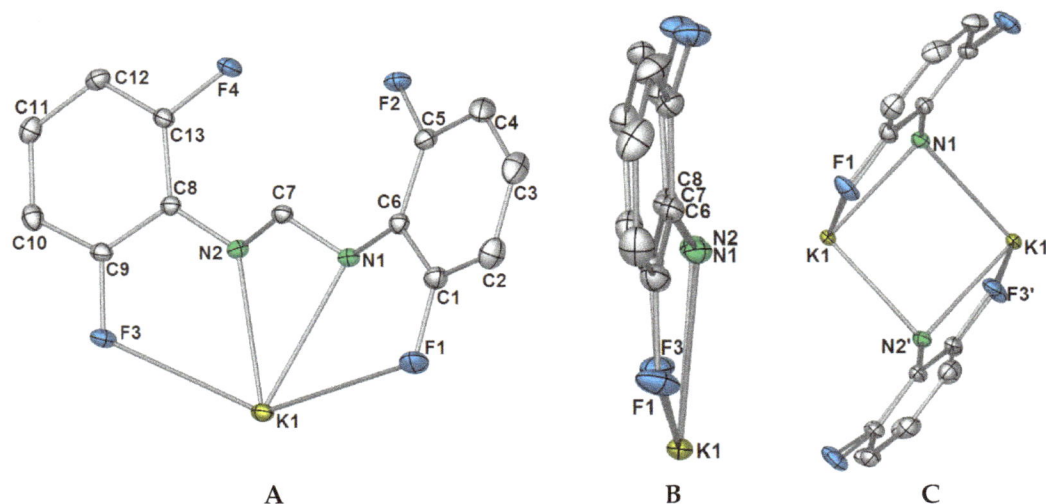

**Figure 1.** (**A**) Asymmetric unit of [K(DFForm)]$\infty$ (**1**). Selected bond lengths (Å) and angles (°): K1–F1: 3.3692(8), K1–N1: 2.8102(9), K1–N2: 2.8057(9), K1–F3: 3.3957(8), C1–F1: 1.3609(13), C9–F3: 1.3581(12), F1–K1–N1: 50.73(2), F3–K1–N2: 50.24(2). Ellipsoids were shown at the 50% probability level, and hydrogen atoms were removed for clarity; (**B**) side view of K(DFForm) showing that the DFForm ligand is not flat; (**C**) simplification of the $\mu$-(*N,N',F,F'*):(*N,N'*) bridging of the DFForm ligand. Selected bond lengths (Å) and angles (°): K1–N1': 2.8102(9), K1–N2': 2.9048(9), K1–K1': 3.4871(4), K1–(N1/N2$_{cent}$)–K1': 84.02(1), N1/2$_{cent}$–K1–N1'/2'$_{cent}$: 95.98(1).

The differences in coordination between the DFForm and FForm ligands to potassium becomes considerably more apparent with expansion of the coordination mode of the ligands through bridging. Initial extension of the coordination of the DFForm ligand in **1** shows that the nitrogen atoms are further bridging to an adjacent potassium ion in a $\mu$-(*N,N'*):(*N,N'*) manner (Figure 1C, also Figure 2A). Such formamidinate bridging is known for other *s*- and *f*-block complexes [16,22,31]. This bridging is mirrored by a symmetry equivalent DFForm ligand, generating a potassium nitrogen based cube of volume: 6.78 Å$^3$ (Figure 1C). In stark contrast, the FForm system shows a twisted $\mu$-($N,C_{ipso}C_{ortho},F$):($N,N',F'$) FForm binding, where the NCHN bite is shared asymmetrically across two anent potassium atoms. It should be further noted that the aromatic group, nitrogen atoms, and backbone/ipso carbon atoms of DFForm are not flat and that the DFForm ligand is tilted (Figure 1B). The two nitrogen atoms coordinate to potassium in an almost symmetrical manner, but C7 is puckered away from the nitrogen atoms (K1–N1/2(cent)–C7: 141.10(8)°), so it is almost in line with the two *ipso* carbon atoms of the phenyl rings (C6–C7–C8: 177.26(6)°, c.f. K(FForm): 168.9(3)°). This puckered nature of the DFForm ligand is typical of other formamidinate complexes which bridge in a $\mu$-($N,N'$):($N,N'$) fashion (e.g., [K(p-TolForm)(dme)]$_\infty$ (K1–N1/2(cent)–C''7": 144.8(3)°, p-TolForm = $N,N'$-bis(4-methylphenyl)formamidinate) [32].

The polymeric network of **1** is complicated. One might expect that, as the DFForm ligand bridges in a $\mu$-($N,N'$):($N,N'$) manner between potassium ions, and that this is the repeating dinuclear unit (e.g., [K$_2$($\mu$-($N,N'$):($N,N'$)-DFForm)$_2$]$_\infty$), but this is not the case. Instead, one dimension of the polymer is generated through aromatic interactions of one 2,6-difluorophenyl group (Figure 2A), where the aromatic ring of N2 binds to K1', and the aromatic ring (but without the *ipso* carbon) of N2' coordinates to K1, both in a $\eta^5$(C2,3,4,5,6) manner. Thus, this direction of the polymeric network has an "A, B" alternating potassium ion arrangement where A = K and B = K' and K" (Figure 2). For the FForm system, the one and only dimension of the polymeric network is generated by additional nitrogen based bonding to two other potassium ions, namely through one ($N,F$) interaction, and one ($N',C_{ispo}'$) interaction, making the overall coordination of each FForm ligand shared across four potassium ions as $\mu_4$-($N,C_{ipso}C_{ortho},F$):($N,N',F'$):($N,F$):($N',C_{ispo}'$). The DFForm ligand is also further bridging to a fourth symmetry equivalent potassium ion, and this binding is completely different from that in the FForm system.

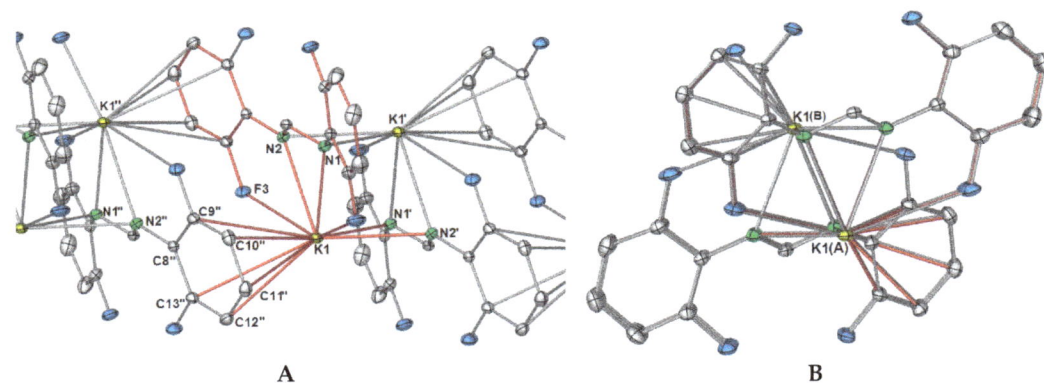

**Figure 2.** Growth of one dimension of the [K(DFForm)]$_\infty$ (**1**) polymer network through aromatic interactions, the red bonds indicate the connectivity to the ASU potassium ion and DFForm ligand. (**A**) View along the side of the polymer; Selected bond lengths (Å) and angles (°): K1–C9': 3.4980(11), K1–C10': 3.3220(11), K1–C11": 3.2417(11), K1–C12": 3.3569(11), K1–C13": 3.5066(10); (**B**) view down the a-axis of the polymeric network of **1**.

As shown in Figure 2A, there is an apparent coordination gap in axial positions of the potassium ions, and it is in this position that the other two fluorine atoms (namely, F2 and F4) of the DFForm ligand become relevant, and expand the one-dimensional polymeric network into a two-dimensional polymer. The further fluorine atoms (F2 and F4) coordinate to an adjacent potassium ion in a $\mu$-($F'',F'''$) manner, generating a ten-membered ring (Figure 3A). Because of this additional coordination, the DFForm ligand is nearly planar across the K and K''' atoms, with the bond angle of K–N1/N2$_{cent}$–K''' being 175.66(1)°. Although the auxiliary fluorine atoms are coordinated at a considerably shorter distance than the K–F1 and K–F3 analogues, there is still only a minor shortening of the C–F bonds from those of DFFormH (C–F: 1.3596(17)–1.3625(18)) [30]).

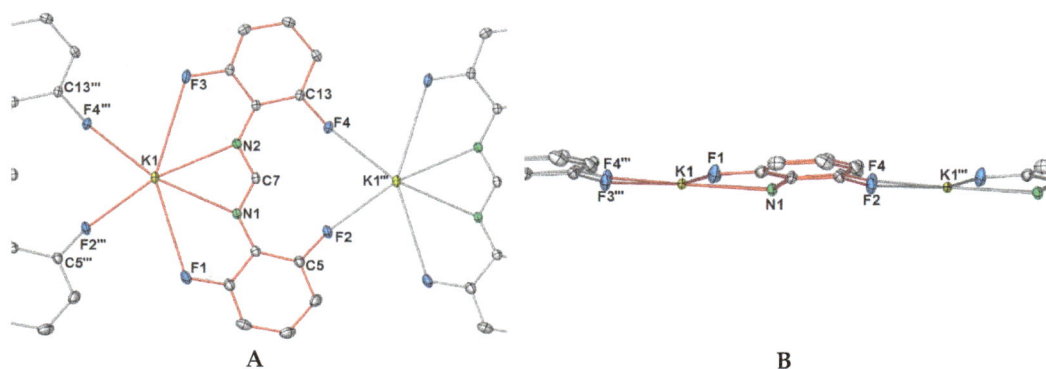

**Figure 3.** Simplified diagram of the bonding of the auxiliary fluorine atoms (F2, F4) to an adjacent potassium ion, expanding the polymeric network into a second direction (across the $b$ axis). (**A**) Top view of bonding showing the formation of a ten-membered ring upon fluorine coordination; (**B**) side view of auxiliary fluorine bonding (or side view of $b$-axis), highlighting the different planes within the DFForm ligand. Selected bond lengths (Å) and angles (°): K1–F2''': 2.7110(8), K1–F4'': 2.7422(7), C5–F2: 1.3641(14), C13–F4: 1.3656(11), K1–K1''': 7.5357(2), K1–N1/2$_{cent}$–K1''': 175.66(1), K1–C7–K1''': 171.37(3), K1–F2/4$_{cent}$–K1''': 176.69(1).

In summation, each DFForm ligand binds four symmetry equivalent potassium ions in a $\mu_4$-1$\kappa$($N,N',F,F'$):2$\kappa$($N,N'$):3$\eta^5$($Ar$-C(2,3,4,5,6)):4$\kappa$($F'',F'''$) manner, giving the potassium ion a coordination number of 11. Such an interesting binding mode exemplifies how the simple addition of other donors to a ligand system can dramatically alter the coordination network. Furthermore, it appears the **1** is the first crystallographically characterised example across all metal classes, where one $N,N'$-bis(aryl)formamidinate ligand generates a two-dimensional polymer network, all other examples are restricted to one dimension [18]. The complete polymeric network of **1** is displayed in Figure 4, showing both how the DFForm bridges across four potassium ions (Figure 4A) and the two dimensions of the polymer (Figure 4B,C).

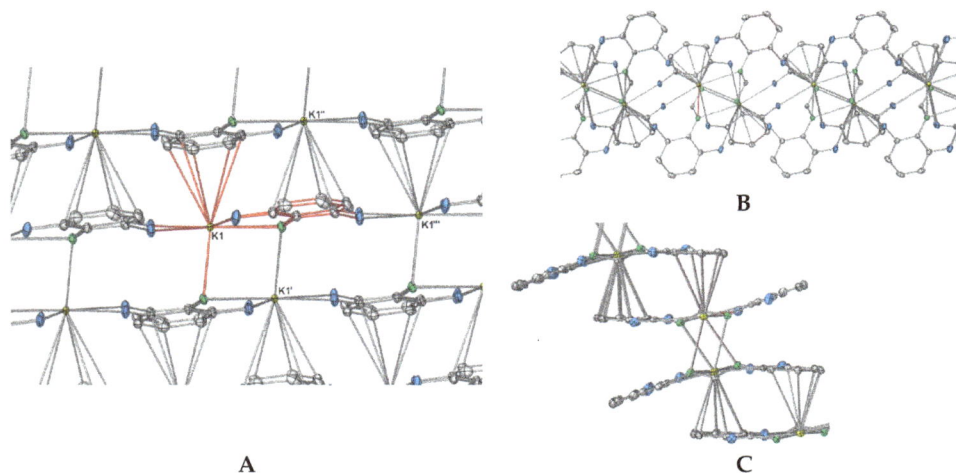

**Figure 4.** Excerpt pictures from the polymeric network of **1**; red-coloured bonds indicate the connectivity to the potassium atom of the ASU and the bonds of the DFForm ligand of the ASU. (**A**) Complete DFForm bonding network across four potassium atoms; (**B,C**) Simplified directions of the polymeric network showing the bridging through fluorine, nitrogen, and aromatic carbon atoms ((**B**) showing nitrogen-based bridging, (**C**) showing aryl group-based bridging).

Crystals of **1** were air- and moisture-sensitive, but under an inert atmosphere the compound appeared stable. Complex **1** was repeatedly obtained by simple exposure of thf solutions of

"[K(DFForm)(thf)$_x$]" to vacuum, giving **1** upon drying. The poor solubility in non-coordinating solvents made analysis by $^1$H NMR and $^{19}$F NMR spectroscopy difficult, giving only broad resonances in both spectra (NC*H*N at 8.88 ppm and F2,6 at −127.2 ppm).All fluorine atoms of the DFForm ligand are equivalent, but clear spectra were generated when **1** was dissolved in thf-d$_8$. In this solvent, the NC*H*N resonance appeared as a pentet, owing to $^5J_{H-F}$ coupling with the *ortho*-fluorine atoms, as the pentet collapsed to a singlet with $^{19}$F decoupling. A broadening of the F resonance (corresponding to F1–F4) was also observed in the $^{19}$F NMR spectrum when it was performed without $^1$H decoupling. It is likely that upon dissolution in thf-d$_8$, the polymeric network is dissociated, and a simpler DFForm coordination mode is adopted e.g., [K(DFForm)(thf)$_x$] (2 < x < 6). Attempts to crystallise a potential monomeric derivative were not successful, but upon concentration of a thf solution of **1**, and storage at −35 °C, crystals of a thf-coordinated species were isolated, namely [K$_2$(DFForm)$_2$(thf)$_2$]$_\infty$ (**2**, Scheme 1ii), identified as a one-dimensional polymer. Complex **2** is probably a transient intermediate between the putative monomeric [K(DFForm)(thf)$_x$] solution species and polymeric **1**.

X-ray data for **2** were solved and refined in the monoclinic space group $P2_1$, with two potassium ions, two DFForm ligands, and two coordinating thf molecules occupying the asymmetric unit (Figure 5A). For the ASU component, the two DFForm ligands bridge between both potassium centres in a $\mu$-1$\kappa$($N,N',F$):2$\kappa$($N,N',F'$) manner, and N1 and N3 coordinate closer to K1, and N2 and N4 coordinate closer to K2. The K$\cdots$F–C coordination in this arrangement is overall shorter than those observed for the tetradentate ($N,N',F,F'$) DFForm coordination in **1**, but longer than the auxiliary fluorine K$\cdots$F–C coordination in **1**. All the C–F bond lengths exhibit only a slight elongation, with the exception of the C13–F4 bond, which is notably longer than the others. An explanation behind the elongation of only C13–F4 is due to the involvement of this fluorine atom in additional bridging to an adjacent potassium atom. This, in conjunction with the two bridging thf ligands, leads to a one-dimensional polymer (Figure 5B).

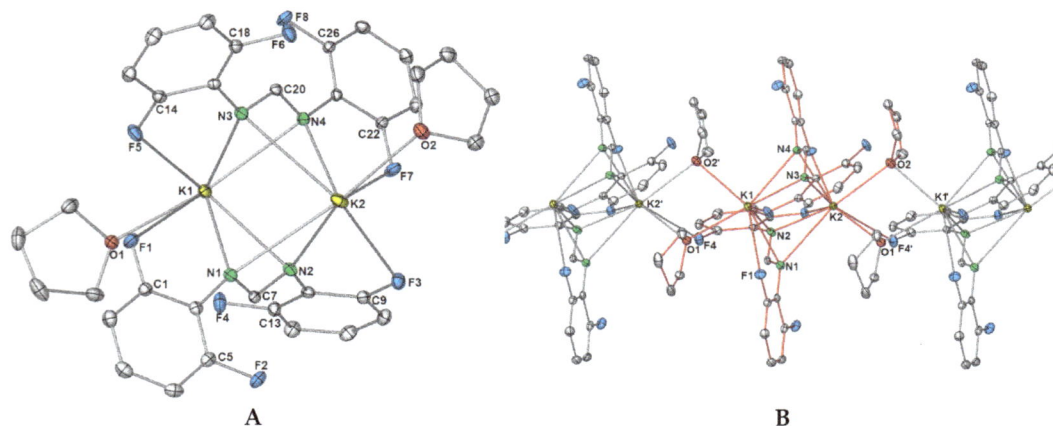

**A**        **B**

**Figure 5.** Molecular structure of [K$_2$(DFForm)$_2$(thf)$_2$]$_\infty$ (**2**). Ellipsoids are shown at 50% probability; hydrogen atoms and lattice solvent were removed for clarity. (**A**) Asymmetric unit. (**B**) Growth of the one-dimensional polymer chain; red bonding indicates the ASU. Selected bond lengths (Å) and angles (°): K1–N1: 2.7921(13), K1–N2: 2.9118(13), K1–N3: 2.7504(13), K1–N4: 3.0472(12), K1–F1: 3.3090(9), K1–F5: 3.1141(10), K1–O1: 2.7830(11), K1–O2′: 2.9351(13), K2–N1: 3.2913(12), K2–N2: 2.7581(13), K2–N3: 3.2957(13), K2–N4: 2.8039(13), K2–F3: 3.3678(10), K2–F4′: 3.2276(11), K2–F7: 2.9353(9), K2–O2: 2.8039(12), K2–O1′: 2.8172(11): C1–F1: 1.3575(19), C5–F2: 1.3644(16), C9–F3: 1.3610(17), C13–F4: 1.3724(14), C14–F5: 1.3608(18), C18–F6: 1.3675(16), C22–F7: 1.3578(16), C26–F8: 1.3640(15), K1–C7–K2: 70.41(3), K1–C20–K2: 68.86(3), O1–K1–K2: 147.97(3), O2–K2–K1: 136.59(3). K1–O1–K2′: 98.05(4), K1–O2′–K2′: 101.41(3).

The thf ligands in **2** bridge in an almost symmetrical manner between the two potassium atoms, though O2′ coordinates closer to K2′ than K1. This type of $\mu$-1$\kappa$($O$):2$\kappa$($O$) bridging of two thf molecules is no stranger to group one chemistry [18], for example in the polymeric sodium diphenyloxidomethanide $(Ph_2CO)^{2-}$ polymer $[Na_2(Ph_2CO)(thf)_2]$ [33], or the potassium 2,4,6-tris(trifluoromethyl)phenolate $(OAr^{CF3})$ complex $[K_2(OAr^{CF3})_2(thf)_4(\mu\text{-}O\text{-}thf)_2]$ [34]. However, examples where the polymeric structure is generated by two thf ligands connecting the dinuclear units is restricted to only one other example in group one chemistry, namely $[K_4(COT)_2(thf)_6]_\infty$ [35]. One difference between the COT (cyclooctatetraenyl) system and **2** is that asymmetric bridging of the thf ligands is more apparent, as both thf ligands favour one metal centre over the other (e.g., K1–O1: 2.839(3), K1–O2: 2.846(5), K2–O1:2.781(3), K2–O2: 2.783(4)). It should also be noted that there are examples where three thf ligands, not two, bridge the monomeric units to create a polymeric network [32,36]. Exposure of crystalline **2** to vacuum immediately causes fracturing of the crystals, giving **1**. Furthermore, when crystals of **2** are isolated and allowed to stand at room temperature, some degree of thf liberation is apparent as the elemental analysis performed on these crystals gave a lower than expected carbon value. The best fit was obtained when the composition was calculated with loss of 0.4 thf molecules from **2**. By examining the structure of **2**, it seems that upon the liberation of bound thf, the DFForm ligand changes from the asymmetric $\mu\text{-}1\kappa(N,N',F):2\kappa(N,N',F')$ coordination to a $\mu\text{-}(N,N',F,F'):(N,N')$ binding mode, and the auxiliary fluorine and aromatic carbon atoms become free to engage with adjacent potassium ions, building the complex polymeric network of **1**. Owing to the rapid loss of thf from **2**, additional characterisation was difficult. Dissolution in $C_6D_6$ gave rapid formation of a powder, presumably **1**, as a large excess of thf was observed in the $^1H$ NMR spectrum.

Although no monomeric $[K(DFForm)(thf)_x]$ species could be obtained from thf, we exploited the well-known affinity of 18-crown-6 for the potassium ion. Treatment of **1** with 18-crown-6 and crystallisation from a $n$-hexane/toluene solution, gave monomeric [K(DFForm)(18-crown-6)] (**3**). The structure was determined by X-ray crystallography, where the data were solved and refined in the monoclinic space group $P2_1/n$, with two molecules occupying the asymmetric unit (only one is depicted in Figure 6). The most surprising feature of this structure is the $\kappa(N,C,F)$ coordination of the DFForm ligand to the potassium centre, as opposed to the expected $\kappa(N,N')$ coordination that is observed in [K($p$-TolForm)(18-crown-6)] [32], and in several other C{NCXN}$^-$C based ligand systems [18], such as [K(pyr)(18-crown-6)] (pyr = 1,3,4,6,7,8-hexahydro-2$H$-pyrimido[1,2-a]pyrimidide) [37]. The *ipso* carbon–potassium bond length (Figure 6) lies in the expected range for such interactions. For example, the *ipso*-carbon potassium interactions observed in the bimetallic 2,6-diphenylphenolate complex $[KCa(OAr^{Ph})_3]$, has a K–$C_{ipso}$ bond lengths of 3.391(6) Å [38] and the K–$C_{ipso}$ bond length in the phenylthiolato complex $[K_2Fe(SPh)_4]$ is 3.477 (5) Å [39]. Despite the non-binding of N2, there is still charge delocalisation across the NCN bite, as there is only a slight shortening of the free C7–N2 bond, making it far too long for a formal double bond (e.g., DFForm(CPh$_3$): C=N: 1.2762(12) Å) [17].

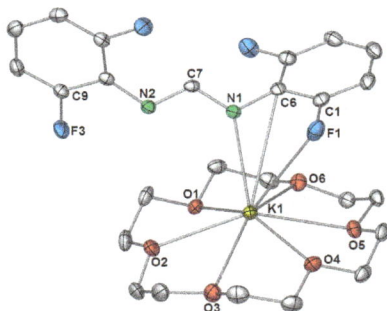

**Figure 6.** Molecular structure of [K(DFForm)(18-crown-6)] (**3**). Ellipsoids shown at 50% probability with hydrogen atoms omitted for clarity. Selected bond lengths (Å): K1–N1: 2.7994(12), K1–C6: 3.4836(14), K1–C1(non-bonding): 3.6886(15), K1–F1: 3.2809(11), K1–N2(non-bonding): 3.9843(13), K1–O(crown): range: 2.8255(11)–2.9525(11), average: 2.89. C1–F1: 1.3584(18), C1–C6:1.401(2), C9–F3: 1.3576(16), C7–N1: 1.3224(18), C7–N2: 1.3161(18).

## 3. Materials and Methods

### 3.1. General Experimental Details

All reactions were undertaken using Schlenk line and glove box techniques. Solvents (thf, toluene, hexane, $C_6D_6$, thf-$d_8$) were purified by distillation over sodium or sodium benzophenone, and were degassed prior to use. NMR experiments were recorded on a Bruker Avance 300 spectrometer or a Bruker AVII+400 machine (Billerica, MA, USA). $^1H$ NMR resonances were referenced to tetramethylsilane by way of the residual $^1H$ resonance of $C_6D_6$ (and $^{19}F$ coupled unless specified otherwise). $^{19}F$-NMR data were $^1H$ decoupled (unless specified otherwise) and referenced to external $CFCl_3$. Microanalyses were performed by the elemental analysis service of London Metropolitan University or by an Elementar Vario Micro cube (Elementar, Langenselbold, Germany) by Wolfgang Bock of Tübingen University. IR spectra were recorded on a Perkin–Elmer 1600 Fourier transform infrared spectrometer ($\bar{v}$ = 4000–500 $cm^{-1}$), as either mulls in sodium-dried Nujol, or a Nicolet 6700 FTIR spectrometer (Thermo Nicolet, Madison, WI, USA) or using a DRIFT chamber with dry KBr/sample mixtures and KBr windows. K[(NSiMe$_3$)$_2$] was purchased from Sigma-Aldrich (St Louis, MO, USA) and used as received. $N,N'$-bis(2,6-difluorophenyl)formamidine (DFFormH) was synthesised by a published procedure [40], 18-crown-6 was purchased from Sigma-Aldrich and used as received.

**[K(DFForm)]$_\infty$ (1)**: K[(NSiMe$_3$)$_2$] (0.25 g, 1.3 mmol) and DFFormH (0.33 g, 1.2 mmol) were each dissolved in toluene and combined with stirring, immediately forming a white, poorly soluble powder. The supernatant solution was decanted and the resulting powder was dried in vacuo. After the addition of thf, the powder dissolved, and then the solution was concentrated and layered with $n$-hexane. Colourless white block crystals grew overnight and were suitable for X-ray diffraction, revealing the composition [K(DFForm)]$_\infty$ (**1**, Yield = 0.30 g, 80%).$^1H$ NMR ($C_6D_6$, 400 MHz, 25 °C): δ 8.88 (br s, 1H, NC$H$N), 6.68 (m, 4 H, Ar-$H$(3,5)), 6.63 (m, 2 H, Ar-$H$(4)).$^{19}F$ NMR ($C_6D_6$, 25 °C): δ −127.2 (br s). $^1H$ NMR (thf-$d_8$, 400 MHz, 25 °C): δ 8.92 (p $^5J_{H-F}$: 3.09 Hz, 1 H, NC$H$N), 6.68 (m, 4 H, Ar-$H$(3,5)), 6.46 (m, 2 H, Ar-$H$(4)). $^1H$ NMR (thf-$d_8$, 400 MHz, 25 °C, $^{19}F$ decoupled): δ 8.89 (s, 1 H, NC$H$N), 6.68 (m, 4 H, Ar-$H$(3,5)), 6.46 (m, 2 H, Ar-$H$(4)).$^{19}F$ NMR (thf-$d_8$, 25 °C): δ −127.8 (s). $^{19}F$ NMR (thf-$d_8$, 25 °C, F–H coupled) −127.8 (br s). IR (DRIFT): ν 1612 (m), 1562 (vs), 1513 (s), 1477 (s), 1464 (s), 1395 (w), 1326 (m), 1287 (w), 1254 (m), 1231 (m), 1199 (s), 1062 (w), 1005 (m), 984 (s), 954 (w), 922 (w), 828 (w), 779 (m), 766 (m). Elemental analysis ($C_{13}H_7F_4KN_2$, 306.31 g·mol$^{-1}$): calcd.: C 50.97, H 2.30, N 9.15, found: C 50.81, H 2.34, N 9.07.

**[K$_2$(DFForm)$_2$(thf)$_2$]$_\infty$ (2)**: **1** (0.10 g, 0.32 mmol) was dissolved in minimal thf and concentrated in vacuo, giving colourless crystals that were not suitable for X-ray diffraction. The concentrated solution was stored at −35 °C, where large colourless block crystals grew of [K$_2$(DFForm)$_2$(thf)$_2$] (**2**), suitable for X-ray diffraction. Upon exposure to vacuum, the crystals fractured and a white powder was obtained, likely consisting of a mixture of **1** and **2**. (Yield = 0.11 g, 89%). $^1H$ NMR ($C_6D_6$, 400 MHz, 25 °C, formation of insoluble white powder upon solvent addition, giving a large excess of thf in solution): δ 1.42 (m, 232 H, thf-β-$CH_2$), 3.57 (m, 232 H, thf-α-$CH_2$), 6.36 (m, 2 H, Ar-$H4$) 6.67 (m, 4 H, Ar-$H$(3,5)), 8.88 (br s, 1 H, NC$H$N). $^{19}F$ NMR ($C_6D_6$, 25 °C): δ −127.1 (br s). IR (DRIFT): ν 1613 (m), 1564 (vs), 1551 (vs), 1514 (m), 1477 (vs), 1464 (vs), 1395 (w), 1325 (s), 1254 (m), 1231 (w), 1200 (s), 1062 (w), 1005 (w), 984.6 (s), 955 (w), 922 (w), 827 (w), 799 (w), 766 (m), 742 (w), 716 (m). Elemental analysis calcd (%) (for $C_{34}H_{30}F_8K_2N_4O_2$, 756.81 g·mol$^{-1}$, pre-dried powder under vacuum). C 53.95, H 4.00, N 7.40, found: C 49.64, H 2.74, N 8.44. When the crystals were dried by slow evaporation in a glove box, a composition of [K$_2$(DFForm)$_2$(thf)$_{1.6}$] was supported, calcd. ($C_{58.4}H_{40.8}F_8K_2N_4O_{1.6}$, 727.96 g·mol$^{-1}$): C 53.46, H 3.71, N 7.69 found: C 53.08, H 3.66, N 7.35.

**[K(DFForm)(18-crown-6)] (3)**: If **1** (~0.10 g, 0.32 mmol) was crystallised from toluene/hexane solutions in the presence of one equivalent of 18-crown-6 (~0.09 g, 0.34 mmol), pale yellow block crystals of [K(DFForm)(18-crown-6)] (**3**) developed. (Yield = ~0.07 g, 34%). $^1H$ NMR ($C_6D_6$, 300 MHz 303.2 K): δ 9.15 (s, 1 H, NC$H$N), 6.84 (m, 4 H, Ar-H(3,5)), 6.42 (m, 2 H, Ar-H(4)), 3.22 (br s, 24 H,

18-crown-6). $^{19}$F NMR (C$_6$D$_6$, 303.2 K): δ = −125.81 (br s). IR (Nujol): $\bar{v}$ = 1588 (vs), 1540 (vs), 1259 (vs), 1193 (m), 1096 (s), 1004 (s), 956 (m), 818 (m). Elemental analysis returned poor C H N values. (C$_{25}$H$_{31}$F$_4$KN$_2$O$_6$, 570.62 g·mol$^{-1}$): calcd. C 52.62, H 5.47, N 4.91, found: C 46.87, H 4.93, N 5.21.

## 3.2. X-ray Crystallography

All compounds were examined on a "Bruker APEX-II CCD" diffractometer at 100.15 or 150.15 K, mounted on a fibre loop in Paratone-N. Absorption corrections were completed using Apex II program suite [41]. Structural solutions were obtained by charge flipping (**1, 2, 3**) [42] methods, and refined using full matrix least squares methods against $F^2$ using SHELX2013 [43], within the OLEX 2 graphical interface [43]. CCDC numbers: **1** (1540263), **2** (1540264), **3** (1540265).

**[K(DFForm)]$_\infty$ (1)**: C$_{13}$H$_7$F$_4$KN$_2$ ($M$ = 306.31 g/mol): triclinic, space group $P$-1 (no. 2), $a$ = 7.4437(2) Å, $b$ = 7.5357(2) Å, $c$ = 11.8891(3) Å, $\alpha$ = 100.6590(10)°, $\beta$ = 101.9020(10)°, $\gamma$ = 101.1070(10)°, $V$ = 622.56(3) Å$^3$, $Z$ = 2, $T$ = 100(2) K, $\mu$(MoK$\alpha$) = 0.466 mm$^{-1}$, $Dcalc$ = 1.634 g/cm$^3$, 10877 reflections measured (5.66° $\leq 2\Theta \leq$ 60.48°), 3657 unique ($R_{int}$ = 0.0152, $R_{sigma}$ = 0.0174) which were used in all calculations. The final $R_1$ was 0.0279 (>2σ(I)) and $wR_2$ was 0.0734 (all data). Note: NCHN hydrogen atom manually assigned from identified Q peak.

**[K$_2$(DFForm)$_2$(thf)$_2$]$_\infty$ (2)**: C$_{34}$H$_{30}$F$_8$K$_2$N$_4$O$_2$ ($M$ = 756.82 g/mol): monoclinic, space group $P2_1$ (no. 4), $a$ = 7.55040(10) Å, $b$ = 19.8885(3) Å, $c$ = 11.6408(2) Å, $\beta$ = 105.4479(6)°, $V$ = 1684.90(4) Å$^3$, $Z$ = 4, $T$ = 100.1 K, $\mu$(MoK$\alpha$) = 0.364 mm$^{-1}$, $Dcalc$ = 1.492 g/cm$^3$, 17983 reflections measured (3.62° $\leq 2\Theta \leq$ 60.66°), 8299 unique ($R_{int}$ = 0.0136, $R_{sigma}$ = 0.0201), which were used in all calculations. The final $R_1$ was 0.0262 (>2σ(I)) and $wR_2$ was 0.0652 (all data).

**2[K(DFForm)(18-crown-6)] (3)**: C$_{50}$H$_{62}$F$_8$K$_2$N$_4$O$_{12}$ ($M$ = 1141.23): note: two molecules present in the asymmetric unit. monoclinic, space group $P2_1/n$ (no. 14), $a$ = 10.9773(4) Å, $b$ = 15.3047(5) Å, $c$ = 31.3860(10) Å, $\beta$ = 93.674(2)°, $V$ = 5262.1(3) Å$^3$, $Z$ = 4, $T$ = 123.15 K, $\mu$(MoK$\alpha$) = 0.273 mm$^{-1}$, $Dcalc$ = 1.441 g/mm$^3$, 78411 reflections measured (2.6 $\leq 2\Theta \leq$ 56.76), 13119 unique ($R_{int}$ = 0.0369, $R_{sigma}$ = 0.0266) which were used in all calculations. The final $R_1$ was 0.0321 (I > 2σ(I)) and $wR_2$ was 0.1146 (all data).

## 4. Conclusions

Complexation of N,N'-bis(2,6-difluorophenyl)formamidinate to potassium generates a species which rapidly liberated coordinated thf, giving a two-dimensional polymeric network [K(DFForm)]$_\infty$ (**1**), based on a complex and unprecedented formamidinate binding mode. This binding was shown to be completely different from the analogous [K(FForm)]$_\infty$ (FForm: N,N'-bis(2-fluorophenyl)formamidinate) mono-directional polymer. With access to two additional auxiliary o-fluorine atoms (namely F2 and F4), a new dimension for the polymeric network could be generated, which was further reinforced by potassium–arene interactions and nitrogen-based bridging of the DFForm ligand. The formation of this network was so favourable that it could be generated by simple n-hexane layering of thf solutions, or the evaporation of thf solutions to dryness, and is the first example of a two-dimensional polymeric N,N'-bis(aryl)formamidinate network. A likely transient species between a monomeric thf solution derivative [K(DFForm)(thf)$_x$] and **1** was also obtained and identified as a one-dimensional polymer with two bridging thf ligands, namely [K$_2$(DFForm)$_2$(thf)$_2$]$_\infty$ (**2**). Complex **2** lost thf in the solid state, slowly forming **1** upon storage at room temperature. A monomeric derivative of **1** was obtained through use of 18-crown-6, giving [K(DFForm)(18-crown-6)] (**3**), which showed a highly unexpected κ(N,C$_{ispo}$,F) coordination. Such examples as these highlight the strong affinity of potassium for donor atoms, especially fluorine, and how the simple addition of more fluorine atoms to a ligand system can expand the coordination network of the ligand, and can generate unexpected structural consequences.

**Acknowledgments:** We thank the ARC (DP16010640) for funding, and Professor Reiner Anwander for use of the elemental analysis and IR services of Tuebingen University.

**Author Contributions:** The project concept was devised equally by all authors. Daniel Werner performed the experiments and analyzed the resulting data. The materials were provided by Glen B. Deacon and Peter C. Junk. The manuscript was written by Daniel Werner and edited by Glen B. Deacon and Peter C. Junk.

## References

1.  Edelmann, F.T. Chapter two—Recent progress in the chemistry of metal amidinates and guanidinates: Syntheses, catalysis and materials. In *Advances in Organometallic Chemistry*; Anthony, F.H., Mark, J.F., Eds.; Academic Press: London, UK, 2013; Volume 61, pp. 55–374.

2.  Edelmann, F.T. Chapter 3 advances in the coordination chemistry of amidinate and guanidinate ligands. In *Advances in Organometallic Chemistry*; Anthony, F.H., Mark, J.F., Eds.; Academic Press: London, UK, 2008; Volume 57, pp. 183–352.

3.  Junk, P.C.; Cole, M.L. Alkali-metal bis(aryl)formamidinates: A study of coordinative versatility. *Chem. Commun.* **2007**, *16*, 1579–1590. [CrossRef] [PubMed]

4.  Cotton, F.A.; Daniels, L.M.; Murillo, C.A. A systematic approach in the preparation of compounds with $\sigma_2$-$\pi_4$ vanadium-to-vanadium triple bonds: Synthesis, reactivity, and structural characterization. *Inorg. Chem.* **1993**, *32*, 2881–2885. [CrossRef]

5.  Deacon, G.B.; Hossain, M.E.; Junk, P.C.; Salehisaki, M. Rare-earth $N,N'$-diarylformamidinate complexes. *Coord. Chem. Rev.* **2017**. [CrossRef]

6.  Cotton, F.A.; Daniels, L.M.; Murillo, C.A. The first complex with a $\sigma_2\pi_4$ triple bond between vanadium atoms in a ligand framework of fourfold symmetry—$[V_2\{(p\text{-}CH_3C_6H_4)NC(H)N(p\text{-}C_6H_4CH_3)\}_4]$. *Angew. Chem. Int. Ed.* **1992**, *31*, 737–738. [CrossRef]

7.  Cotton, F.A.; Daniels, L.M.; Maloney, D.J.; Matonic, J.H.; Murillo, C.A. Divalent metal chloride formamidine complexes, $M_{11}$ = Fe, Co and Pt. Syntheses and structural characterization. *Polyhedron* **1994**, *13*, 815–823. [CrossRef]

8.  Lyhs, B.; Bläser, D.; Wölper, C.; Schulz, S. Syntheses and X-ray crystal structures of organoantimony diazides. *Chem. Eur. J.* **2011**, *17*, 4914–4920. [CrossRef] [PubMed]

9.  Deacon, G.B.; Junk, P.C.; Wang, J.; Werner, D. Reactivity of bulky formamidinatosamarium(II or III) complexes with C=O and C=S bonds. *Inorg. Chem.* **2014**, *53*, 12553–12563. [CrossRef] [PubMed]

10. Jones, C.; Mills, D.P.; Rivard, E.; Stasch, A.; Woodul, W.D. Synthesis and crystal structures of anionic gallium(II) and gallium(III) heterocyclic compounds derived from a gallium(I) *n*-heterocyclic carbene analogue. *J. Chem. Crystallogr.* **2010**, *40*, 965–969. [CrossRef]

11. Cole, M.L.; Davies, A.J.; Jones, C.; Junk, P.C. Persistent $\pi$-arene interactions in bulky formamidinate complexes of potassium. *J. Organomet. Chem.* **2007**, *692*, 2508–2518. [CrossRef]

12. Cole, M.L.; Junk, P.C. Potassium complexes of the 'super' formamidine $(2,6\text{-}pr_2{}^iC_6H_3)NC(H)NH(2,6\text{-}pr_2{}^iC_6H_3)$, Hdippform.: Synthesis and molecular structure of $[\{K(DippForm)_2K(thf)_2\}_n]\cdot n$thf and $[K(DippForm)(thf)_3]\cdot$HDippForm. *J. Organomet. Chem.* **2003**, *666*, 55–62. [CrossRef]

13. Baldamus, J.; Berghof, C.; Cole, M.L.; Evans, D.J.; Hey-Hawkins, E.; Junk, P.C. Attenuation of reactivity by product solvation: Synthesis and molecular structure of $[K\{(\eta^6\text{-Mes})NC(H)N(Mes)\}\{(\eta^6\text{-Mes})NHC(H)N(Mes)\}]$, the first formamidinate complex of potassium. *J. Chem. Soc. Dalton Trans.* **2002**, *14*, 2802–2804. [CrossRef]

14. Hamidi, S.; Jende, L.N.; Martin Dietrich, H.; Maichle-Mössmer, C.; Törnroos, K.W.; Deacon, G.B.; Junk, P.C.; Anwander, R. C–H bond activation and isoprene polymerization by rare-earth-metal tetramethylaluminate complexes bearing formamidinato *n*-ancillary ligands. *Organometallics* **2013**, *32*, 1209–1223. [CrossRef]

15. Kulkarni, N.V.; Elkin, T.; Tumaniskii, B.; Botoshansky, M.; Shimon, L.J.W.; Eisen, M.S. Asymmetric bis(formamidinate) group 4 complexes: Synthesis, structure and their reactivity in the polymerization of $\alpha$-olefins. *Organometallics* **2014**, *33*, 3119–3136. [CrossRef]

16. Deacon, G.B.; Junk, P.C.; Werner, D. Enhancing the value of free metals in the synthesis of lanthanoid formamidinates: Is a co-oxidant needed? *Chem. Eur. J.* **2016**, *22*, 160–173. [CrossRef] [PubMed]

17. Werner, D.; Deacon, G.B.; Junk, P.C.; Anwander, R. Cerium(III/IV)formamidinate chemistry, and a stable cerium(IV) diolate. *Chem. Eur. J.* **2014**, *20*, 4426–4438. [CrossRef] [PubMed]

18. Allen, F.H. The cambridge structural database: A quarter of a million crystal structures and rising. *Acta Crystallogr. Sec. B* **2002**, *58*, 380–388. [CrossRef]

19. Cole, M.L.; Evans, D.J.; Junk, P.C.; Smith, M.K. Structural studies of N,N'-di(ortho-fluorophenyl)formamidine group 1 metallation. *Chem. Eur. J.* **2003**, *9*, 415–424. [CrossRef] [PubMed]

20. Cotton, F.A.; Murillo, C.A.; Pascual, I. Quadruply bonded dichromium complexes with variously fluorinated formamidinate ligands. *Inorg. Chem.* **1999**, *38*, 2182–2187. [CrossRef] [PubMed]

21. Cole, M.L.; Davies, A.J.; Jones, C.; Junk, P.C. Lithium and sodium N,N'-di(2,6-dialkylphenyl)formamidinate complexes. *J. Organomet. Chem.* **2004**, *689*, 3093–3107. [CrossRef]

22. Cole, M.L.; Junk, P.C.; Louis, L.M. Synthesis and structural characterisation of some novel lithium and sodium N,N'-di(para-tolyl)formamidinate complexes. *J. Chem. Soc. Dalton Trans.* **2002**, 3906–3914. [CrossRef]

23. Cole, M.L.; Davies, A.J.; Jones, C.; Junk, P.C. Mononuclear formamidinate complexes of lithium and sodium. *Z. Anorg. Allg. Chem.* **2011**, *637*, 50–55. [CrossRef]

24. Cole, M.L.; Deacon, G.B.; Forsyth, C.M.; Junk, P.C.; Konstas, K.; Wang, J.; Bittig, H.; Werner, D. Synthesis, structures and reactivity of lanthanoid(II)formamidinates of varying steric bulk. *Chem. Eur. J.* **2013**, *19*, 1410–1420. [CrossRef] [PubMed]

25. Werner, D.; Zhao, X.; Best, S.P.; Maron, L.; Junk, P.C.; Deacon, G.B. Bulky ytterbium formamidinates stabilise complexes with radical ligands, and related samarium "tetracyclone" chemistry. *Chem. Eur. J.* **2017**, *23*, 2084–2102. [CrossRef] [PubMed]

26. Deacon, G.B.; Junk, P.C.; Werner, D. The synthesis and structures of rare earth 2-fluorophenyl- and 2,3,4,5-tetrafluorophenyl-N,N'-bis(aryl)formamidinate complexes. *Polyhedron* **2016**, *A103*, 178–186. [CrossRef]

27. Cole, M.L.; Deacon, G.B.; Forsyth, C.M.; Junk, P.C.; Konstas, K.; Wang, J. Steric modulation of coordination number and reactivity in the synthesis of lanthanoid(III)formamidinates. *Chem. Eur. J.* **2007**, *13*, 8092–8110. [CrossRef] [PubMed]

28. Deacon, G.B.; Junk, P.C.; Werner, D. Lanthanoid induced C–F activation of all fluorine atoms of one $CF_3$ group. *Eur. J. Inorg. Chem.* **2015**, *9*, 1484–1489. [CrossRef]

29. Shannon, R.D. Revised effective ionic radii and systematic studies of interatomic distances in halides and chalcogenides. *Acta Cryst.* **1976**, *A32*, 155–169. [CrossRef]

30. Krackl, S.; Inoue, S.; Driess, M.; Enthaler, S. Intermolecular hydrogen–fluorine interaction in dimolybdenum triply bonded complexes modified by fluorinated formamidine ligands for the construction of 2D- and 3D-networks. *Eur. J. Inorg. Chem.* **2011**, *13*, 2103–2111. [CrossRef]

31. Deacon, G.B.; Junk, P.C.; Macreadie, L.K.; Werner, D. Structural and reactivity consequences of reducing steric bulk of N,N'-diarylformamidinates coordinated to lanthanoid ions. *Eur. J. Inorg. Chem.* **2014**, 5240–5250. [CrossRef]

32. Baldamus, J.; Berghof, C.; Cole, M.L.; Evans, D.J.; Hey-Hawkins, E.; Junk, P.C. N,N'-di(tolyl)formamidinate complexes of potassium: Studies of ancillary donor imposed molecular and supramolecular structure. *J. Chem. Soc. Dalton Trans.* **2002**, *22*, 4185–4192. [CrossRef]

33. Geier, J.; Ruegger, H.; Grutzmacher, H. Sodium compounds of the benzophenone dianion (diphenyloxidomethanide). *Dalton Trans.* **2006**, *1*, 129–136. [CrossRef] [PubMed]

34. Brooker, S.; Edelmann, F.T.; Kottke, T.; Roesky, H.W.; Sheldrick, G.M.; Stalke, D.; Whitmire, K.H. Comparison of the X-ray crystal structures of the sodium and potassium 2,4,6-tris(trifluoromethyl)phenoxides (RO-) and 2,4,6-tris(trifluoromethyl)benzenethiolates (RS-); [Na(OR)(thf)$_2$]$_2$, [k(OR)(thf)$_2$($\mu$-thf)]$_2$, [Na(SR)(thf)$_2$·0.25thf] and [K(SR)(thf)](thf = tetrahydrofuran). *J. Chem. Soc. Chem. Commun.* **1991**, *3*, 144–146.

35. Hu, N.; Gong, L.; Jin, Z.; Chen, W. Crystal structure of cyclooctatetraenylpotassium, $C_8H_8K_2\cdot(OC_4H_8)_3$. *J. Organomet. Chem.* **1988**, *352*, 61–66. [CrossRef]

36. Antolini, F.; Hitchcock, P.B.; Khvostov, A.V.; Lappert, M.F. Synthesis and structures of alkali metal amides derived from the ligands [N(SiMe$_2$Ph)(SiMe$_3$)]$^-$, [N($t$bu)(SiMe$_3$)]$^-$, [N(Ph)(2-C$_5$H$_4$N)]$^-$, and [N(2-C$_5$H$_4$N)$_2$]$^-$. *Eur. J. Inorg. Chem.* **2003**, *18*, 3391–3400. [CrossRef]

37. Coles, M.P.; Hitchcock, P.B. Bicyclic guanidinates in mono- and di-valent metal complexes, including group 1/2 and group 1/12 heterometallic systems. *Aus. J. Chem.* **2013**, *66*, 1124–1130. [CrossRef]

38.   Zuniga, M.F.; Deacon, G.B.; Ruhlandt-Senge, K. Developments in heterobimetallic s-block systems: Synthesis and structural survey of molecular M/AE (M = Li, Na, K, Cs; AE = Ca, Sr) aryloxo complexes. *Inorg. Chem.* **2008**, *47*, 4669–4681. [CrossRef] [PubMed]

39.   Yu, X.-Y.; Jin, G.-X.; Weng, L.-H. Phenylthiolate as a σ and π donor ligand: Synthesis of a 3-D organometallic coordination polymer $[K_2Fe(SPh)_4]_n$. *Chem. Commun.* **2004**, *13*, 1542–1543. [CrossRef] [PubMed]

40.   Roberts, R.M. Acid Catalyzed Reaction of Diarylformamidines with Ethyl Orthoformate. *J. Am. Chem. Soc.* **1949**, *71*, 3848–3849. [CrossRef]

41.   Sheldrick, G.M. *SADABS*; program for empirical absorption correction; University of Gottingen: Gottingen, Germany, 1996.

42.   Dolomanov, O.V.; Bourhis, L.J.; Gildea, R.J.; Howard, J.A.K.; Puschmann, H. *OLEX2*: A complete structure solution, refinement and analysis program. *J. Appl. Cryst.* **2009**, *42*, 339–341. [CrossRef]

43.   Sheldrick, G.M. A short history of *SHELX*. *Acta. Crystallogr.* **2008**, *A64*, 112. [CrossRef] [PubMed]

# 17

# Hydrogen Storage Stability of Nanoconfined MgH$_2$ upon Cycling

Priscilla Huen [1], Mark Paskevicius [2] (iD), Bo Richter [1], Dorthe B. Ravnsbæk [3] and Torben R. Jensen [1,*]

[1]   Center for Materials Crystallography, Interdisciplinary Nanoscience Center and Department of Chemistry, Aarhus University, Langelandsgade 140, 8000 Aarhus C, Denmark; priscilla.huen@inano.au.dk (P.H.); richter@chem.au.dk (B.R.)

[2]   Department of Physics and Astronomy, Fuels and Energy Technology Institute, Curtin University, Kent Street, Bentley, WA 6102, Australia; M.Paskevicius@curtin.edu.au

[3]   Department of Physics, Chemistry and Pharmacy, University of Southern Denmark, Campusvej 55, 5230 Odense M, Denmark; dbra@sdu.dk

*   Correspondence: trj@chem.au.dk

**Abstract:** It is of utmost importance to optimise and stabilise hydrogen storage capacity during multiple cycles of hydrogen release and uptake to realise a hydrogen-based energy system. Here, the direct solvent-based synthesis of magnesium hydride, MgH$_2$, from dibutyl magnesium, MgBu$_2$, in four different carbon aerogels with different porosities, i.e., pore sizes, $15 < D_{avg} < 26$ nm, surface area $800 < S_{BET} < 2100$ m$^2$/g, and total pore volume, $1.3 < V_{tot} < 2.5$ cm$^3$/g, is investigated. Three independent infiltrations of MgBu$_2$, each with three individual hydrogenations, are conducted for each scaffold. The volumetric and gravimetric loading of MgH$_2$ is in the range 17 to 20 vol % and 24 to 40 wt %, which is only slightly larger as compared to the first infiltration assigned to the large difference in molar volume of MgH$_2$ and MgBu$_2$. Despite the rigorous infiltration and sample preparation techniques, particular issues are highlighted relating to the presence of unwanted gaseous by-products, Mg/MgH$_2$ containment within the scaffold, and the purity of the carbon aerogel scaffold. The results presented provide a research path for future researchers to improve the nanoconfinement process for hydrogen storage applications.

**Keywords:** hydride; nanoconfinement; carbon scaffold

---

## 1. Introduction

The development of a cleaner and more sustainable energy system is urgently needed to meet our increasing energy demand, and to avoid global warming and environmental pollution due to increasing levels of carbon dioxide and other toxic gases. Hydrogen is considered a potential energy carrier, since it is an abundant, non-greenhouse gas and can be produced by the electrolysis of water [1–4]. However, gaseous hydrogen at ambient conditions has a low density of 0.082 g/L, which is a disadvantage for mobile applications, even with compression [5]. Therefore, the solid state storage of hydrogen in a metal hydride has been investigated [3,4,6,7]. Magnesium hydride, MgH$_2$, as one of the most extensively studied hydride materials, has a moderately high theoretical gravimetric H$_2$ density of $\rho_m(\text{MgH}_2) = 7.6$ wt % H$_2$, and a volumetric H$_2$ density of $\rho_v(\text{MgH}_2) = 110$ g H$_2$/L [8]. However, the practical application of an MgH$_2$-based system is hindered from the unfavourable thermodynamics and the typically slow kinetics of the hydrogen release and uptake [9,10].

To improve the hydrogen storage properties of MgH$_2$, nanoconfinement in porous materials can be considered [11–17]. Preparing nanosized MgH$_2$ from this bottom-up approach can reduce the hydrogen diffusion distance and increase the amount of hydrogen in the grain boundaries, leading to improved kinetics of hydrogenation/dehydrogenation [18]. Nanoconfinement has also been employed for other hydride materials (e.g., $NaAlH_4$, $LiBH_4$, and $NH_3BH_3$) and demonstrates an improvement in gas release properties [15,19–22]. Nanoconfined $MgH_2$ in mesoporous scaffolds can be prepared through an Mg melt infiltration process followed by hydrogenation, or by a direct synthesis route using a precursor (e.g., dibutyl magnesium, $MgBu_2$) [11–13,23–25]. The loading of $MgH_2$ in the scaffold is between 3.6 wt % and 22.0 wt % [19].

Previous work reveals that smaller pore sizes within resorcinol-formaldehyde carbon aerogel (CA) scaffolds lead to improved hydrogen release kinetics of nanoconfined $MgH_2$, by reducing the particle size and increasing the surface area of $MgH_2$ [11]. However, mainly the first hydrogen release cycle has been investigated up to now. Therefore, this present study includes multiple cycles of hydrogen release and uptake. Thermal treatment of the CA scaffold in a gas flow (often $CO_2$) can increase the surface area, up to >2000 $m^2/g$, and the total pore volume to 2–3 mL/g, but has almost no effect on the pore size distribution. This procedure is often denoted scaffold "activation". Therefore, CA scaffolds are considered very customisable and may possess a wide range of porosity parameters. Previous investigations of sodium aluminium hydride, $NaAlH_4$, nanoconfined in activated scaffolds reveal that more material can be infiltrated onto an activated scaffold, i.e., there is a larger hydrogen storage capacity due to a larger pore volume, but these materials show slower kinetics for hydrogen release as compared to nonactivated scaffold [26]. Nanoconfined hydrides are mostly shown to exhibit improved kinetics of hydrogen release and uptake, but a change in thermodynamics is only observed when the scaffolds have pore sizes smaller than 2–3 nm [27].

There are a number of studies that have investigated the effect of nanoconfinement on the dehydrogenation properties of $MgH_2$, but there is little information about the reversible hydrogen storage capacity of nanoconfined $MgH_2$ upon cycling (hydrogen release and uptake). In addition, it has been discovered that butane gas is released (in conjunction with hydrogen) in the thermal treatment of nanoconfined $MgH_2$, which may refer to the incomplete hydrogenation of $MgBu_2$ after infiltration [28]. Here, we maximize the hydrogen storage capacity of nanoconfined $MgH_2$ through multiple infiltrations and use a variety of carbon aerogel scaffolds with different pore networks. The properties of nanoconfined $MgH_2$ samples are then compared with a focus on their hydrogen storage capacity after multiple hydrogen release and uptake cycles.

## 2. Results and Discussion

### 2.1. Porosity of the Nanoporous Scaffolds and Confinement of Magnesium Hydride

Magnesium hydride, $MgH_2$, was nanoconfined in four different carbon aerogel scaffolds with different texture properties as shown in Table 1. The porosities of the as-synthesised scaffolds X1 and X2 are similar except for the average pore sizes, $D_{max}$, of 16.6 ± 0.5 and 27.1 ± 2.7 nm, respectively. The surface area, $S_{BET}$, and total pore volume, $V_{tot}$, of the activated scaffolds CX1 and CX2 increase significantly after heat treatment in a flow of carbon dioxide, but $D_{max}$ remains almost constant.

**Table 1.** Texture properties of the carbon aerogel scaffolds and amount of magnesium hydride present after three infiltrations.

| Carbon Aerogels | $S_{BET}$ ($m^2/g$) | $D_{avg}$ (nm) | $V_{micro}$ ($cm^3/g$) | $V_{meso}$ ($cm^3/g$) | $V_{tot}$ ($cm^3/g$) | $MgH_2$ (wt %) [a] | $MgH_2$ (vol %) [b] |
|---|---|---|---|---|---|---|---|
| X1 | 829 ± 16 | 16.6 ± 0.5 | 0.23 ± 0.01 | 1.13 ± 0.03 | 1.32 ± 0.04 | 24.8 | 17.3 |
| X2 | 801 ± 16 | 27.1 ± 2.7 | 0.25 ± 0.01 | 1.11 ± 0.08 | 1.32 ± 0.10 | 24.3 | 16.7 |
| CX1 | 1940 ± 131 | 14.7 ± 0.6 | 0.54 ± 0.07 | 1.85 ± 0.06 | 2.37 ± 0.12 | 37.1 | 17.1 |
| CX2 | 1803 ± 30 | 25.0 ± 0.8 | 0.56 ± 0.02 | 1.89 ± 0.04 | 2.38 ± 0.05 | 40.3 | 19.6 |

[a] Calculated stoichiometrically from the uptake of di-*n*-butylmagnesium; [b] Calculated from the volume of the empty scaffold and the bulk density of $MgH_2$.

The procedures for the direct synthesis of nanoconfined magnesium hydride, $MgH_2$ utilised in this investigation are a new modification of a previously described approach using monoliths of carbon aerogel scaffold [11,12]. The aim of this investigation is to explore new approaches to prepare high hydrogen capacity materials based on nanoconfined magnesium hydride. A total of three dibutyl magnesium infiltrations, each with three hydrogenations, were conducted in order to increase the loading of $MgH_2$ in the porous scaffolds, with details provided in Table 1 and Table S1. The infiltrated amount of dibutyl magnesium, $MgBu_2$, is measured gravimetrically after mechanically removing excess dibutyl magnesium that was crystallised on the surface of the scaffolds. Scaffolds X2 and CX2 show decreasing amounts of infiltrated $MgBu_2$ for each consecutive cycle of infiltration, see Table S1, assigned to increasing amounts successfully infiltrated in each cycle. In contrast, the amount of infiltrated $MgBu_2$ in X1 and CX1 vary more so, possibly due to difficulties in efficiently removing $MgBu_2$ from the surface. A graphical presentation of the results from the infiltrations is presented in Figure 1. Dibutyl magnesium is assumed to be completely converted to $MgH_2$ following the reaction Scheme (1):

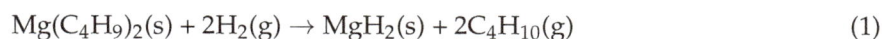

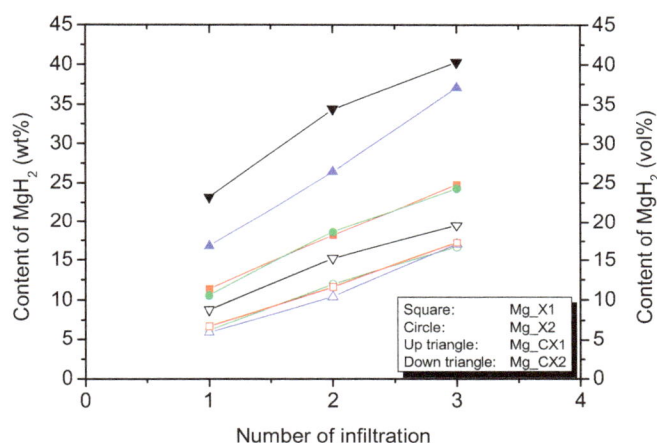

$$Mg(C_4H_9)_2(s) + 2H_2(g) \rightarrow MgH_2(s) + 2C_4H_{10}(g) \tag{1}$$

**Figure 1.** The cumulative gravimetric (solid symbols) and volumetric (open symbols) infiltration of magnesium hydride $MgH_2$ after each infiltration procedure in the four different carbon aerogel (CA) scaffolds.

The gravimetric and volumetric quantity of infiltrated magnesium hydride is calculated using the mass of scaffold, total pore volume, and bulk density of $MgH_2$. The volumetric loading of $MgH_2$ in the three scaffolds X1, X2, and CX1 are similar, ~17 vol %, whereas CX2 is slightly larger, ~20 vol %. However, the gravimetric hydride content varies more significantly, ~24 wt % for X1 and X2, ~37 wt % for CX1, and ~40 wt % for CX2. Recall that three independent infiltrations of $MgBu_2$ were conducted in this work, each with three individual hydrogenations. However, this work reveals that only a moderate increase in the infiltrated amount of $MgH_2$ is obtained after three infiltrations as compared to 12 vol % $MgH_2$ after one infiltration in a previous work [11]. That is mainly assigned to the large difference in molar volume of $MgH_2$ (18.2 cm$^3$/mol) and $MgBu_2$ (188.2 cm$^3$/mol). As such, $MgBu_2$ takes up a large volume after the infiltration, and only one-tenth of this volume is converted to $MgH_2$. This is similar to the utilisation of butyllithium for the direct synthesis of nanoconfined LiH, where loadings in the range of 12–17 wt % were obtained [29]. Secondly, $MgH_2$ may have a tendency to block the pores and stop further infiltration, which may hamper the full infiltration of the smaller pores.

## 2.2. Hydrogen Storage Capacity upon Cycling

Reversible hydrogen storage properties were investigated for five cycles of hydrogen release ($T$ = 355 °C, $t$ = 15 h in vacuum) and uptake ($T$ = 355 °C, $t$ = 15 h in $p(H_2)$ = 50 bar), i.e., $\Delta p(H_2)$ = 50 bar, denoted *condition* 1, for the four nanoconfined $MgH_2$ samples (see Figure 2). In the first decomposition, Mg_CX1 released 3.1 wt % $H_2$, which is slightly higher than the calculated hydrogen content of the sample based on the calculated quantity of $MgH_2$, 2.82 wt % (see Table 2). The observed hydrogen release from Mg_X1, 1.8 wt % $H_2$, is in accordance with the calculated value (1.88 wt % $H_2$). Samples Mg_X2 and Mg_CX2, with larger average pore sizes, release a lower quantity of gas, 1.3 and 2.2 wt % $H_2$, which corresponds to 68% and 71% of the calculated hydrogen content, respectively. For the following cycles, Table 2 and Figure 2 reveal a general stabilisation of the hydrogen storage capacity after the second desorption cycle.

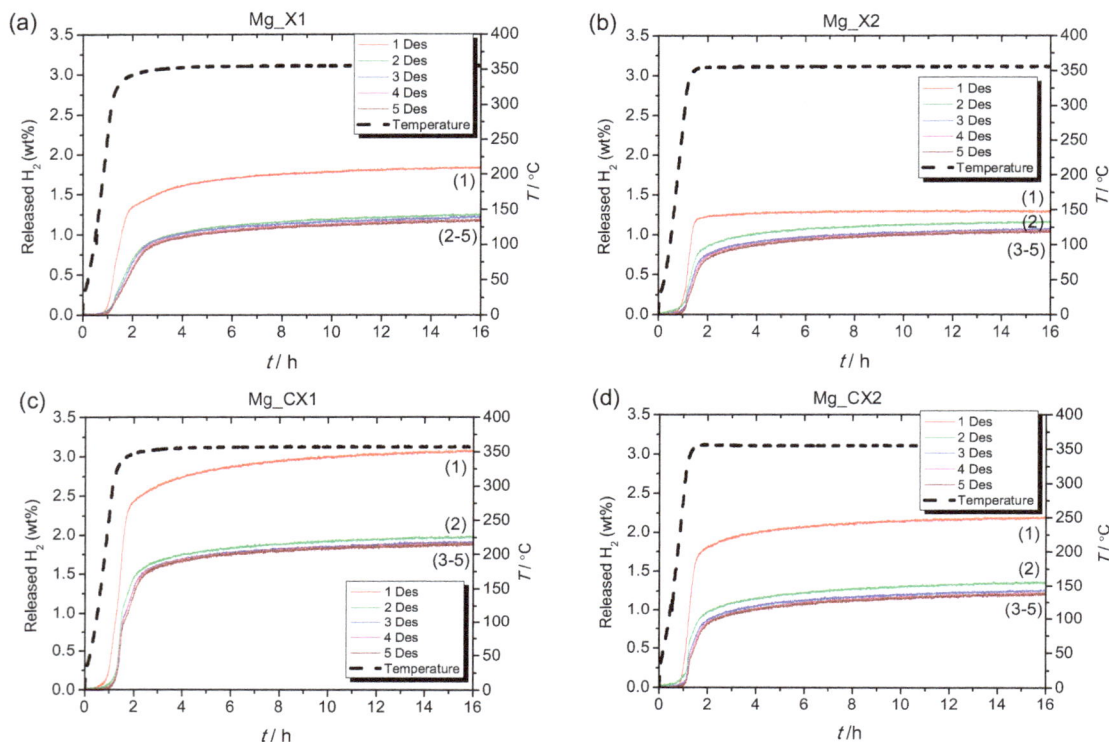

**Figure 2.** Sievert's measurements of (**a**) Mg_X1; (**b**) Mg_X2; (**c**) Mg_CX1; (**d**) Mg_CX2 under *condition* 1. Samples heated in vacuum from room temperature to 355 °C ($\Delta T / \Delta t = 5$ °C/min) for 15 h and reabsorbed for 15 h under $p(H_2) = 50$ bar.

**Table 2.** Calculated hydrogen content in infiltrated carbon aerogels and the hydrogen release measured by Sievert's method in desorption one (Des1) to five (Des5) using *condition* 1. The percentages in parentheses correspond to the retained hydrogen storage capacity compared to the initial values in Des1.

| Sample | $D_{max}$ (nm) | $\rho_m$ $(H_2)/(wt \%)$ | Des1 ($H_2$ wt %) | Des2 ($H_2$ wt %) | Des3 ($H_2$ wt %) | Des4 ($H_2$ wt %) | Des5 ($H_2$ wt %) |
|---|---|---|---|---|---|---|---|
| Mg_X1 | 17 | 1.88 | 1.8 (100%) | 1.3 (72%) | 1.2 (67%) | 1.2 (67%) | 1.2 (67%) |
| Mg_X2 | 26 | 1.85 | 1.3 (100%) | 1.2 (92%) | 1.1 (85%) | 1.1 (85%) | 1.0 (77%) |
| Mg_CX1 | 15 | 2.82 | 3.1 (100%) | 2.0 (67%) | 1.9 (61%) | 1.9 (61%) | 1.9 (61%) |
| Mg_CX2 | 25 | 3.06 | 2.2 (100%) | 1.4 (64%) | 1.3 (59%) | 1.2 (55%) | 1.2 (55%) |

For all four samples, the hydrogen release temperature is lower for the first cycle in comparison to further cycles. This may indicate that other reactions, besides the release of hydrogen, mainly occur in the first cycle. A thermal analysis using mass spectroscopy revealed that butane release occurs in addition to hydrogen release. This is unexpected due to the rigorous infiltration procedure, where a total of nine hydrogenation and evacuation steps are undertaken. In fact, butane is typically released at lower temperatures than hydrogen, generally in the range of 100 to 350 °C (this is further discussed later).

A similar investigation of the reversible hydrogen storage properties of the four nanoconfined samples was conducted using *condition* 2, i.e., the same temperature and time but a higher back-pressure for hydrogen release ($p(H_2)$ = 4–5 bar) and lower hydrogen pressure for uptake ($p(H_2)$ = 12 bar), i.e., $\Delta p(H_2)$ ~7.5 bar. Figure S1 shows a dramatic difference in the hydrogen release properties in comparison to Figure 2, where hydrogen desorption was conducted under vacuum and hydrogen absorption was conducted under 50 bar.

## 2.2.1. Thermodynamic Considerations

*Conditions* 1 and 2 for hydrogen release and uptake were selected so that *condition* 2 was just above/below the thermodynamic equilibrium pressure for hydrogen absorption/release of Mg/MgH$_2$ at 355 °C, i.e., $p_{eq}(H_2)$ = 6.4 bar [30], whereas *condition* 1 operates at a considerable "over-pressure". The hydrogen release data is presented in Figure 2 and Figure S1, respectively, showing dramatically different hydrogen release properties. Specifically, Figure S1 displays much lower gravimetric hydrogen release (i.e., 0.4 wt % vs. 1.8 wt % for the same sample and same cycle).

For *condition* 2, hydrogen is absorbed at $p(H_2)$ = 11–12 bar and desorbed at $p(H_2)$ < 5.2 bar, which is well above/below the thermodynamically limiting equilibrium pressure of $p_{eq}(H_2)$ = 6.4 bar [30]. Thus, from a thermodynamic point of view, *conditions* 1 and 2 should provide the same hydrogen storage properties, including hydrogen capacity. The hydrogen release profiles of *conditions* 1 and 2 (Figure 2 and Figure S1) are similar, which suggest that hydrogen release kinetics are similar and the majority of hydrogen release is within the first 3 h in all cases. However, the amount of hydrogen release is much lower in *condition* 2.

The very different pressures during hydrogenation, 50 or 12 bar for *conditions* 1 and 2, may lead to large differences in the degree of hydrogenation for several reasons: (i) Hydrogen is known to have slow diffusion in bulk Mg and MgH$_2$; (ii) The larger molar volume of magnesium hydride, $\rho_{mol}(MgH_2)$ = 18.15 cm$^3$/mol as compared to magnesium $\rho_{mol}(Mg)$ = 13.98 cm$^3$/mol may lead to core/shell formation during the hydrogenation of magnesium particles. Thus, a magnesium hydride layer may retard further hydrogenation; (iii) The material expansion of Mg to MgH$_2$ could lead to the blocking of the smaller pores in the scaffold, which may also retard further hydrogenation. A larger "over-pressure" as applied in *condition* 1 may limit the above mentioned drawbacks, (i) to (iii), and lead to complete hydrogenation of the samples.

## 2.2.2. Kinetics of Hydrogen Release of Nanoconfined MgH$_2$

For all the nanoconfined magnesium hydride samples, the majority of hydrogen is desorbed during heating from room temperature to 355 °C. Furthermore, in all cases, the first H$_2$ release profile is significantly different to the following ones, whereas the second is similar to the third, and then the H$_2$ release profiles become almost identical. This is clearly observed in Figure 2. For all four samples, the first decomposition has faster kinetics for hydrogen release and also a lower onset temperature. The initial 10 to 50% H$_2$ for the first cycle is released at a rate of 0.024, 0.030, 0.046, and 0.046 wt % H$_2$/min for the samples Mg_X1, Mg_X2, Mg_CX1 and Mg_CX2, respectively. The later hydrogen release profiles, cycle no. 2 to 5, consist of two regimes, see Figure S2. Initially, the hydrogen release rate appears to increase exponentially and then linearly at higher temperatures (see Figure S2). This suggests that the hydrogen release mechanism consists of more than one process, which is also observed for Mg$_{1-x}$Ti$_x$H$_2$ nanoparticles [31]. Here, we assume that the individual hydrogen release processes are independent and are due to differences in particle size, location in small or large pores or being located outside the scaffold, or consisting of Mg/MgH$_2$/MgBu$_2$ core–shell particles [32]. Assuming independent individual processes for hydrogen release, then the fastest process would occur at lower temperatures.

The data presented here for hydrogen release is not measured under isothermal conditions, which makes the kinetic analysis more challenging. The overall hydrogen release profile has a distorted sigmoidal shape, which cannot be modelled using Avrami-type kinetic equations, which have previously successfully been used to evaluate hydrogen release from Mg–Al–H, Mg–Cu–H, and Mg–Ni–H systems [33–35]. The first exponentially increasing hydrogen release does not match a power law, but the linear part of the profile can be fitted to a linear equation of the type, $\alpha(t) = b + kt$, where k is assigned an apparent rate constant. Apparent kinetic data is useful to compare similar samples in a more quantitative way. The degree of hydrogen release, $\alpha(t)$, from the normalised hydrogen release profiles (see Figure 3) also expresses the degree of magnesium formation. For the two as-synthesised scaffolds, the linear part of the curve is approximately in the range $0.3 < \alpha(t) < 0.6$. The apparent rate constants for these two samples, Mg_X1 and Mg_X2, are $k_1 = 1.33(4) \times 10^{-4}$ s$^{-1}$ and $k_2 = 2.3(1) \times 10^{-4}$ s$^{-1}$, respectively. The carbon dioxide activated sample, Mg_CX2, is somewhat similar, $0.37 < \alpha(t) < 0.62$, with $k_4 = 1.65(7) \times 10^{-4}$ s$^{-1}$, whereas the linear hydrogen release profile occurs at a higher degree of formation, $0.50 < \alpha(t) < 0.75$, for Mg_CX1, with $k_3 = 1.23(2) \times 10^{-4}$ s$^{-1}$. The linear regime for the hydrogen release rates have onsets in the temperature range 300 to ~330 °C and in some cases continue into the isothermal heating at $T = 355$ °C. We note that the calculated values for the apparent rate constants have the same order of magnitude as the values for bulk- and nickel-doped magnesium hydride, i.e., $1.0 < k < 5.3 \times 10^{-4}$ s$^{-1}$, but at significantly higher temperatures, 370 to 390 °C [35].

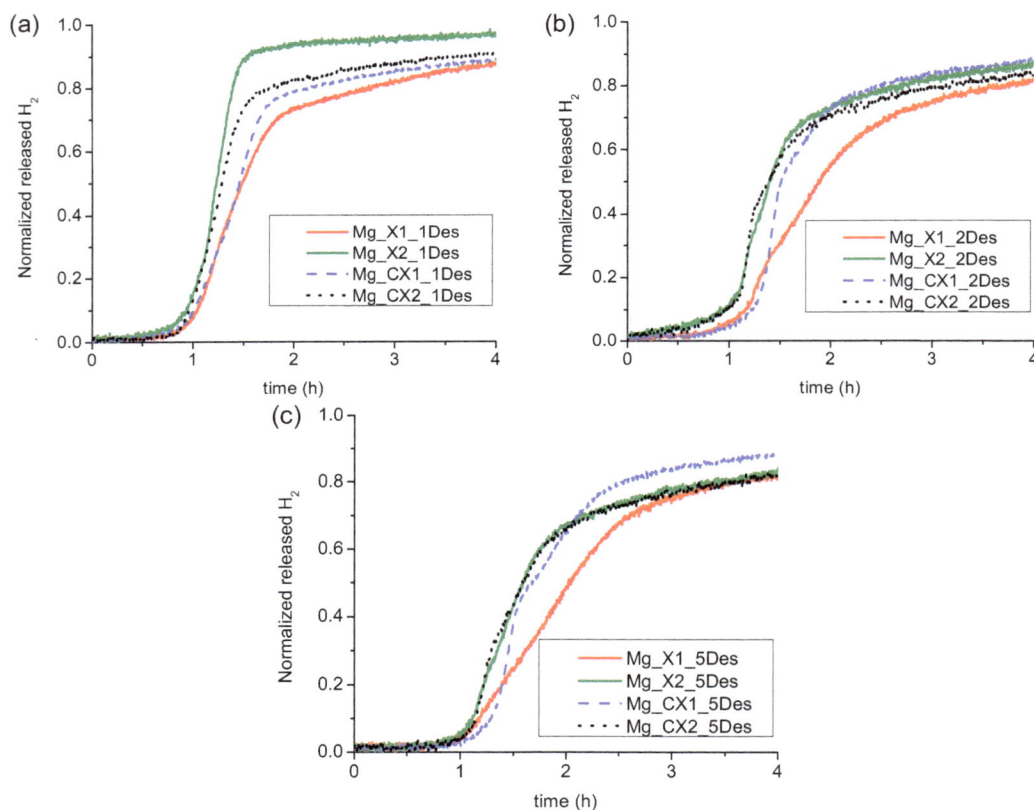

**Figure 3.** Normalized Sieverts gas release profiles of the four samples of nanoconfined magnesium hydride, (**a**) first desorption cycle; (**b**) second desorption cycle; and (**c**) fifth desorption cycle.

The first gas release with an exponential increasing rate is assigned to MgH$_2$ confined in the smaller pores, whereas hydrogen release at higher temperatures in the linear regime is assigned to MgH$_2$ confined in the larger cavities or outside the scaffold. Clearly, the rate of hydrogen release is lower for the larger particles as compared to the initial hydrogen release for the smaller in all cases, despite the significantly higher temperatures in the linear regime, which is illustrated in Figure S2. Accordingly, the four samples have similar apparent rate constants. However, the hydrogen storage capacities for the nanoconfined samples presented in Table 2 are significantly lower as compared to well-known magnesium hydride–metal oxide systems, which may also show fast kinetics, e.g., $MgH_2$–$Nb_2O_5$ [36,37]. However, this is due to a reduction of the metal and the formation of a solid solution, $Mg_xNb_{1-x}O$ [38].

### 2.3. Analysis of the Released Gases and Samples after Cycling

TGA-MS reveals that nanoconfined $MgH_2$ samples release hydrogen in accordance with Sievert's measurements (see Figure 4). However, there is also a significant quantity of butane gas that is also released, not just in the first cycle but also small but still detectable amounts on the fifth desorption cycle. However, after five desorption/absorption cycles under *condition 1*, the amount of butane released by Mg_X2 is about 100 times less compared to the as-prepared Mg_X2. It should again be reiterated that the sample preparation in this study was meticulous in pre-cycling a hydrogen reduction step three times in an attempt to completely transform the $MgBu_2$ precursor, but the release gas stream is still contaminated with butane. The conversion of the $MgBu_2$ precursor to $MgH_2$ was conducted at $T = 150$ °C during sample preparation. This treatment appears more efficient for scaffolds with larger pores, which release less butane. Scaffolds with smaller pores may more effectively contain and isolate $MgBu_2$, preventing it from hydrogenating during activation. This leads to butane release in the later hydrogenation cycles. In terms of hydrogen release, the temperature of maximum hydrogen release shifts to a higher temperature due to the particle growth of $MgH_2$, as revealed by powder X-ray diffraction (see Section 2.4).

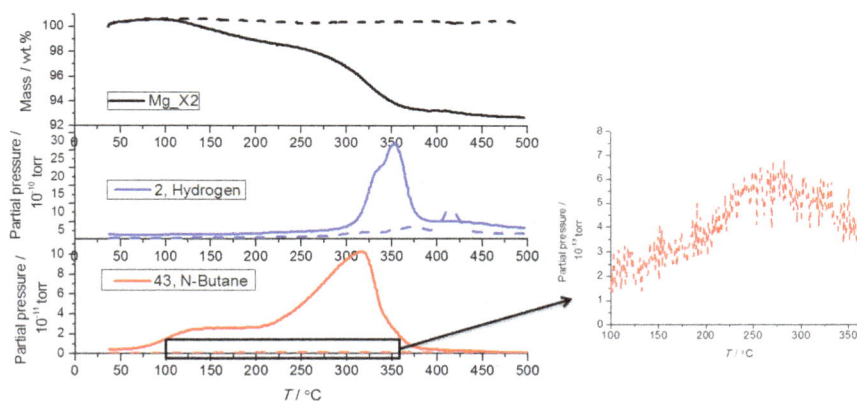

**Figure 4.** Thermogravimetric and mass spectroscopic analysis of the hydrogen and butane release from as-prepared Mg_X2 (solid line) and cycled Mg_X2 (dash line) during constant heating from room temperature (RT) to 500 °C ($\Delta T/\Delta t = 5$ °C/min).

The minor increase in the measured mass at low temperature is caused by buoyancy. The total mass loss of the as-prepared Mg_X2 upon decomposition was 7.3 wt %, which is significantly higher than the calculated hydrogen content (1.85 wt %). Larger than expected mass loss is also observed for other samples. In addition to hydrogen and butane, other types of gas (e.g., observed as $m/z$ ratio = 28, 36, and 38) are also released from the samples in the first decomposition (see Figure S3). The impurities may come from the organic solvent or from the scaffolds above 250 °C [11]. In the first decomposition cycles, impurities in the as-prepared samples vaporize. Thus, in further cycles, the gas stream is more pure hydrogen whilst other gases are absent and do not contribute to extra mass loss.

### 2.4. Comparison of As-Prepared and Cycled Nanoconfined MgH₂

The four nanoconfined magnesium hydride samples were examined by powder X-ray diffraction (PXD) before and after five cycles of hydrogen release and uptake. Figure 5 reveals that the as-prepared nanoconfined sample and the five-times cycled sample Mg_CX1 contain crystalline $MgH_2$ and MgO. Figure 5 also reveals an extreme difference in the diffraction peak width for $MgH_2$ in the two samples. All the diffraction data was analysed quantitatively for the composition of the crystalline fraction of the sample and the average crystallite sizes using Rietveld refinement (see Table 3). In the as-prepared samples, the crystallite size of $MgH_2$ is significantly smaller than the average pore size of the scaffold. This is due to the relatively low temperature for conversion of $MgBu_2$ to $MgH_2$ (150 °C), and the fact that the molar volume of $MgBu_2$ is a factor ten larger than that of $MgH_2$. However, only 38% to 48% of the crystalline fraction is $MgH_2$; the major part is nanocrystalline MgO.

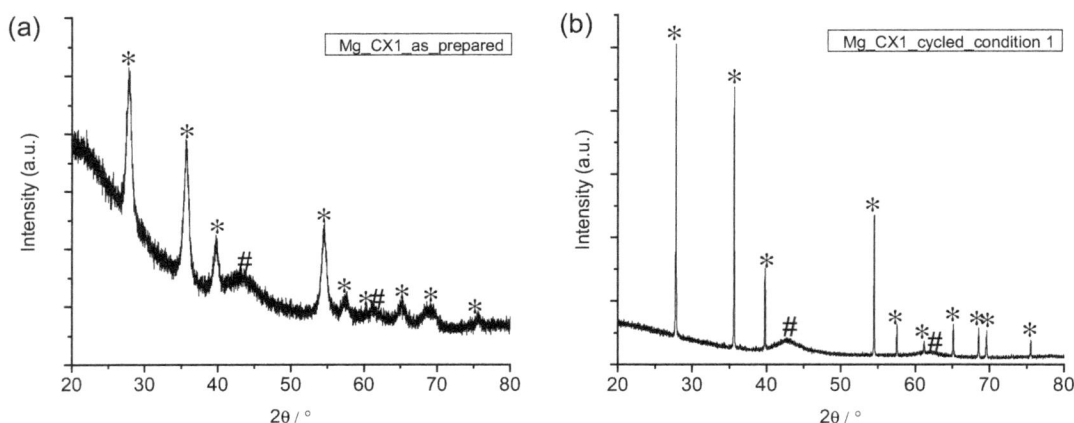

**Figure 5.** Powder X-ray diffraction (PXD) of Mg_CX1 (**a**) before; and (**b**) after five desorption/absorption cycles. * $MgH_2$; # MgO.

**Table 3.** Calculated average $MgH_2$ crystallite size and crystalline weight fraction from PXD as infiltrated and after five desorption cycles using *condition* 1. The remaining crystalline weight fraction is from MgO, which in all cases exists with ~1 nm crystallites.

| Sample | $D_{max}$ (nm) | As Infiltrated | | After Five Cycles | |
|---|---|---|---|---|---|
| | | $MgH_2$ Cryst. Size (nm) | $MgH_2$ Cryst. wt % | $MgH_2$ Cryst. Size (nm) | $MgH_2$ Cryst. wt % |
| Mg_X1 | 17 | 13 | 0.38 | 210 | 0.21 |
| Mg_X2 | 26 | 10 | 0.40 | 248 | 0.19 |
| Mg_CX1 | 15 | 8 | 0.48 | 300 | 0.35 |
| Mg_CX2 | 25 | 13 | 0.23 | 95 | 0.12 |

For sample Mg_CX2, the distribution of $MgH_2$ and MgO is 23% and 77%, respectively. This decrease in active hydrogen storage material is in accordance with the decrease in hydrogen storage capacity measured by Sievert's method (see Figures 2 and 3). For all investigated samples, magnesium oxide is present as stable nanocrystallites (~1 nm). This can be ascribed to the fact that MgO is a much more refractory material, which does not take part in any reactions at temperatures used in the present study. The presence of oxygen is obviously a significant problem for the long term stability of nanoconfined $MgH_2$. The primary source of oxygen appears to be the "inert" carbon aerogel scaffold. It has been found that a carbon aerogel synthesised by a variety of routes has a significant oxygen content (C–O and C=O) [39]. Typically, the oxygen content is a few percent, with much higher oxygen content reported on the surface (~10%). Magnesium is an excellent oxygen scavenger, and the results here show that it strongly reacts with the oxygen within the carbon aerogel scaffold during synthesis and hydrogen cycling at an elevated temperature.

After five cycles of hydrogen release/uptake, the Bragg peaks of $MgH_2$ are much sharper, revealing an average crystallite size that is one order of magnitude or two orders of magnitude greater than in the as-prepared samples (Table 3). These average crystallite sizes are also much larger than the average pore sizes in the scaffolds, which demonstrates the high mobility of $Mg/MgH_2$ during cycling (hydrogen release and uptake) at 350 °C. Thus, $Mg/MgH_2$ tends to migrate or agglomerate in larger pore voids or outside of the scaffold. Particle growth contributes to increasing temperatures for hydrogen release due to hindered kinetics. Nanoparticles have a well-known tendency to grow to larger particles. Previous work demonstrates that sodium alanate, $NaAlH_4$, prefers to crystallise in the larger pores in CA scaffolds [26], and may also migrate out of the scaffold upon cycling [40].

The infiltrated scaffolds, before and after hydrogen cycling, were investigated by transmission electron microscopy (TEM) (see Figure 6). After infiltration, the $MgH_2$ is well-dispersed in the carbon scaffold (<25 nm). After five desorption/absorption cycles, $MgH_2$ particles appear to form larger agglomerations (~100 nm). However, it is difficult to determine if the agglomerates of $MgH_2$ are still within the scaffold or on the surface from the TEM data given that it is a transmission-based technique. Given the average carbon aerogel pore size of 25 nm, it seems likely that $Mg/MgH_2$ has migrated to the surface of the scaffold outside of the pore network.

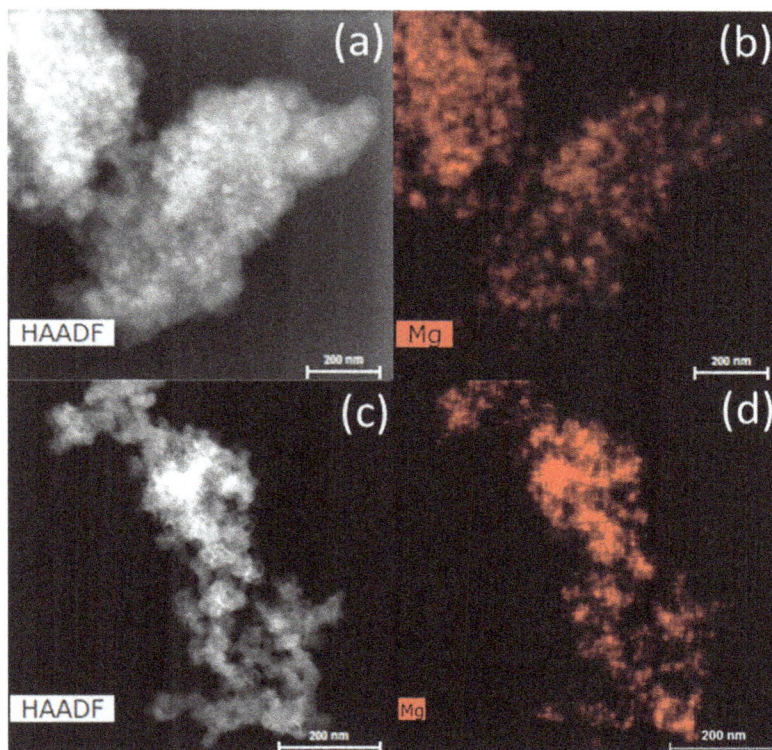

**Figure 6.** Scanning transmission electron microscope-high-angle annular dark-field (STEM-HAADF) images and elemental mapping of the as-prepared Mg_CX2 (**a,b**) and Mg_CX2 after five desorption/absorption cycles (**c,d**).

## 3. Materials and Methods

### 3.1. Synthesis of Carbon Scaffolds

Two batches (denoted X1 and X2) of resorcinol-formaldehyde carbon aerogel were synthesized as described previously [11,41]. Resorcinol (41.3 g, Sigma-Aldrich, Brøndby, Denmark, ≥99.0%) and formaldehyde (56.9 mL, 37 wt % in $H_2O$, stabilized by 10–15% methanol, Sigma-Aldrich) were added to deionized water (56.6 mL) under stirring. Sodium carbonate, $Na_2CO_3$ (65 mg, Sigma-Aldrich, 99.999%) was added to the synthesis of X1 (pH = 6.47) and 40 mg to that of X2 (pH = 6.20). The mixtures were kept in sealed containers at room temperature for 24 h, then at 50 °C for 24 h, and finally at 90 °C for 72 h. The depth of the solution in the sealed containers was less than 0.5 cm to ensure the homogeneity of the carbon aerogel. After cooling, the solid gels were immersed in an acetone bath to exchange all the water inside the pores. The solid gels were then cut into small pieces with average dimension 1 cm × 0.5 cm × 0.4 cm and pyrolysed at 800 °C ($\Delta T / \Delta t$ = 3 °C/min) in $N_2$ for 6 h. A portion of both samples X1 and X2 underwent further heat treatment from room temperature (RT) to 950 °C ($\Delta T / \Delta t$ = 6 °C/min) followed by an isothermal step at 950 °C for 5 h in a constant $CO_2$ flow in order to increase the surface area ($S_{BET}$) and total pore volume ($V_{tot}$) [42]. These samples are denoted CX1 and CX2. The average dimension of the monoliths decreased significantly to only 10–20% of their initial volume. All the synthesized carbon aerogels were degassed in vacuum at 350 °C for several hours and stored inside an argon-filled glovebox.

### 3.2. Direct Synthesis of Nanoconfined Magnesium Hydride

Monoliths of carbon aerogel with an average volume of 0.2 $cm^3$ were immersed in 1 M di-$n$-butylmagnesium, $Mg(CH_2CH_2CH_2CH_3)_2$, denoted $MgBu_2$ (~5 mL, in ether and hexanes, Sigma-Aldrich) for two days. The solvent was removed using Schlenk techniques and the monoliths were dried for several hours in an inert argon atmosphere. Excess white $MgBu_2$ on the surface of the black scaffold was removed mechanically. The amount of infiltrated dibutyl magnesium was determined from the weight gain of the monoliths before and after each infiltration. Afterwards, the infiltrated monoliths were placed in an autoclave (Swagelok, Esbjerg, Denmark) and heated to 150 °C ($\Delta T / \Delta t$ = 5 °C/min) under $p(H_2)$ = 100 bar and kept at 150 °C for 1 h to convert $MgBu_2$ to $MgH_2$ and butane. The autoclave was then evacuated and kept in dynamic vacuum for 30 min to remove the released butane gas. The hydrogenation and evacuation procedures were repeated two further times at 150 °C to ensure a high conversion of $MgBu_2$ to $MgH_2$. Finally, the samples were cooled to room temperature under hydrogen pressure. These $MgBu_2$ infiltration and consequent hydrogenation procedures were repeated three times (3×) for each of the four monolithic samples, and finally the prepared samples were hand ground into powder for further characterisation. The infiltrated volumetric quantity of hydrogen storage material, $MgH_2$, is calculated from the weight gain of the scaffold and the bulk densities $\rho(MgH_2)$ = 1.45 g/$cm^3$ and $\rho(MgBu_2)$ = 0.736 g/$cm^3$. Table S1 provide details about the amounts of $MgBu_2$ infiltrated in each procedure and the total amounts of magnesium hydride in each scaffold. The magnesium hydride-containing scaffolds are denoted Mg_X1, Mg_X2, etc. The samples were stored and handled inside an argon-filled glovebox with $H_2O/O_2$ levels below 1 ppm.

### 3.3. Characterisation

The porosity analysis was performed using a Nova 2200e surface area and pore size analyser (Quantachrome Instruments, Odelzhausen, Germany). The properties of the carbon aerogels were deduced from $N_2$ adsorption/desorption measurements at 77 K. The surface area ($S_{BET}$) was measured using the Brunauer–Emmett–Teller (BET) method, and the micropore volume ($V_{micro}$) was determined by the $t$-plot method [43,44]. The average pore size ($D_{max}$) and mesopore volume ($V_{meso}$) were recorded by the Barrett–Joyner–Halenda (BJH) method during desorption [45]. The total pore volume ($V_{tot}$) of the scaffold was obtained from the point at maximum $p/p_0 \sim 1$.

The thermal properties of nanoconfined $MgH_2$ before and after the desorption/absorption cycles were studied by thermogravimetric analysis (TGA) coupled with mass spectroscopy (MS). TGA was carried out using a STA 6000 (Perkin Elmer, Skovlunde, Denmark), and the evolved gases were detected by a HPR-20 QMS Mass Spectrometer (Hiden Analytical, Warrington, UK). A few milligrams of sample was placed in an aluminium crucible and heated ($\Delta T/\Delta t = 5\ °C/min$) in an argon flow of 40 mL/min.

The stability of the hydrogen storage capacity of nanoconfined $MgH_2$ samples was investigated over five cycles of hydrogen release and uptake by Sievert's measurements using an in-house custom apparatus [30]. Approximately 100 mg of sample was sealed in an autoclave and studied for five desorption and absorption cycles under two different conditions. For *condition* 1, the samples were heated in vacuum from room temperature to 355 °C ($\Delta T/\Delta t = 5\ °C/min$) and kept isothermal for 15 h during hydrogen release. Then, hydrogen absorption was conducted at $p(H_2) = 50$ bar for 15 h at 355 °C, i.e., $\Delta p(H_2) = 50$ bar. The sample was then cooled to room temperature under the same hydrogen pressure. For *condition* 2, the samples were heated to 355 °C ($\Delta T/\Delta t = 5\ °C/min$) and kept at 355 °C for five cycles. Hydrogen release was conducted at $p(H_2) = 4$–5 bar for 15 h at 355 °C and hydrogen absorption at $p(H_2) = 12$ bar for 15 h at 355 °C, i.e., $\Delta p(H_2) \sim 7.5$ bar. The hydrogen equilibrium pressure for $Mg/MgH_2$ at 355 °C is $p_{eq}(H_2) = 6.4$ bar [30].

Powder X-ray diffraction was conducted to characterize the nanoconfined $MgH_2$ samples before and after five desorption/absorption cycles. This was done by using a SmartLab diffractometer (Cu K$\alpha_1$ source, $\lambda = 1.5406$ Å, Rigaku, Ettlingen, Germany). The samples were mounted in 0.5 mm-diameter Lindemann glass capillaries, and the diffraction patterns were collected with an angular step of 3° per minute. The Rietveld analysis was performed in Topas (Bruker, Cambridge, UK) along with crystallite size refinement using fundamental parameters after an instrument calibration using $LaB_6$. The crystallite size was calculated using the LVol-IB method (volume averaged column height calculated from the integral breadth), which provides a measure of the volume-weighted crystallite size.

The distribution of $MgH_2$ in the samples before and after five desorption/absorption cycles was studied using an Talos F200X (S)TEM-microscope (FEI, Copenhagen, Denmark) equipped with an advanced energy dispersive X-ray spectroscopy (EDS) system operated at 200 kV. Samples were dispersed on a copper grid coated in a holey carbon film after suspension in (dry) cyclohexane. Sample grids were attached to the TEM sample holder in ambient conditions, i.e., exposing the sample to air for several minutes.

## 4. Conclusions

$MgH_2$ was infiltrated into four different carbon aerogel scaffolds using a comprehensive activation process. Multiple infiltrations showed a limited increase in the amount of $MgH_2$ (18.2 $cm^3$/mol) due to the large molar volume of $MgBu_2$ (188.2 $cm^3$/mol). The volumetric loading of $MgH_2$ after three loading steps was 17–20 vol % in the various scaffolds. Despite the vigilant infiltration and activation procedure, hydrogen cycling resulted in the production of butane from the conversion of residual $MgBu_2$ in the scaffold. It appears as though batch-wise hydrogenation of $MgBu_2$ is inefficient in fully converting it to $MgH_2$, and future studies may benefit from high pressure flow-through hydrogenation to decrease the $MgBu_2$ content. The nanoconfined $MgH_2$ samples also displayed significant hydrogen capacity loss after cycling that appears to be due to the formation of large quantities of MgO from interactions between $MgH_2$ and the carbon aerogel scaffold. Carbon aerogel scaffolds are not pure carbon, and can contain C–O and C=O groups that could be reduced by Mg at high temperature. Overall, we observe hydrogen release of 1.3 to 3.1 wt % in the first cycle, which for some samples is higher than previously reported ref. [11–13], and 1.0 to 1.9 in the fifth cycle, which may be slightly lower. Further work must be directed towards further purifying carbon aerogel scaffolds or finding alternative, less reactive scaffolds. Hydrogen kinetics was also found to decrease due to $Mg/MgH_2$ growth after cycling at high temperature. It is likely that Mg is able to migrate out of the pore network under vacuum (or low pressure) at high temperature. Other nanoconfinement studies should focus on unreactive scaffold design, improved flow-through $MgH_2$ activation procedures, and work towards understanding the migration of active metal hydride material within the scaffold at high temperature.

**Supplementary Materials:**
Table S1: The mass of scaffold, initial pore volume, and gain in mass of each infiltration of $MgBu_2$ for the four nanoconfined samples, Figure S1: Sievert's measurements of nanoconfined samples under *condition* 2, Figure S2: Sievert's measurements of the first 3 h of nanoconfined samples under *condition* 1, Figure S3: Mass spectroscopic analysis of the gas release from the as-prepared Mg_X1 at 348 °C, Figure S4: Rietveld refinement and difference plots of as-prepared and cycled Mg_CX1.

**Acknowledgments:** This research project received funding from the People Program (Marie Curie Actions) of the European Union's Seventh Framework Program FP7/2007–2013/ under REA grants agreement No. 607040 (Marie Curie ITN ECOSTORE). Furthermore, the work was supported by the Danish National Research Foundation, Center for Materials Crystallography (DNRF93), The Innovation Fund Denmark (project HyFill-Fast), and by the Danish Research Council for Nature and Universe (Danscatt). We are grateful to the Carlsberg Foundation.

**Author Contributions:** Priscilla Huen was involved in all stages of the work, including planning, conducting experiments, and analyzing the data; Mark Paskevicius conducted part of the Sievert's measurements and the Rietveld refinement of diffraction patterns; Bo Richter performed the TEM-EDS experiments; Dorthe B. Ravnsbæk acted as co-supervisor and was involved in the discussion of the results and work planning; Torben R. Jensen acted as the main supervisor and helped with the data analysis and work planning; Priscilla Huen, Mark Paskevicius, and Torben R. Jensen wrote the paper; and all the authors contributed to the revision of the paper.

# References

1. Mazloomi, K.; Gomes, C. Hydrogen as an energy carrier: Prospects and challenges. *Renew. Sustain. Energy Rev.* **2012**, *16*, 3024–3033. [CrossRef]

2. Holladay, J.D.; Hu, J.; King, D.L.; Wang, Y. An overview of hydrogen production technologies. *Catal. Today* **2009**, *139*, 244–260. [CrossRef]

3. Ley, M.B.; Jepsen, L.H.; Lee, Y.-S.; Cho, Y.W.; Bellosta von Colbe, J.M.; Dornheim, M.; Rokni, M.; Jensen, J.O.; Sloth, M.; Filinchuk, Y.; et al. Complex hydrides for hydrogen storage—New perspectives. *Mater. Today* **2014**, *17*, 122–128. [CrossRef]

4. Møller, K.T.; Jensen, T.R.; Akiba, E.; Li, H. Hydrogen—A sustainable energy carrier. *Prog. Nat. Sci. Mater. Int.* **2017**, *27*, 34–40. [CrossRef]

5. Haynes, W.M. *CRC Handbook of Chemistry and Physics*, 95th ed.; CRC Press: Boca Raton, FL, USA, 2014; ISBN 9781482208689.

6. Lai, Q.; Paskevicius, M.; Sheppard, D.A.; Buckley, C.E.; Thornton, A.W.; Hill, M.R.; Gu, Q.; Mao, J.; Huang, Z.; Liu, H.K.; et al. Hydrogen Storage Materials for Mobile and Stationary Applications: Current State of the Art. *ChemSusChem* **2015**, *8*, 2789–2825. [CrossRef] [PubMed]

7. Paskevicius, M.; Jepsen, L.H.; Schouwink, P.; Černý, R.; Ravnsbæk, D.B.; Filinchuk, Y.; Dornheim, M.; Besenbacher, F.; Jensen, T.R. Metal borohydrides and derivatives—Synthesis, structure and properties. *Chem. Soc. Rev.* **2017**, *46*, 1565–1634. [CrossRef] [PubMed]

8. Webb, C.J. A review of catalyst-enhanced magnesium hydride as a hydrogen storage material. *J. Phys. Chem. Solids* **2015**, *84*, 96–106. [CrossRef]

9. Crivello, J.-C.; Denys, R.V.; Dornheim, M.; Felderhoff, M.; Grant, D.M.; Huot, J.; Jensen, T.R.; de Jongh, P.; Latroche, M.; Walker, G.S.; et al. Mg-based compounds for hydrogen and energy storage. *Appl. Phys. A* **2016**, *122*, 85. [CrossRef]

10. Crivello, J.-C.; Dam, B.; Denys, R.V.; Dornheim, M.; Grant, D.M.; Huot, J.; Jensen, T.R.; de Jongh, P.; Latroche, M.; Milanese, C.; et al. Review of magnesium hydride-based materials: Development and optimisation. *Appl. Phys. A* **2016**, *122*, 97. [CrossRef]

11. Nielsen, T.K.; Manickam, K.; Hirscher, M.; Besenbacher, F.; Jensen, T.R. Confinement of $MgH_2$ Nanoclusters within Nanoporous Aerogel Scaffold Materials. *ACS Nano* **2009**, *3*, 3521–3528. [CrossRef] [PubMed]

12. Zhang, S.; Gross, A.F.; Van Atta, S.L.; Lopez, M.; Liu, P.; Ahn, C.C.; Vajo, J.J.; Jensen, C.M. The synthesis and hydrogen storage properties of a $MgH_2$ incorporated carbon aerogel scaffold. *Nanotechnology* **2009**, *20*, 204027. [CrossRef] [PubMed]

13. Gross, A.F.; Ahn, C.C.; Van Atta, S.L.; Liu, P.; Vajo, J.J. Fabrication and hydrogen sorption behaviour of nanoparticulate $MgH_2$ incorporated in a porous carbon host. *Nanotechnology* **2009**, *20*, 204005. [CrossRef] [PubMed]

14. Jia, Y.; Sun, C.; Cheng, L.; Abdul Wahab, M.; Cui, J.; Zou, J.; Zhu, M.; Yao, X. Destabilization of Mg–H bonding through nano-interfacial confinement by unsaturated carbon for hydrogen desorption from MgH$_2$. *Phys. Chem. Chem. Phys.* **2013**, *15*, 5814. [CrossRef] [PubMed]

15. Nielsen, T.K.; Javadian, P.; Polanski, M.; Besenbacher, F.; Bystrzycki, J.; Jensen, T.R. Nanoconfined NaAlH$_4$: Determination of Distinct Prolific Effects from Pore Size, Crystallite Size, and Surface Interactions. *J. Phys. Chem. C* **2012**, *116*, 21046–21051. [CrossRef]

16. De Jongh, P.E.; Adelhelm, P. Nanosizing and Nanoconfinement: New Strategies Towards Meeting Hydrogen Storage Goals. *ChemSusChem* **2010**, *3*, 1332–1348. [CrossRef] [PubMed]

17. Gosalawit-Utke, R.; Thiangviriya, S.; Javadian, P.; Laipple, D.; Pistidda, C.; Bergemann, N.; Horstmann, C.; Jensen, T.R.; Klassen, T.; Dornheim, M. Effective nanoconfinement of 2LiBH$_4$–MgH$_2$ via simply MgH$_2$ premilling for reversible hydrogen storages. *Int. J. Hydrogen Energy* **2014**, *39*, 15614–15626. [CrossRef]

18. Bérubé, V.; Radtke, G.; Dresselhaus, M.; Chen, G. Size effects on the hydrogen storage properties of nanostructured metal hydrides: A review. *Int. J. Energy Res.* **2007**, *31*, 637–663. [CrossRef]

19. Nielsen, T.K.; Besenbacher, F.; Jensen, T.R. Nanoconfined hydrides for energy storage. *Nanoscale* **2011**, *3*, 2086–2098. [CrossRef] [PubMed]

20. Gutowska, A.; Li, L.; Shin, Y.; Wang, C.M.; Li, X.S.; Linehan, J.C.; Smith, R.S.; Kay, B.D.; Schmid, B.; Shaw, W.; et al. Nanoscaffold Mediates Hydrogen Release and the Reactivity of Ammonia Borane. *Angew. Chem.* **2005**, *117*, 3644–3648. [CrossRef]

21. Ngene, P.; van Zwienen, M.; de Jongh, P.E. Reversibility of the hydrogen desorption from LiBH$_4$: A synergetic effect of nanoconfinement and Ni addition. *Chem. Commun.* **2010**, *46*, 8201. [CrossRef] [PubMed]

22. Paskevicius, M.; Filsø, U.; Karimi, F.; Puszkiel, J.; Pranzas, P.K.; Pistidda, C.; Hoell, A.; Welter, E.; Schreyer, A.; Klassen, T.; et al. Cyclic stability and structure of nanoconfined Ti-doped NaAlH$_4$. *Int. J. Hydrogen Energy* **2016**, *41*, 4159–4167. [CrossRef]

23. Zhao-Karger, Z.; Hu, J.; Roth, A.; Wang, D.; Kübel, C.; Lohstroh, W.; Fichtner, M. Altered thermodynamic and kinetic properties of MgH$_2$ infiltrated in microporous scaffold. *Chem. Commun.* **2010**, *46*, 8353. [CrossRef] [PubMed]

24. De Jongh, P.E.; Wagemans, R.W.P.; Eggenhuisen, T.M.; Dauvillier, B.S.; Radstake, P.B.; Meeldijk, J.D.; Geus, J.W.; de Jong, K.P. The Preparation of Carbon-Supported Magnesium Nanoparticles using Melt Infiltration. *Chem. Mater.* **2007**, *19*, 6052–6057. [CrossRef]

25. Utke, R.; Thiangviriya, S.; Javadian, P.; Jensen, T.R.; Milanese, C.; Klassen, T.; Dornheim, M. 2LiBH$_4$–MgH$_2$ nanoconfined into carbon aerogel scaffold impregnated with ZrCl$_4$ for reversible hydrogen storage. *Mater. Chem. Phys.* **2016**, *169*, 136–141. [CrossRef]

26. Nielsen, T.K.; Javadian, P.; Polanski, M.; Besenbacher, F.; Bystrzycki, J.; Skibsted, J.; Jensen, T.R. Nanoconfined NaAlH$_4$: Prolific effects from increased surface area and pore volume. *Nanoscale* **2014**, *6*, 599–607. [CrossRef] [PubMed]

27. Fichtner, M. Nanoconfinement effects in energy storage materials. *Phys. Chem. Chem. Phys.* **2011**, *13*, 21186. [CrossRef] [PubMed]

28. Roedern, E.; Hansen, B.R.S.; Ley, M.B.; Jensen, T.R. Effect of Eutectic Melting, Reactive Hydride Composites, and Nanoconfinement on Decomposition and Reversibility of LiBH$_4$–KBH$_4$. *J. Phys. Chem. C* **2015**, *119*, 25818–25825. [CrossRef]

29. Bramwell, P.L.; Ngene, P.; de Jongh, P.E. Carbon supported lithium hydride nanoparticles: Impact of preparation conditions on particle size and hydrogen sorption. *Int. J. Hydrogen Energy* **2017**, *42*, 5188–5198. [CrossRef]

30. Paskevicius, M.; Sheppard, D.A.; Buckley, C.E. Thermodynamic Changes in Mechanochemically Synthesized Magnesium Hydride Nanoparticles. *J. Am. Chem. Soc.* **2010**, *132*, 5077–5083. [CrossRef] [PubMed]

31. Cuevas, F.; Korablov, D.; Latroche, M. Synthesis, structural and hydrogenation properties of Mg-rich MgH$_2$–TiH$_2$ nanocomposites prepared by reactive ball milling under hydrogen gas. *Phys. Chem. Chem. Phys.* **2012**, *14*, 1200–1211. [CrossRef] [PubMed]

32. Pasquini, L.; Boscherini, F.; Callini, E.; Maurizio, C.; Pasquali, L.; Montecchi, M.; Bonetti, E. Local structure at interfaces between hydride-forming metals: A case study of Mg-Pd nanoparticles by X-ray spectroscopy. *Phys. Rev. B* **2011**, *83*, 184111. [CrossRef]

33.  Andreasen, A.; Sørensen, M.B.; Burkarl, R.; Møller, B.; Molenbroek, A.M.; Pedersen, A.S.; Vegge, T.; Jensen, T.R. Dehydrogenation kinetics of air-exposed $MgH_2/Mg_2Cu$ and $MgH_2/MgCu_2$ studied with in situ X-ray powder diffraction. *Appl. Phys. A* **2006**, *82*, 515–521. [CrossRef]

34.  Andreasen, A.; Sørensen, M.B.; Burkarl, R.; Møller, B.; Molenbroek, A.M.; Pedersen, A.S.; Andreasen, J.W.; Nielsen, M.M.; Jensen, T.R. Interaction of hydrogen with an Mg–Al alloy. *J. Alloys Compd.* **2005**, *404–406*, 323–326. [CrossRef]

35.  Jensen, T.; Andreasen, A.; Vegge, T.; Andreasen, J.; Stahl, K.; Pedersen, A.; Nielsen, M.; Molenbroek, A.; Besenbacher, F. Dehydrogenation kinetics of pure and nickel-doped magnesium hydride investigated by in situ time-resolved powder X-ray diffraction. *Int. J. Hydrogen Energy* **2006**, *31*, 2052–2062. [CrossRef]

36.  Dornheim, M.; Eigen, N.; Barkhordarian, G.; Klassen, T.; Bormann, R. Tailoring Hydrogen Storage Materials Towards Application. *Adv. Eng. Mater.* **2006**, *8*, 377–385. [CrossRef]

37.  Barkhordarian, G.; Klassen, T.; Bormann, R. Catalytic Mechanism of Transition-Metal Compounds on Mg Hydrogen Sorption Reaction. *J. Phys. Chem. B* **2006**, *110*, 11020–11024. [CrossRef] [PubMed]

38.  Nielsen, T.K.; Jensen, T.R. $MgH_2$–$Nb_2O_5$ investigated by in situ synchrotron X-ray diffraction. *Int. J. Hydrogen Energy* **2012**, *37*, 13409–13416. [CrossRef]

39.  Alegre, C.; Sebastián, D.; Baquedano, E.; Gálvez, M.E.; Moliner, R.; Lázaro, M. Tailoring Synthesis Conditions of Carbon Xerogels towards Their Utilization as Pt-Catalyst Supports for Oxygen Reduction Reaction (ORR). *Catalysts* **2012**, *2*, 466–489. [CrossRef]

40.  Chumphongphan, S.; Filsø, U.; Paskevicius, M.; Sheppard, D.A.; Jensen, T.R.; Buckley, C.E. Nanoconfinement degradation in $NaAlH_4$/CMK-1. *Int. J. Hydrogen Energy* **2014**, *39*, 11103–11109. [CrossRef]

41.  Li, W.-C.; Lu, A.-H.; Weidenthaler, C.; Schüth, F. Hard-Templating Pathway to Create Mesoporous Magnesium Oxide. *Chem. Mater.* **2004**, *16*, 5676–5681. [CrossRef]

42.  Lin, C.; Ritter, J.A. Carbonization and activation of sol-gel derived carbon xerogels. *Carbon* **2000**, *38*, 849–861. [CrossRef]

43.  Brunauer, S.; Emmett, P.H.; Teller, E. Adsorption of Gases in Multimolecular Layers. *J. Am. Chem. Soc.* **1938**, *60*, 309–319. [CrossRef]

44.  Deboer, J. Studies on pore systems in catalysts VII. Description of the pore dimensions of carbon blacks by the t method. *J. Catal.* **1965**, *4*, 649–653. [CrossRef]

45.  Barrett, E.P.; Joyner, L.G.; Halenda, P.P. The Determination of Pore Volume and Area Distributions in Porous Substances. I. Computations from Nitrogen Isotherms. *J. Am. Chem. Soc.* **1951**, *73*, 373–380. [CrossRef]

# Permissions

All chapters in this book were first published in INORGANICS, by MDPI AG; hereby published with permission under the Creative Commons Attribution License or equivalent. Every chapter published in this book has been scrutinized by our experts. Their significance has been extensively debated. The topics covered herein carry significant findings which will fuel the growth of the discipline. They may even be implemented as practical applications or may be referred to as a beginning point for another development.

The contributors of this book come from diverse backgrounds, making this book a truly international effort. This book will bring forth new frontiers with its revolutionizing research information and detailed analysis of the nascent developments around the world.

We would like to thank all the contributing authors for lending their expertise to make the book truly unique. They have played a crucial role in the development of this book. Without their invaluable contributions this book wouldn't have been possible. They have made vital efforts to compile up to date information on the varied aspects of this subject to make this book a valuable addition to the collection of many professionals and students.

This book was conceptualized with the vision of imparting up-to-date information and advanced data in this field. To ensure the same, a matchless editorial board was set up. Every individual on the board went through rigorous rounds of assessment to prove their worth. After which they invested a large part of their time researching and compiling the most relevant data for our readers.

The editorial board has been involved in producing this book since its inception. They have spent rigorous hours researching and exploring the diverse topics which have resulted in the successful publishing of this book. They have passed on their knowledge of decades through this book. To expedite this challenging task, the publisher supported the team at every step. A small team of assistant editors was also appointed to further simplify the editing procedure and attain best results for the readers.

Apart from the editorial board, the designing team has also invested a significant amount of their time in understanding the subject and creating the most relevant covers. They scrutinized every image to scout for the most suitable representation of the subject and create an appropriate cover for the book.

The publishing team has been an ardent support to the editorial, designing and production team. Their endless efforts to recruit the best for this project, has resulted in the accomplishment of this book. They are a veteran in the field of academics and their pool of knowledge is as vast as their experience in printing. Their expertise and guidance has proved useful at every step. Their uncompromising quality standards have made this book an exceptional effort. Their encouragement from time to time has been an inspiration for everyone.

The publisher and the editorial board hope that this book will prove to be a valuable piece of knowledge for researchers, students, practitioners and scholars across the globe.

# List of Contributors

**Stefano Nuzzo, Michelle P. Browne, Brendan Twamley, Michael E. G. Lyons and Robert J. Baker**
School of Chemistry, Trinity College, University of Dublin, 2 Dublin, Ireland

**Zineb Mouline**
Department of Frontier Materials, Graduate School of Engineering, Nagoya Institute of Technology, Gokiso-cho, Showa-ku, Nagoya 466-8555, Japan

**Mohd Nazri Mohd Sokri**
Department of Frontier Materials, Graduate School of Engineering, Nagoya Institute of Technology, Gokiso-cho, Showa-ku, Nagoya 466-8555, Japan;
Department of Energy Engineering and Advanced Membrane Technology Research Centre, Faculty of Chemical and Energy Engineering, Universiti Teknologi Malaysia (UTM), Johor Bahru 81310, Malaysia

**Sawao Honda, Yusuke Daiko and Yuji Iwamoto**
Department of Frontier Materials, Graduate School of Engineering, Nagoya Institute of Technology, Gokiso-cho, Showa-ku, Nagoya 466-8555, Japan;
Core Research for Evolutional Science and Technology (CREST), Japan Science and Technology Agency, Gokiso-cho, Showa-ku, Nagoya 466-8555, Japan

**Hiroki Noma and Keishi Ohara**
Graduate School of Science and Engineering, Ehime University, 2-5, Bunkyo-cho, Matsuyama 790-8577, Japan

**Toshio Naito**
Graduate School of Science and Engineering, Ehime University, 2-5, Bunkyo-cho, Matsuyama 790-8577, Japan
Division of Material Science, Advanced Research Support Center (ADRES), Ehime University, 2-5, Bunkyo-cho, Matsuyama 790-8577, Japan

**Eriko Miyamae**
Department of Chemistry, Faculty of Science, Kanagawa University, Hiratsuka, Kanagawa 259-1293, Japan

**Satoshi Matsunaga, Takuya Otaki, Yusuke Inoue, Kohei Mihara and Kenji Nomiya**
Department of Chemistry, Faculty of Science, Kanagawa University, Hiratsuka, Kanagawa 259-1293, Japan

**Wojciech I. Dzik**
Van't Hoff Institute for Molecular Sciences (HIMS), Homogeneous, Supramolecular and Bio-Inspired Catalysis, Universiteit van Amsterdam, GS Amsterdam, The Netherlands

**Martin Valldor**
Physics of Correlated Matter, Max Planck Institute for Chemical Physics of Solids, Nöthnitzer Str. 40, 01187 Dresden, Germany

**Victoria K. Greenacre and Ian R. Crossley**
Department of Chemistry, University of Sussex, Brighton BN1 9QJ, UK

**Theresia M. M. Richter, Sabine Strobel and Rainer Niewa**
Institute of Inorganic Chemistry, Universität Stuttgart, Pfaffenwaldring 55, 70569 Stuttgart, Germany

**Nicolas S. A. Alt and Eberhard Schlücker**
Institute of Process Machinery and Systems Engineering, University of Erlangen-Nuremberg, Cauerstraße 4, 91058 Erlangen, Germany

**Shin-ichi Kawaguchi**
Center for Education and Research in Agricultural Innovation, Faculty of Agriculture, Saga University, 152-1 Shonan-cho Karatsu, Saga 847-0021, Japan

**Yuki Sato, Yoshiaki Minamida, Akihiro Nomoto and Akiya Ogawa**
Department of Applied Chemistry, Graduate School of Engineering, Osaka Prefecture University, 1-1 Gakuen-cho, Nakaku, Sakai, Osaka 599-8531, Japan

**Yuta Saga**
Department of Applied Chemistry, Graduate School of Engineering, Osaka Prefecture University, 1-1 Gakuen-cho, Nakaku, Sakai, Osaka 599-8531, Japan
Katayama Chemical Industries Co., Ltd., 26-22, 3-Chome, Higasinaniwa-cho, Amagasaki, Hyogo 660-0892, Japan

**Arno L. Görne, Janine George and Jan van Leusen**
Institute of Inorganic Chemistry, RWTH Aachen University, Landoltweg 1, 52056 Aachen, Germany

**Richard Dronskowski**
Institute of Inorganic Chemistry, RWTH Aachen University, Landoltweg 1, 52056 Aachen, Germany

Jülich-Aachen Research Alliance, JARA-HPC, RWTH Aachen University, 52056 Aachen, Germany

**Kirsten Reuter, Fabian Dankert, Carsten Donsbach and Carsten von Hänisch**
Fachbereich Chemie and Wissenschaftliches Zentrum für Materialwissenschaften (WZMW),
Philipps-Universität Marburg, Hans-Meerwein Straße 4, D-35032 Marburg, Germany

**Nina V. Kosova and Daria O. Rezepova**
Institute of Solid State Chemistry and Mechanochemistry SB RAS, 18 Kutateladze, Novosibirsk 630128, Russia

**Olga S. Bokareva, Tobias Möhle, Sergey I. Bokarev, Stefan Lochbrunner and Oliver Kühn**
Institut für Physik, Universität Rostock, Albert-Einstein-Str. 23-24, 18059 Rostock, Germany

**Antje Neubauer**
Institut für Physik, Universität Rostock, Albert-Einstein-Str. 23-24, 18059 Rostock, Germany;
Becker & Hickl GmbH, Nahmitzer Damm 30, 12277 Berlin, Germany

**Emmanuel Kottelat and Fabio Zobi**
Department of Chemistry, University of Fribourg, Chemin du Musée 9, CH-1700 Fribourg, Switzerland

**Daniel Werner**
Institut für Anorganische Chemie, University of Tübingen (EKUT) Auf der Morgenstelle 18, 72076 Tübingen, Germany

**Glen B. Deacon**
School of Chemistry, Monash University, Clayton, Victoria 3800, Australia

**Peter C. Junk**
College of Science & Engineering, James Cook University, Townsville, Queensland 4811, Australia

**Priscilla Huen, Bo Richter and Torben R. Jensen**
Center for Materials Crystallography, Interdisciplinary Nanoscience Center and Department of Chemistry, Aarhus University, Langelandsgade 140, 8000 Aarhus C, Denmark

**Mark Paskevicius**
Department of Physics and Astronomy, Fuels and Energy Technology Institute, Curtin University, Kent Street, Bentley, WA 6102, Australia

**Dorthe B. Ravnsbæk**
Department of Physics, Chemistry and Pharmacy, University of Southern Denmark, Campusvej 55, 5230 Odense M, Denmark

# Index